THE INORGANIC CHEMISTRY
OF
BIOLOGICAL PROCESSES

THE

INORGANIC CHEMISTRY

OF

BIOLOGICAL PROCESSES
Second Edition

M. N. Hughes

Reader in Chemistry, Queen Elizabeth College,
University of London

JOHN WILEY & SONS
Chichester · New York · Brisbane · Toronto

British Library Cataloguing in Publication Data:

Hughes, Martin Neville
   The inorganic chemistry of biological processes.
   2nd ed.
   1. Metals in the body
   2. Metal ions
   I. Title
   574.1′9214   QP532   80–40499

ISBN 0 471 27815 7

Photosetting by Thomson Press (India) Limited, New Delhi and Printed in the United States of America.

CONTENTS

PREFACE TO THE FIRST EDITION

The overlap region between inorganic chemistry and the biological sciences is one where exciting and significant developments are taking place. It is now appreciated that metal ions control a vast range of processes in biology; that life is really as dependent upon inorganic chemistry as organic chemistry. New developments in instrumental techniques have further accelerated the growth of 'inorganic biochemistry' so that it is now probably true to say that this subject involves one of the most rapidly expanding areas in the chemical and biochemical sciences.

This book is intended to present an introduction to this most important field. It has its origins in third year undergraduate courses at Queen Elizabeth College and is written primarily for chemists, particularly inorganic chemists. The material presented includes a survey of the occurrence and role of the metal ions of biological importance and shows how the function of these ions may be studied experimentally. While most topics of current interest are discussed, the coverage is not intended to be exhaustive. The book does not depend upon a prior knowledge of biological subjects, some relevant material is summarized in Chapter 1. It is also hoped that this book may be of interest to workers in the biological sciences, and so, primarily for this purpose, a brief survey of the relevant properties of transition metal complexes is presented in Chapter 2, together with an account of the mechanisms of their reactions in solution.

I am happy to acknowledge the assistance of a number of colleagues and friends; in particular Dr. K. J. Rutt for his helpful comments on the early chapters, Dr. C. W. Bird for his encouragement throughout all stages of writing this book, and Miss Jane Cooper for her excellent typing of the manuscript. I am grateful to Professors J. Brachet, J. Coleman, and S. Lindskog for permission to reproduce Figures 1.1, 4.3, 4.4 and 4.5 respectively and also to the appropriate Editors as specified in the text.

The reprinting of this book has given an opportunity to rewrite part of the section on the iron–sulphur proteins. Significant developments have taken place since 1972, largely as a result of the publication of the structures of several of these proteins. I have also included a short section at the end of Chapter 10 on the use of platinum complexes in cancer chemotherapy.

PREFACE TO THE SECOND EDITION

The years since 1972 have witnessed a dramatic expansion of research activity and interest in Inorganic Biochemistry. Many advances have been made, partly as a result of the availability of new instrumental techniques. Some apparently simple questions have remained unanswered, however, and these doubtless will serve as a stimulus for further effort.

The general arrangement of the book remains unchanged, although most of the chapters have been almost completely rewritten. Two new chapters have been included: one on the transport and storage of iron, and a short one on non-metals. The edition has been written in the light of journals received in the United Kingdom by mid-1979.

I am indebted to a number of colleagues for helpful suggestions and encouragement. The early stages were written at the University Chemical Laboratory, Cambridge, while on sabbatical leave, and I am grateful for the hospitality shown to me there, and for useful discussions with Dr. L. F. Lindoy and Dr. J. R. Thornback. Finally, I must pay tribute to the very considerable contribution made by my wife Jenny to the preparation of the manuscript. She has my deepest appreciation and thanks.

February 1980

CHAPTER ONE

INTRODUCTION

Metal ions play a vital role in a vast number of widely differing biological processes. Increasing knowledge will almost certainly serve to demonstrate this fact more effectively. Some of these processes are quite specific in their metal ion requirements in that only certain metal ions, in specified oxidation states, can fulfil the necessary catalytic or structural requirement, while other processes are much less specific, and it is possible to replace one metal ion by another, although the activity may be reduced.

Metal ion dependent processes are found throughout the Life Sciences and vary tremendously in their function and complexity. Three examples, from biochemistry, physiology, and cytochemistry respectively, are given to illustrate this point. Thus the metal ions potassium, magnesium, manganese, iron, cobalt, copper, molybdenum, and zinc are all important catalysts of a variety of enzyme reactions such as, for example, group transfer, redox or hydrolytic processes. Not only, however, are these metal ions involved in such processes, but, in certain cases, there are other protein systems involved in storing and controlling the concentration of the metal ion, and then in transporting it to the appropriate site for incorporation into the necessary enzyme system. Sodium, potassium, and calcium, on the other hand, are heavily involved in certain physiological control and trigger mechanisms, while potassium, calcium, and magnesium ions are all important in maintaining the structure and controlling the function of cell walls. The metals cited in these examples are not the only ones involved in biological processes, others although quantitatively less important, also have biological functions. Of the cited metal ions, Na^+, K^+, Mg^{2+}, and Ca^{2+} are present in much greater amounts than the heavy metal ions. Thus in the human body these four cations constitute some 99% of the total metal ion content.

The physiological and biochemical functions of the metal ion in all these processes are obviously a matter of fundamental importance, but the difficulties involved in attempting to clarify their role should not be minimized. Such a study presents many difficult problems, the solutions of which often require an overlap of disciplines. For the inorganic chemist, without doubt, the field which holds out most hope of effective exploitation is that of the role of metal ions in enzyme and similar systems. It should be stressed that here there are often additional advantages associated with the presence of the metal ion which contribute very markedly to an understanding of the system. This is particularly true of transition metal ions. Thus as a result of the electronic properties of the metal, a variety of powerful instrumental techniques may be brought into play. Again the presence of the metal ion provides an extra guide in elucidating the mechanism of the enzyme action, if only in providing an extra check on its correctness in terms of correlating the specificity of the system for that ion in the light of modern awareness of the preferred environment, stereochemistry, and electronic properties of the ion.

Developments in inorganic and organometallic chemistry have resulted in a very significantly increased understanding of the bonding, structure, and reactivity of coordination compounds. This has been reflected in certain areas of inorganic chemistry, for example the activity of small molecules such as CO, H_2, and olefins on coordination to a metal, and this is leading to an increased understanding of certain important catalytic processes. Equally, however, the border area between inorganic chemistry and the Life Sciences should present a challenge to the inorganic chemist to apply his increased understanding to the design of model systems that throw light on the behaviour of metal ions in biological processes, and ultimately to look more closely at these processes themselves. These developments in inorganic chemistry have been matched by developments in biochemistry in that it is now possible from the biochemical point of view to consider processes at the molecular level. Certain progress in the field of metal ion dependent enzymes has been made, particularly in the case of metalloenzymes where the metal is firmly bound to protein. Most metal cations in living organisms will be associated with proteins and so the subject of metal–protein binding is a fundamental one in the overall context of this book.

BACKGROUND MATERIAL

It is necessary at this stage to introduce some background material in order to explain the terms and concepts used in later chapters. Much of this material is presented at an elementary level.

Amino acids, peptides, and proteins

The proteins are macromolecules of great biological importance. They are made up of α-amino acids of L configuration, linked together *via* peptide bonds —CONH—. By suitable treatment they can be degraded to smaller peptides and finally to the constituent amino acids. Some twenty of these amino acids are found in nature, together with the α-amino acids proline and hydroxyproline. These are listed in Table 1.1. Each naturally occurring polypeptide or protein involves a specific sequence of amino acid residues, which may be determined by chemical and biochemical methods. The sequence of amino acid residues will determine the physical and chemical properties of the protein in terms of the chemical and physical interactions occurring between the side chains R in $NH_2CH(R)COOH$. The important side chains in this connection are those involving aromatic groups, sulphur-containing groups, and —NH_2, —OH, and —COOH groups. The nature of the residues may generate hydrophobic or hydrophilic environments in certain regions of the protein chain.

Each amino acid has at least two ionizable groups, the amino and carboxyl groups. The —NH_3 group is less acidic than the carboxyl group and so, between pH 4–9, the amino acid exists as a zwitterion, $H_3N^+CH(R)COO^-$. The

TABLE 1.1 Naturally occurring amino acids $NH_2CH(R)COOH$

R		R	
H—	Glycine	$HOCH_2$—	Serine
CH_3—	Alanine	$CH_3CH(OH)$—	Threonine
$(CH_3)_2CH$—	Valine	$^-OOCCH_2$—	Aspartic acid
$(CH_3)_2CHCH_2$—	Leucine	$^-OOCCH_2CH_2$—	Glutamic acid
$CH_3CH_2CH(CH_3)$—	Isoleucine	$NH_2C(O)CH_2CH_2$—	Glutamine
$\overset{+}{N}H_3(CH_2)_3CH_2$—	Lysine	$NH_2C(O)CH_2$—	Asparagine
$\overset{+}{N}H_3CH(OH)(CH_2)_2CH_2$—	Hydroxylysine		
$SHCH_2$—	Cysteine		
$CH_3SCH_2CH_2$—	Methionine		

$$\overset{+}{N}H_3 \diagdown \atop ^-OOC \diagup CHCH_2SSCH_2- \quad Cystine$$

HO—⟨benzene ring⟩—CH_2—   Tyrosine

⟨benzene ring⟩—CH_2—   Phenylalanine

$$\begin{array}{c} H_2N \diagdown \\ C \!\!=\!\! NH(CH_2)_2CH_2 - Arginine \\ H_2N \diagup + \end{array}$$

⟨indole ring⟩CH_2—   Tryptophan

⟨imidazole ring⟩CH_2—   Histidine

α-carboxyl and amino functions are involved in the formation of the peptide link, and so each peptide has terminal —NH_2 and —COOH groups, together with peptide links and side chains. These are all possible metal-binding sites. Certain low molecular weight peptides are biologically important molecules.

The molecular weights of proteins are in the range of 10^4–10^6. The determination of the chemical structure and the spatial configuration of proteins is a very important matter, as this is bound up with the biological function of the proteins. At the present time an increasing number of protein structures have been determined to a high degree of resolution by X-ray diffraction techniques. The structure of proteins is discussed in terms of primary, secondary, tertiary, and quaternary structure.

Primary structure

This is the sequence of amino acid residues in the chain. For a particular enzyme some residues are less important than others and may in enzymes from different species be replaced by other residues.

Secondary structure

This is concerned with the configuration of the protein chain that results largely from hydrogen bonding between peptide links. Pauling and Corey began their

$$\diagdown C=O \cdots HN \diagup$$

classic study of this problem by determining the structures of a range of simpler compounds, including amides. This has demonstrated that for a stable secondary structure the peptide link is always planar; that the carbon atoms each side of the peptide link are *trans* to each other, so lowering repulsive forces; and finally that there is a maximum amount of hydrogen bonding between carbonyl oxygen and amide nitrogen atoms.

This hydrogen bonding may either be intramolecular or intermolecular. In the first case, it gives rise to an α-helical structure and in the latter case to a pleated sheet structure. X-ray studies have confirmed the α-helix structure and have shown that there are 3.7 residues per turn of the helix. Helical structures occur in globular and fibrous protein. Pleated sheet structures may involve peptide chains having all N-termini at one end or with every other N-terminus at one end. These are the parallel and antiparallel forms respectively.

Secondary structure is also dependent upon the nature of the side chains in that interactions between these may, for example, lower the stability of the α-helix through repulsion.

Tertiary structure

This is concerned with the way in which the protein chain (with its secondary structure) folds upon itself, and the resulting shape of the molecule. Thus it may be globular or rod-like. This results from the interactions between side chains, and may reflect the formation of S—S bonds between cysteine residues, hydrogen bonding between side chains, so called 'salt' or ionic linkages between oppositely charged groups such as $-\overset{+}{N}H_3$ and $-COO^-$, and hydrophobic interactions between aromatic residues. It appears that the last named of these is most important.

The overall shape of the protein molecule may be studied in a number of ways. The most powerful technique is of course that of X-ray diffraction, but other techniques include the measurement of the intrinsic viscosity and the light scattering properties of the macromolecule. X-ray diffraction will also confirm or determine the primary structure.

A schematic representation of the structure of myoglobin is given in Figure 1.1. This globular protein is made up of eight helical segments separated by regions of random coil. It has a high α-helix content of 77 per cent, and has no S—S bridges. It is rather atypical in this.

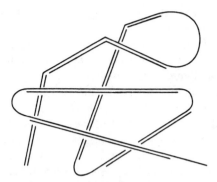

Figure 1.1 Schematic representation of the structure of myoglobin (the double lines indicate α-helical segments).

Quaternary structure

Many proteins are made up of linked subunits held together by other than covalent bonds. The quaternary structure reflects the way in which these subunits come together. Hemoglobin is made up of four such units, and substantial changes in interaction between these occur when dioxygen is taken up or lost.

Proteins in solution, denaturation and other topics

One important question is the extent to which protein structures, determined in the solid state, change in solution. Only if there is good reason for assuming that little change has occurred can we extrapolate the results of X-ray structure determinations to the problems of enzyme mechanisms in solution. In the past it has been assumed that solution and crystal structures are similar. Thus the alkylation of side chains in myoglobin proceeds readily for those residues which the X-ray determinations show to be exposed and does not proceed for those deep inside the molecule. However this assumption must be viewed with caution, in view of results on the enzyme carboxypeptidase.

It is clear from the preceding paragraphs that the secondary and tertiary structure of the protein is dependent upon a variety of interactions between various groups in the protein molecule. On heating or change of pH some of these interactions may be affected. The addition of solvents may break down some hydrogen bonding, while treatment with reducing agents may break S—S linkages. Clearly, therefore, proteins are readily subject to changes which convert an ordered structure to a disordered one. This is termed denaturation. Sometimes these changes are reversible and sometimes they are irreversible. Denaturation results in a number of changes in the chemical and physical properties of a protein. The sensitivity of proteins to denaturation is a major difficulty in experimental work.

Much work has been carried out on synthetic polypeptides such as poly-L-alanine and this has provided a basis for the study of naturally occurring proteins; for example, the value of certain physical techniques in measuring the α-helical content of proteins.

The determination of the helical content of proteins is an important measurement. Optical rotatory dispersion (ORD) studies have been widely used, as the helix itself contributes to this in addition to the constituent amino acids. Other approaches have involved the use of infrared dichroism studies using plane polarized infrared light and hydrogen–deuterium exchange. Standard texts should be consulted for a full account of these techniques.

The use of models and the study of synthetic polypeptides with only one type of amino acid residue have also contributed to an understanding of the secondary structure of proteins. Thus it has been demonstrated that a mixture of D- and L-amino acid residues is not compatible with the α-helix. Similarly the requirements of the α-helix structure cannot be met by proline residues and so these must terminate the α-helix configuration. van der Waals interactions between substituents such as in valine and isoleucine will also lower the stability of the helix.

One important property of proteins is the fact that they have a number of ionizable groups that may be involved in acid–base equilibria. These are the side chains of aspartic and glutamic acids, arginine, cysteine, histidine, lysine, and tyrosine. As a result of the large number of titratable groups present in each protein the interpretation of protein titration curves is complex. In addition, the environment of the protein (hydrogen bonding, medium effects) may well affect the pK_a of the amino acid residue so that it is greater or less than the value for the free acid by up to one pK unit. Certain residues may also be buried within the interior of the protein molecule and so not be accessible during the titration. By carrying out a back titration it is possible to discover if denaturation has occurred, resulting in inaccessible groups now being available. However, in favoured cases the analysis of titration curves has allowed the estimation of the number and type of the protonation sites available.

Enzymes

These are proteins which by reason of their particular three-dimensional structure are able to act as highly specific biochemical catalysts. The catalytic effect is considerable. Thus sometimes rate constants for enzyme-catalysed reactions and model reactions differ by as much as a factor of 10^{12}. A number of explanations have been suggested for this, and we shall consider some later. Because the enzymes are proteins the general problems of protein stability hold here, and therefore pH and temperature must be carefully controlled.

In general the cell requires a different enzyme for each of its reactions, although a limited number will catalyse reactions of a general type. The esterases will thus catalyse ester hydrolysis. The need for such high specificity can readily be seen. Life processes usually consist of a series of interrelated complex reactions. Often the product of one reaction becomes the starting material for another. The reactions must be very specific in order to avoid complications from other simultaneously occurring systems.

The need for extra factors in an enzyme reaction can often be shown by the process of dialysis, in which the enzyme solution, in a cellophane container, is suspended in distilled water. The cellophane pores allow the low molecular weight components to diffuse through into the surrounding water, leaving the protein molecules in the cellophane container. If the enzyme depends upon any of the low molecular weight substances it will fail to function until they are added back.

The enzyme activators can be metal ions or complex organic molecules such as nucleotides or certain B vitamins. These are termed coenzymes and are bound to the enzyme protein, being removed on prolonged dialysis. Sometimes a coenzyme is bound so firmly that it is not removed by dialysis, in which case it is termed a prosthetic group.

The mechanism of enzyme action

An old established analogy is that of the lock and key. This represents the complex three-dimensional relationship between an enzyme and the substrate on which it acts, a relationship which may require the incorporation of certain cofactors or activators before it is complete. The result of this is to activate the substrate so that it is able to react in the required way. This analogy is still useful in that X-ray studies on enzymes have shown, in all cases, the presence of a cleft at the active site into which the substrate must fit. The lock and key analogy is, however, inadequate in that it does not indicate the other effects which occur; for example, the substantial conformational changes that result when the substrate is bound to carboxypeptidase. The active site itself, that portion of the protein chain involved in the interaction with substrate, will only involve a few residues, although of course they may be from well-separated parts of the protein chain as a result of protein folding. It is the configuration of the protein around this portion that provides the cleft into which the substrate must fit to become activated. This explains the specificity of the enzyme. Why, for example, an enzyme may only be able to attack certain optical isomers. Thus malate dehydrogenase oxidizes L-malate exclusively to oxaloacetate in the presence of coenzyme 1 (NAD), while D-malate is unaffected, thus providing an effective method for obtaining the D- isomer from the DL form. The role of inhibitors can also be understood in a general sense, in terms of their modifying the overall molecular shape at the active site so that the substrate may not fit into it, or by their competing with the substrate for the active centre of the enzyme.

However, the portion of the enzyme known as the active site may not only act by generating a certain steric specificity. The nature of the side chains is also important in terms of their hydrophobic or hydrophilic properties. It appears that the active site is surrounded always by non-polar residues, thus providing an environment at the active site which is of lower dielectric constant than that of the aqueous solution in which the enzyme is found. This means that a number of interactions are different from what might have been expected; pK_a values are affected, ionic interactions will be stronger. All of these will be necessary for the correct functioning of the enzyme. Again, the active site provides the directional hydrogen bonding and van der Waals interactions to bind the substrate to the enzyme. In a number of cases X-ray studies have suggested that the initial interaction between substrate and enzyme induces further interactions that cause other parts of the enzyme to close in upon the substrate.

The actual source of catalytic power has been ascribed to a number of causes such as proximity and orientation effects and electron push–pull effects. It is, however, difficult to put these on a quantitative basis. One recent suggestion which has aroused some controversy is that of orbital steering, i.e. an orientation effect involving the orientation of orbitals in the reacting atoms of enzyme and substrate.

Nucleosides, nucleotides, and nucleic acids

Nucleosides are composed of a pyrimidine or purine base attached to the sugar ribose *via* the N-1 and N-9 atoms respectively. In nucleotides the sugar is linked to a phosphate group, which may be a mono-, di-, or triphosphate species. The structure of the important nucleotide adenosine triphosphate or ATP is given below to illustrate these structural features. This contains the purine, adenine.

The nucleic acids are polymers, built up from nucleotides *via* phosphodiester bond formation between the 3′-hydroxyl of one nucleotide and the 5′-hydroxyl group of the adjacent nucleotide. The sequence of nucleotides is extremely important as it constitutes the genetic code in DNA. Different nucleotides vary in the nature of the purine or pyrimidine base. The four important bases in RNA (ribonucleic acid) are the purines, adenine and guanine, and the pyrimidines, cytosine and uracil. In deoxyribonucleic acid (DNA), which is a polymer of 2′-deoxyribose nucleotides, 5-methyluracil (thymine) is present instead of uracil. These five bases are given below.

Adenine   Guanine   Cytosine   Uracil   Thymine

The biological roles of the nucleic acids and certain nucleotides are dependent on metal ions. In view of this, the interaction of nucleosides and nucleotides with metal ions has been much studied, particularly by NMR techniques. Base, phosphate and ribose groups are all potential ligands, but the sugar hydroxyl groups will be very weak donors. Phosphate will be a good ligand, while the bases will contain in appropriate cases both oxygen and nitrogen donors. Complexes of adenine nucleotides have been particularly well studied. Mg^{2+} and Ca^{2+} bind only to the β- and γ-phosphate of ATP, while Zn^{2+}, Mn^{2+}, Cu^{2+}, and Ni^{2+} also bind to the N-7 nitrogen atom of the base in addition to the phosphate. Recently the first example of a nucleoside coordinated to a metal *via* the sugar group has been reported, Cu_3 (2'-deoxyguanosine)$_2$ (OH)$_4$ 4H$_2$O.*

RNA is a single strand polymer, but DNA consists of two intertwined helices with each heterocyclic base hydrogen bonded to a base on the other strand in the famous 'double helix' structure. *Replication* of DNA involves unwinding into separate strands, with formation of complementary strands for each of the original two, so that two new DNA double helices are formed. The DNA acts as a template for the synthesis of messenger RNA in the process of *transcription*. This messenger RNA thus carries the genetic code, which is conveyed through trinucleotide sequences or 'codons', which code for a specific amino acid, e.g. the group CUU codes for phenylalanine. These codons associate with trinucleotide sequences in transfer RNA which contain the heterocyclic bases that are complementary to those in the codon. This takes place on the surface of ribosomes. The transfer RNA also binds the specific amino acid corresponding to the original codon. In the example given above, the transfer RNA for phenylalanine contains the nucleotide sequence GAA, which is complementary for CUU. Thus the code in the messenger RNA is *translated* into the correct sequence of amino acids, which is followed by the formation of peptide links. These essential processes of replication, transcription, and translation are dependent on metal ions.

The cell

The complexity of the cell is illustrated by the fact that not only are there many-celled organisms, but there are also one-celled species, such as bacteria, viruses, and moulds whose behaviour must be entirely accounted for by the activities occurring within that single cell. The following comments apply, in general terms, to animal cells and unicellular plants (green plants will also contain chloroplasts, associated with photosynthesis) but not to bacterial cells which differ in many respects, e.g. they do not possess a nucleus and therefore belong to the prokaryotic class of cells.

Cells, in general, are encased by a membrane, the function of which is extremely important and which is dependent on metal ions. This membrane is selectively permeable to different metal ions (and other species) and this is readily

*H. C. Nelson and J. F. Villa, *Inorg. Chem.*, **18**, 1725 (1979).

10

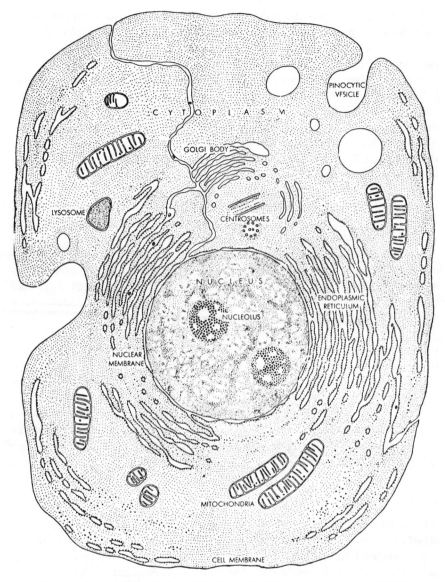

Figure 1.2 A typical cell (from *The Living Cell*, Brachet. Copyright © 1961 by Scientific American, Inc. All rights reserved).

associated with the function and distribution of, for example, the s block metal cations. Thus Mg^{2+} and K^+ are concentrated in the cell by the action of the membrane, while Na^+ and Ca^{2+} are rejected by it. This is associated with the utilization of Ca^{2+} as a structural factor in teeth, bones, and shells, and as an activator of extracellular enzymes, while Mg^{2+} and K^+ are associated with intracellular processes, both as structural stabilizers and also as enzyme activators. Plant cells are surrounded by an additional wall of cellulose giving extra rigidity.

The complex internal structure of the cell has been demonstrated by electron microscopy. Figure 1.2 shows a typical cell. In the cytoplasm of the cell are a number of structures, whose existence must be correlated with the various degradative and synthetic pathways involved in the working of the cell. All cells have a nucleus which contains the cell DNA, a little RNA, and certain associated proteins, together with relevant enzyme systems such as RNA and DNA polymerases and those associated with membrane synthesis. The nucleus occupies an appreciable part of the cell volume and is surrounded by its own membrane. The remaining cytoplasm contains a large number of particles, the mitochondria, which are smaller than the nucleus by a factor of up to a hundred, while the general body of the cell contains the tubules of the endoplasmic reticulum and small groups of particles, the ribosomes, which contain most of the cell RNA.

Many types of cells also contain a large number of hydrolytic enzymes, such as DNA-ases, phosphatases, and esterases, which are capable of destroying the cell components, but which are kept isolated by a membrane. Relatively large concentrations of the s block elements are associated with structural and functional aspects of the living cell, together with smaller quantities of the trace elements which are involved in enzyme activation. Other inorganic ions are also found in the cytoplasm, as are other enzyme systems.

The distribution of materials in the cell, the control of the influx of reactants and the efflux of products is dependent on the properties of the membranes of the cell and of the cell components. The properties of cell membranes have received much attention and are the subject of some controversy.

Membrane structure

The main constituents of membranes are lipids and proteins. For membrane preparations from red blood cells the protein constitutes some 65% of the total dry weight. For many years the membrane was thought to consist of a bimolecular lipid layer sandwiched between two protein layers, the total being some 60 Å thick. Lipid molecules have a polar head group (consisting of a phosphate and some other group), which will interact with the protein, and long hydrophobic, hydrocarbon tails. This gives membranes a central section of low dielectric constant which then acts as a barrier to the transport of ionic and hydrophilic molecules, for which specific transport pathways are then necessary.

The proteins of membranes are associated with important functions such as the transport of ions and small molecules, and electron transport and oxidative phosphorylation. The question of the localization of the protein has attracted much attention, and the model described above has now been replaced by one in which globular proteins are embedded in the lipid bilayer. This 'Fluid Mosaic' model is represented in Figure 1.3, which serves also to demonstrate the tail-to-tail structure of the bilayer.

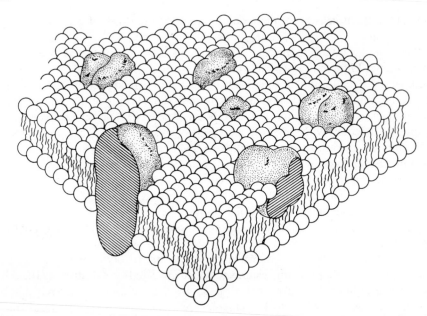

Figure 1.3 The Fluid Mosaic Model for the membrane (Reproduced by permission from *Science*, **175**, 720 (1972). Copyright (1972) by the American Association for the Advancement of Science).

These proteins are postulated either to span the lipid layer or to be embedded in it and so be accessible from one side only. These 'integral' proteins will always have hydrophilic and hydrophobic sections, so that the hydrophobic section is in contact with the lipid, while the hydrophilic layer would be exposed to the aqueous medium outside the membrane. Proteins that span the membrane would have two hydrophilic segments separated by a section of hydrophobic character which is long enough to span the membrane. Such a 'spanning' protein would thus provide a hydrophilic, water-filled channel through the membrane, entry to which could then be controlled to give specificity. Another class of membrane proteins is 'peripheral' proteins, which would be bound to integral proteins protruding from the membrane. These may be dissociated from it readily. There is evidence to show that cytochrome *c* is attached to the mitochondrial inner membrane at sites on both cytochrome *c* reductase and cytochrome oxidase proteins, both of which may well be integral proteins that protrude from the surface of the inner membrane.

The ordering of the lipid bilayers has been much studied by magnetic resonance techniques. The fluidity of such structures has been demonstrated by EPR, with lipids labelled with the free-radical nitroxide group (a spin label), and by NMR. These properties are temperature dependent. Thus the artificial bilayer dipalmitoyl lecithin shows a transition at 43 °C, above which temperature the hydrocarbon chains are in a liquid condition, and below which the chains are crystalline and motion is very restricted.

The cell components

It is obviously of interest to separate all the components of the cell and to observe their specific function. The biochemist attempts to do this by rupturing the cell membrane and applying the technique of centrifugation, the various particles thus being separated in turn. The separated cell fractions can then be examined independently for their enzymatic activity.

Of great biochemical interest are the mitochondria. They contain several sets of enzymes and cofactors not found elsewhere in the cell. In this way, in the mitochondria, all the enzymes and cofactors for certain important cycles, such as the breakdown of pyruvic acid by the citric acid cycle, are kept together in an highly organized system. Each mitochondrion is surrounded by an outer membrane. There is also an inner membrane with an irregular folded-type structure. Along this inner membrane is distributed a complex system involved in oxidation and phosphorylation. The complexity and interlinking of various biochemical processes certainly implies, in this case, a highly organized system of enzymes, coenzymes, and electron carriers. This is provided by attaching these systems to the mitochondrial membrane. The structure of the internal membrane increases the capacity of each mitochondrion for this ordered system. The result is a rigid system ideally arranged, for example, for hydrogen carrying, one that would not be achieved anywhere near so ideally if enzymes and cofactors were mixed in solution. The outer mitochondrial membrane is associated with such enzymes as succinic dehydrogenase and NAD dehydrogenase.

The control of the mammalian cell

The behaviour of the cell has to be regulated. As we have noted, the membrane plays a very important role in this by its selectivity towards outside species. However, there are external stimuli, such as nerve impulses and certain special chemicals (the hormones) secreted by the glands, which will affect the behaviour of membranes. Several of the hormones are important in controlling the concentrations of metal cations and other inorganic groups retained or disposed of through the urine *via* the kidneys. Thus the mineralo-corticoids, a group of adrenal cortical steroid hormones, are involved in increasing the loss of K^+ and decreasing the loss of Na^+. The parathyroid hormone stimulates the excretion of calcium and phosphate at the expense of bone calcium phosphate, while calcitonin inhibits the release of calcium.

The chemical reactions undergone in the cell are catalysed by the enzymes. The requirement for the appropriate coenzymes and cofactors is another controlling factor.

METALLOPROTEINS AND METAL–PROTEIN COMPLEXES

Most naturally occurring metal ions are bound by proteins. The question of the nature and extent of metal ion–protein interaction is therefore a most important one. It has been the subject of much work using amides, amino acids, and small peptides as model compounds, as discussed in Chapter 3.

It is possible to classify metal–protein systems under two headings. The classification is not an absolute one but is nevertheless a useful general guide to the nature of metal–protein interaction.

Metalloproteins (including metalloenzymes). Here the metal ion and protein are firmly linked together so that the metal ion can be regarded as an integral part of the protein structure, from which it cannot be separated except by extreme chemical attack. The activity of the metalloprotein is usually lost if the metal is replaced by another.

Metal-activated proteins or metal–protein complexes (including metal-activated enzymes). This group involves those examples where the metal ion is combined reversibly with the protein. Systems such as these are much less amenable to study than the metalloproteins for which it is usually possible to say something about the site symmetry of the metal and the nature of the binding groups that hold it to the protein. The metal ion–binding group interaction for this second class will of course be much less than that involved in the first class. This will be an important factor in the mechanisms of the reactions catalysed by these metal-activated enzymes. A suitable example is the role of Mg^{2+} in the hydrolysis of ATP by the appropriate enzyme. The binding of metal ions to free ATP is through the phosphate groups. The hydrolysis of ATP does not change considerably from metal to metal in model systems. However, the state of affairs is very different in the enzymatic reaction. Cd(II), Ni(II), and Co(II) have an inhibitory effect while Mg(II) and Mn(II) have a catalytic effect. This phenomenon demonstrates, first of all, the importance of the role of the protein. Catalysis is a reciprocal effect, not just one due to the presence of the metal ion. Secondly, it is probably safe to say that the inhibition by certain metals is due to their blocking of a certain site on the enzyme which is necessary for catalysis, while the weaker binding of the metals magnesium and manganese has resulted in this site being free.

Specific tests can be applied to ascertain to which type any particular protein–metal system belongs. For metalloproteins (and metalloenzymes) the ratio of metal to protein tends to a constant value as the purity of the system is increased, finally becoming independent of further purification when in a highly pure state. This ratio will then be the stoichiometric ratio of metal to protein molecule. The ratio may vary from one mole of metal ion per mole (such as zinc in carbonic anhydrase) up to, for example, eight moles of metal ion per mole, such as is the case for copper in ascorbic acid oxidase. Often for copper the metal ions occur in pairs, while enzymes such as aldehyde oxidase, and xanthine oxidase involve two different metal ions, in this case eight moles of iron and two moles of molybdenum. Another example is that of ferredoxin, an iron–sulphur protein, where there are eight iron atoms per mole of protein.

In the case of the metal-activated proteins, metal ion analysis will vary as the metal can be removed very easily, e.g. by dialysis. This will result in reduced activity, but it is usually readily recovered on adding the metal ion again. The specificity in these cases, towards the metal ion, is usually much lower than that of the metalloproteins. Metal-activated enzymes show a similar lower metal ion specificity and are usually easily inhibited by adding certain reagents. This in itself can provide useful mechanistic clues.

It is only possible to inhibit the action of a metalloenzyme by the addition of ligands of high activity for the metal ion and for its particular stereochemical environment. This is reversible when the metal still remains partly bonded to the protein molecule but it is often irreversible when the metal is removed completely from the protein. The resulting 'apoenzyme' (protein minus metal) can, however, in certain cases be reconstituted by adding the metal previously removed from the protein, but it can be very difficult to reconstitute with full activity.

The binding of metals in metalloproteins is a difficult question to resolve with certainty. There are many alternative donor sites, the side chains, peptide and terminal $-NH_2$ and $-COOH$ groups. It is reasonable to suppose, however, that certain groups will have particularly enhanced basic properties and tend to dominate the competition among the various potential binding sites for the metal. Two common amino acids have such outstanding binding properties—these are histidine, through its imidazole ring, and cysteine through

$$\begin{array}{cc} \overset{\displaystyle CH_2CH-COOH}{\underset{\displaystyle NH_2}{\mid}} & \overset{\displaystyle HSCH_2CH-COOH}{\underset{\displaystyle NH_2}{\mid}} \\ HN \quad N & \\ \text{Histidine} & \text{Cysteine} \end{array}$$

the thiol group. So the overall picture is probably less complicated than it might seem. There are many potential donor sites, but the cation must bind quite strongly to particular sites and only weakly to other sites to avoid having a number of alternative co-ordination compounds formed. Different metal ions would, of course, not necessarily be bound to the same ligand groups on the protein. For certain cases the full amino acid sequence of the protein molecule has been determined and the metal-binding groups pinpointed with some certainty. Again, X-ray analysis is most important. In some cases it is possible through the techniques of transition metal chemistry to make reasonable predictions regarding the metal environment. These methods are discussed in Chapter 3.

THE ROLE OF THE METAL IN METAL–PROTEIN SYSTEMS. SOME EXAMPLES

The first point that must be stressed is that the relationship between metal and protein is a reciprocal one. The presence of the metal ion can influence the electronic and structural arrangement of the protein and so affect its reactivity. But the very fact that a complex protein rather than a simple structure is required indicates that the protein is equally important. So the protein can enforce unusual stereochemistries upon the metal ion in that the protein internal structural requirements do not allow it to provide a normal symmetrical binding site, or even normal metal–ligand distances, in some cases. This, in turn, can affect the reactivity of the metal. Thus the irregular stereochemistry forced on the iron atom in myoglobin makes the addition of oxygen to the iron atom a more energetically attractive process in that the iron is already in the transition state structure for a certain type of ligand substitution reaction.

Metal ions can act in a number of different ways,

(a) in trigger and control mechanisms,

(b) in a structural context, including 'template' reactions,

(c) as Lewis acids,

(d) as redox catalysts.

We shall briefly treat these in turn, prior to a full treatment in Chapter 3. It should be emphasized that one should not always attempt to limit the behaviour of any one metal ion in an enzyme reaction to one of these roles. It should also be noted that there are a number of metal proteins which appear at present to have no known biological function. It could well be that these are storage proteins, there being many such systems having this role. The storage, control, and supply of certain factors in life processes is a critical one. Such proteins as conalbumin (in egg white) and transferrin (in blood plasma) reversibly bind and transport iron, while ceruloplasmin is a copper-carrying protein. The protein ferritin can store iron and then transport it to an appropriate site to aid in the biosynthesis of other molecules involving this particular metal. Other proteins control the concentrations of calcium, magnesium, zinc, and possibly other metals.

Certain other metal-containing macromolecules are coenzymes. The best known examples are those of the cobalt-containing vitamin B_{12} series. Here cobalt is bound in the centre of a tetrapyrrole ring, this being the basic structure of the cobamide family of coenzymes.

(a) Trigger and control mechanisms

The cations Na^+, K^+, Mg^{2+}, and Ca^{2+} are associated with a number of trigger and control mechanisms. The selective distribution of these cations inside and outside the cell means that a breakdown of this selectivity can act as a trigger for some biochemical event. Thus calcium, which is normally excluded from the cell cytoplasm, can enter the cell in response to an appropriate stimulus and trigger off a range of processes such as morphological changes, muscle contraction, and secretion of hormones, defence chemicals, neurotransmitters etc.

Nerve impulses can be described as electrical impulses conducted along the membrane. This is associated with the rapid influx of Na^+ into the nerve cell as a result of the temporary breakdown of the membrane's selectivity against Na^+.

(b) Structural influences

The binding of metal ions may control the conformation of biological macromolecules and so affect their chemical and biological properties. Thus certain metal ions stabilize the DNA double helix, as shown by the fact that the temperature at which DNA unwinds into single strands increases with metal ion concentration. Unwinding of the helix normally arises from the repulsion between negatively charged phosphate groups. Binding of cations to the phosphate neutralizes the charge and stabilizes the double helix. Divalent metal ions, particularly Mg^{2+}, are most effective in stabilizing the structure.

In contrast, Cu^{2+} favours the unwinding of the double helix, as Cu^{2+} binds to the base as well as the phosphate. Binding to the base prevents hydrogen bonding and so the double helix will be destabilized. Such unwinding of DNA by Cu^{2+} only occurs at low ionic strength. Thus addition of solid electrolytes can result in the reformation of DNA after unwinding by Cu^{2+}. The fact that the double helix can be regenerated under these circumstances reflects the fact that the two single strands are still cross-linked through binding to Cu^{2+} at some points so that the two strands are kept in close proximity to each other. Zn(II) is also able to bring about the reversible unwinding and rewinding of DNA on heating and cooling.

Metal ions are known to stabilize enzyme configurations. Thus Ca^{2+}, Mg^{2+}, and Mn^{2+} can influence the equilibrium between native and reversibly denatured protein. Again certain bacteria can function for short time periods at remarkably high temperatures. In some cases here it is suggested that the presence of high metal concentrations prevents denaturation of enzyme protein. Sometimes binding of metal ions induces conformational changes in a protein that give rise to the active site. Potassium ions appear to be important in this context, and many examples of K^+-activation of enzymes will probably involve such conformational changes. A further example involves glutamine synthetase which binds twenty-four moles of Mn^{2+} per mole of protein. The binding of the first twelve cations results in conformational changes that lead to the formation of twelve new sites for the binding of the remaining twelve cations, which then have a functional role.

The simplest function for a metal ion in an enzyme is purely that of a template to bring the reacting groups into the correct relative orientation for reaction. In this case the metal will normally be bonded to both protein and substrate, rather than either only to the protein or only to the substrate, i.e. E−M−S rather than M−E−S or E−S−M. The use of metal ions as templates in the synthesis of certain complex organic compounds has been demonstrated. Thus the two halves of corrin were synthesized separately and then cyclized by use of a nickel(II) template (Figure 1.4).

If metal ion cofactors in enzyme reactions have this type of role then it ought to be possible to understand the specificity of the metal ion in terms of the stereochemistry of non-enzymatic model compounds and the binding strength of different metals for certain groups. The similarity between Mg^{2+} and Mn^{2+} as enzyme activators can be understood on such a basis.

Metal ions are also important in maintaining the structure of cell walls. Calcium and magnesium are both metal ions whose concentrations are higher than that expected if their role was only that of enzyme activator. These divalent metal ions could produce a stiffening mechanism for lipoprotein membranes by

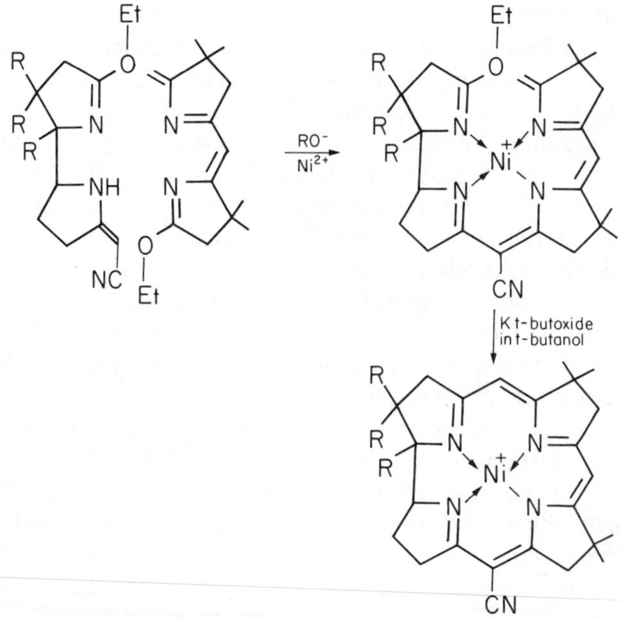

Figure 1.4 The use of nickel(II) as a template ion.*

bridging neighbouring carboxylate groups. Ca^{2+} is particularly well suited for this role.

(c) Lewis acid behaviour

Metal ions can accept electron pairs and so act as Lewis acids. On coordination to the metal the ligand will be polarized with increase in its reactivity. Different metal ions have different Lewis acid strengths, which will clearly increase with the charge on the metal and with decreasing ionic radius. For the transition metals other factors are also important, and so for a series of divalent ions the following order generally holds:

$$Mn^{2+} < Fe^{2+} < Co^{2+} < Ni^{2+} < Cu^{2+} > Zn^{2+}$$

This order of activity is observed for the reactions of most model compounds involving Lewis acid catalysis by metal ions, but does not hold completely for metal ion catalysis of certain enzyme reactions. The breakdown of this order is significant and will be discussed later.

The mechanistic role of the metal ion then is one of general acid catalysis, but differs from proton catalysis in that (a) the metal ion can coordinate to several ligands simultaneously, and (b) metal ion catalysis is possible in pH ranges where proton catalysis would be ineffective.

* A. Eschenmoser, R. Scheffold, E. Bertelle, M. Pesaro, and H. Gschwind, *Proc. Roy. Soc. A,* **288**,306 (1965).

An example of an organic reaction catalysed by a metal ion in this way is the decarboxylation of dimethyl oxaloacetic acid. This is catalysed by ferric ion. The scheme is given below:

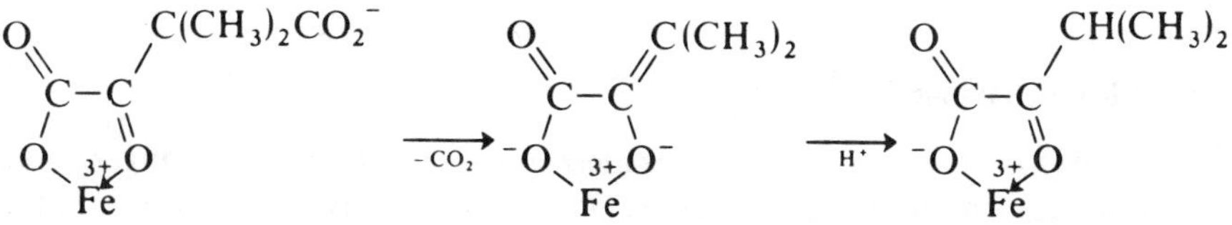

The enol chelation is confirmed by the appearance of a blue colour, due to a charge-transfer band, that is characteristic of ferric enolate complexes. The role of the metal ion can be easily understood, the Lewis acid behaviour of Fe^{3+} allowing the CO_2 group to leave more easily.

Zinc and cobalt are good examples of strong Lewis acid catalysts, for example, in the hydrolysis of phosphates (by phosphatases) and esters (by esterases), while Mg^{2+}, Ca^{2+}, Mn^{2+}, as has already been seen, are good catalysts for substrates involving weaker base centres, such as polyphosphates.

Magnesium is particularly well known as an activator of those enzymes associated with phosphate systems. The formation of a phosphate ester is an important biochemical reaction and is often the first step in a complex synthetic sequence. Mono-, di-, and triphosphate esters can all be formed.

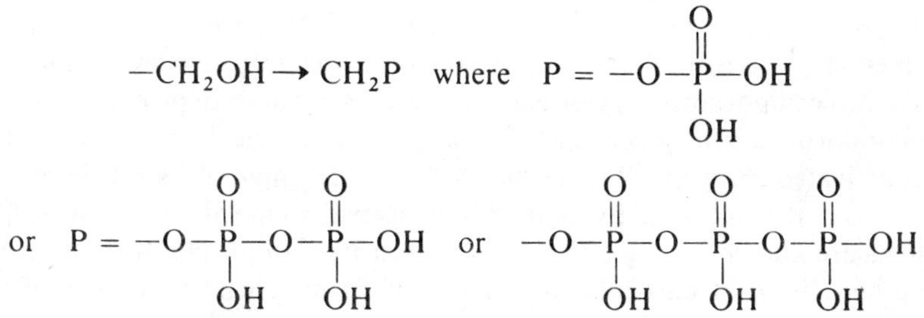

$$-CH_2OH \rightarrow CH_2P \quad \text{where} \quad P = -O-\overset{\overset{\displaystyle O}{\|}}{\underset{\underset{\displaystyle OH}{|}}{P}}-OH$$

$$\text{or} \quad P = -O-\overset{\overset{\displaystyle O}{\|}}{\underset{\underset{\displaystyle OH}{|}}{P}}-O-\overset{\overset{\displaystyle O}{\|}}{\underset{\underset{\displaystyle OH}{|}}{P}}-OH \quad \text{or} \quad -O-\overset{\overset{\displaystyle O}{\|}}{\underset{\underset{\displaystyle OH}{|}}{P}}-O-\overset{\overset{\displaystyle O}{\|}}{\underset{\underset{\displaystyle OH}{|}}{P}}-O-\overset{\overset{\displaystyle O}{\|}}{\underset{\underset{\displaystyle OH}{|}}{P}}-OH$$

Almost all enzymes that are involved in phosphate transfer require a metal, but magnesium is the most important. It is possible that the role of the metal ion may also be one of charge neutralization. Adenosine triphosphate is usually the substrate.

Examples of ATPases which require magnesium as a cofactor are myosin, which cleaves ATP to ADP (adenosine diphosphate) and inorganic phosphate, and hexokinase which is involved in the phosphorylation of glucose.

$$\text{glucose} + \text{ATP} + \text{hexokinase} \xrightarrow{\quad Mg^{2+} \quad} \text{glucose-6-phosphate} + \text{ADP}.$$

Zinc is also associated with hydrolytic enzymes, in particular with phosphatases, peptidases, and esterases. Zinc-containing systems have been studied with some success because it is often possible to replace zinc by a range of transition metal ions and hence obtain useful comparative information. Detailed reference will be made in Chapter 4 to the role of zinc in certain systems such as carbonic anhydrase in order to illustrate the usefulness of this approach. Carbonic anhydrase is vital for respiration in animals as it catalyses the normally slow carbonic acid–carbon dioxide reaction.

(d) Redox behaviour

In these reactions redox changes in the metal ions catalyse valence changes in the substrates, such as in the nitrogen cycle. Here the specificity of the enzyme for the metal tends to be much higher. The transition metal ions are involved in a wide range of catalytic functions, while zinc also has a role as a catalyst of hydride transfer. A characteristic feature of these redox metalloenzymes is the irregular stereochemistry associated with the metal. This is very important in explaining their function. The redox potentials of the metal ions in these metalloenzymes are also of interest and may be related to the irregularity of the metal environment.

The more important metals in biological redox processes are iron, copper, and cobalt with molybdenum involved to a lesser degree. The processes vary from electron transfer, oxygen atom and hydroxyl group incorporation to hydrogen atom and hydride ion removal.

Copper and iron are extremely important metals in biology. Both are involved in respiratory processes; iron in hemoglobin and copper in hemocyanin are oxygen carriers, for example. Iron is a component of various cytochromes, peroxidases, and catalase, which are all porphyrin enzymes. There are also iron transport enzymes (ferritin and transferrin) and concentration-controlling enzymes (transferrin). Ferredoxin is a non-heme iron–protein complex of considerable importance, having eight iron atoms per mole of protein. It is found in the chloroplasts of green plants and is the initial electron acceptor of the photoactivated chlorophyll molecule. A further example of its role in electron transfer is in the biological fixation of dinitrogen. A range of other iron–sulphur proteins are known having one, two, and four iron atoms per mole of protein. These have been of particular interest to inorganic chemists in view of their interesting structural and magnetic properties.

Copper proteins are widespread. In addition to its role in hemocyanin, copper is involved in cytochrome c oxidase in the respiratory chain, in blue copper proteins that catalyse electron transfer, in blue oxidases that catalyse the reduction of O_2 to H_2O, and in non-blue oxidases that catalyse the reduction of O_2 to H_2O_2. The most well-studied copper protein is probably ceruloplasmin, whose exact physiological function is still not known with confidence. It shows important oxidase properties, and it has also been suggested that it is associated with the incorporation of iron into apotransferrin. Alternatively it may be a copper transport protein. Another copper-containing protein of current interest is superoxide dismutase, once known as erythrocuprein. It is now suggested that the physiological function of this protein is to catalyse the disproportionation of superoxide ion, a supposedly toxic species, but this view has been challenged.

Molybdenum is involved in some half dozen redox enzymes. It is particularly well known for its involvement in the biochemistry of the Nitrogen cycle, both in fixation of dinitrogen and in the reduction of nitrate ion. In many cases it is noteworthy that molybdoenzymes catalyse the transfer of oxo groups (in $NO_3^- \rightleftharpoons NO_2^-$ and $SO_3^{2-} \rightleftharpoons SO_4^{2-}$, for example). This is interesting in the light of the fact that a dominant feature of the chemistry of molybdenum is its tendency to form oxo complexes. This could be the reason for the selection of molybdenum in these enzymes, as it is the only second row transition element to have a major biological role.

Another transition element, manganese, does not in fact appear to be involved in redox reactions. As manganese(II) it is essential for the activity of many degradative enzymes. Manganese(II) is very similar to magnesium(II) and often can replace that metal. This has been used with some profit as manganese(II) can be studied by EPR and other techniques. Manganese, and possibly copper, are involved in photosynthesis in that they appear to be essential for the biosynthesis of chlorophyll.

Even though it is not a transition metal, zinc is also found in certain dehydrogenases. Its role here is probably one of catalysis of hydride transfer. It is known in organic chemistry that pyridinium compounds react as if the *para*-carbon were positively charged; thus they can add hydride ions. In biological systems the most common pyridinium compound which carries out this reaction is the pyridine nucleotide coenzyme NAD^+ (nicotinamide adenine dinucleotide).

$$\text{NAD}^+ \qquad\qquad \text{NADH}$$

The best studied enzyme is the alcohol dehydrogenase. In the past, the detailed role of the zinc has been unclear, and schemes have been suggested in which the NADH was bound to the zinc while others have postulated the binding of the alcohol substrate. More recently conclusive evidence has been put forward for the latter proposal, in which the zinc promotes hydride transfer to NAD^+ *via* the formation of a zinc(II) alcoholate complex.

THE ADVANTAGES ASSOCIATED WITH THE PRESENCE OF A TRANSITION METAL ION

The presence of a transition metal ion in a metalloenzyme enables a wide range of physical techniques to be applied, giving information that cannot be obtained for metal–enzymes where the metal is magnesium, calcium, zinc, etc. Because of this it is very worthwhile to replace these metals by transition metals. Thus zinc can usually be replaced by cobalt and magnesium by manganese. In these cases the new metal ion probably occupies a similar site to the original metal. But the more metals that can be substituted, the more the comparative information that can be obtained. The presence of a transition metal, with its unfilled d orbitals, allows the standard techniques of transition metal chemistry to be applied to the problem of the site symmetry and the type of binding groups utilized. These are the techniques of magnetochemistry and electronic spectra, together with various other more specialized spectroscopic techniques. Certain non-transition metals may also be used as probes; for example, thallium(I) is an NMR probe for K^+.

The replacement of one metal by another has to be considered carefully, as, apart from the examples cited, there is always the possibility that the new metal may bind to different groups or tend to favour different stereochemistries. Again, the strength of the metal–ligand bond could be very important in determining the reaction pathway. This could vary with the metal.

It is very important that there are quantitative assessments of the interactions between metal ions and potential ligands. This is measured by stability or formation constants and these are most important in the context of biological inorganic chemistry. There are now extensive compilations of formation constant data which provide some very useful information.

FURTHER READING

Many excellent articles will be found in the following series:
Advances in Protein Chemistry, Academic Press, New York.
The Proteins (Ed. H. Neurath), (2nd end.) Academic Press, New York.
The Enzymes (Ed. P. D. Boyer), (3rd edn.) Academic Press, New York.

Detailed discussion of topics of bioinorganic interest will be found in the following texts and series:
G. L. Eichhorn (Ed.), *Inorganic Biochemistry*, Elsevier, 1973.
H. Sigel (Ed.), *Metal Ions in Biological Systems*, Vol. 1–10, Dekker.
C. A. McAuliffe (Ed.), *Techniques and Topics in Bioinorganic Chemistry*, Macmillan, 1975.
D. R. Williams (Ed.), *An Introduction to Bioinorganic Chemistry*, C. C. Thomas, Ipp., 1976.
K. N. Raymond (Ed.), *Bioinorganic Chemistry II*, Advances in Chemistry Series, No. 162, 1977.
R. J. P. Williams and J. R. R. F. da Silva, *New Trends in Bioinorganic Chemistry*, Academic Press, London, 1978.
S. J. Singer and G. L. Nicholson, *Science*, **175**, 720 (1972) (The Fluid Mosaic Model of Membrane Structure). Also S. J. Singer, *J. Coll. Interfac. Sci.*, **58**, 452 (1977).

The series *Progress in Inorganic Chemistry, Structure and bonding*, and *Coordination Chemistry Reviews* include reviews of bioinorganic topics, while a Chemical Society Specialist Periodical Report on *Inorganic Biochemistry* is now available (Ed. H. A. O. Hill, 1979, Vol. 1). Another new series is *Advances in Inorganic Biochemistry*, G. L. Eichhorn and L. G. Marzilli (Eds), Elsevier-North Holland, **1** (1980).

CHAPTER TWO

PROPERTIES OF TRANSITION METAL IONS

The important transition elements in biological processes are the redox catalysts iron, copper, cobalt, and molybdenum. Manganese is also important, although its function is rather different. The study of metalloproteins containing these ions is made easier by the presence of the transition metal ion, as its characteristic properties allow the study of the biological metal-binding site by the use of instrumental techniques. In this chapter it is hoped to summarize the aspects of transition metal chemistry that are relevant to the study of metalloenzymes, that is those properties that give information on the symmetry of the metal-binding site, the nature of the binding groups, and the electronic state of the metal. Some comments on the mechanisms of reaction of metal ions and complexes will also be included.

Coordination numbers and stereochemistry

The transition elements are those whose ions have incompletely filled d orbitals. Their characteristic properties include the formation of coloured complexes, a range of oxidation states, and the formation of paramagnetic compounds, indicating the presence of unpaired electrons. The elements of the first transition series, in which the $3d$ orbitals are being filled, tend to have a maximum

TABLE 2.1 Stereochemistries and oxidation states of some $3d$ elements

Oxid. state	C.N.	Stereochemistry	Oxid. state	C.N.	Stereochemistry
$Cu^I(d^{10})$	2	Linear	$Co^{II}(d^7)$	4*	Tetrahedral
	3	Planar		4	Square planar
	4*	Tetrahedral		5	Trigonal bipyramidal
$Cu^{II}(d^9)$	4*	Square planar		5	Square pyramidal
	4	Distorted tetrahedral		6*	Octahedral
	5	Square pyramidal	$Co^{III}(d^6)$	4	Tetrahedral
	5	Trigonal bipyramidal		5	Square pyramidal
	6*	Distorted octahedral		6*	Octahedral
$Ni^{II}(d^8)$	4*	Square planar	$Fe^{II}(d^6)$	4	Tetrahedral
	4*	Tetrahedral		6*	Octahedral
	5	Trigonal bipyramidal	$Fe^{III}(d^5)$	4	Tetrahedral
	6*	Octahedral		6*	Octahedral
$Co^I(d^8)$	4	Square planar	$Mn^{II}(d^5)$	4	Tetrahedral
	5	Trigonal bipyramidal		4	Square planar
	6	Octahedral		6*	Octahedral

*Common states.

coordination number of six. The most common coordination numbers are four and six, although five is quite often encountered. The second and third row transition elements can increase their coordination numbers beyond six.

Four-cordinate complexes may have either tetrahedral or square-planar stereochemistry, while six-coordinate complexes are usually octahedral. In many cases, and often in biological systems, these structures will be distorted. Five-coordinate compounds may have either a square-pyramidal or a trigonal-bipyramidal structure. In some compounds there is a rapid interconversion between these two structures, while in others the structure may lie between these two extremes. Table 2.1 lists the stereochemistries observed for the biologically important elements of the first transition series.

BONDING IN TRANSITION METAL COMPLEXES

The coordinate bond may be simply described in terms of electron-pair donation to the metal ion from the coordinating group or ligand. A more detailed description of the bonding is put forward in a number of bonding theories that will account semiquantitatively or quantitatively for the magnetic and spectroscopic properties of complexes.

Crystal field theory

This approach neglects covalent bonding between metal ion and ligand and assumes that the interaction is only electrostatic, the ligands being treated as point charges. While this assumption in incorrect this theory accounts semiquantitatively for many aspects of transition metal chemistry.

Figure 2.1 schematically represents the orientation of the five d orbitals in space. It may be seen that these are not all spatially equivalent. Three orbitals, the d_{xy}, d_{yz}, and the d_{zx} orbitals forms one group, with their electron density distributed between the axes, while the other group, made up of the $d_{x^2-y^2}$ and d_{z^2} orbitals, has the electron density lying along the axes.

In the formation of complexes, as the negative (or polarized) ligand approaches the metal ion along the cartesian axes, so the orbitals of the metal ion will rise in energy. Not all the orbitals will be affected in the same way. If we consider an octahedral complex, then it may readily be seen that the more stable of these two sets of orbitals will be the one involving the d_{xy}, d_{yz}, and d_{zx} orbitals (termed the t_{2g} orbitals) because the electrons in these orbitals will experience less repulsion from the electrons of the ligand. The other orbitals, the $d_{x^2-y^2}$ and $d_{z^2}(e_g)$ lying along the axes, will be more destabilized. Thus the crystal field causes the splitting of the degenerate d orbitals of the free ion into two groups, one a set of triply degenerate orbitals and the other a set of doubly degenerate orbitals. The splitting between these sets of orbitals is known as the crystal field splitting (Δ_o or $10Dq$) and may be determined from electronic spectra. The splitting of d orbitals by an octahedral field is shown in Figure 2.2, where the energy of the unsplit orbitals is that corresponding to the hypothetical situation in which the

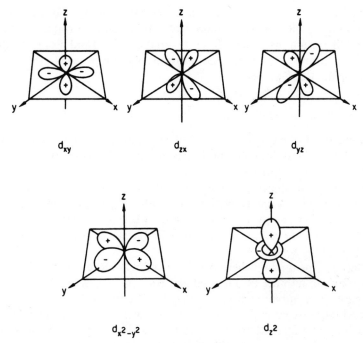

d_{xy} d_{zx} d_{yz}

$d_{x^2-y^2}$ d_{z^2}

Figure 2.1 The d orbitals.

ion is surrounded by a spherically symmetrical crystal field. Figure 2.2 also shows the splitting of the d orbitals by a tetrahedral field. In this case the splitting is reversed compared to the octahedral situation, and the value of Δ_t is much smaller ($\Delta_t = -4/9\Delta_0$).

Regular stereochemistries are not always observed. In octahedral complexes two *trans* ligands may be nearer or further away from the metal ion than the other four, so giving a tetragonal distortion. In the case when the two ligands on the z axis are further away than the other four, the degeneracy of the e_g orbitals

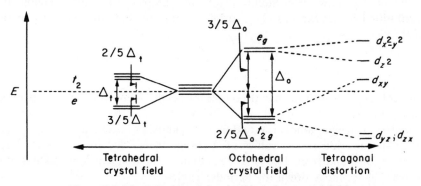

Figure 2.2 The splitting of the d orbitals by crystal fields.

is removed as the d_{z^2} orbital becomes more stable than the $d_{x^2-y^2}$ orbital. Similarly the d_{yz} and d_{zx} orbitals become more stable than the d_{xy} orbital, as shown in Figure 2.2. In certain cases the d_{z^2} level crosses the d_{xy} level and becomes almost as stable as the d_{yz} and d_{zx} pair. Severe distortions of this type lead to the formation of square-planar complexes.

The number of unpaired electrons

The arrangement of electrons in an unfilled shell is governed by Hund's rules, which state that the stable configuration is the one with the maximum number of unpaired electrons, arranged with parallel spins. However, this state of affairs is complicated by the splitting of the d orbitals, as we now have to consider the possibility that it is energetically more favourable to pair electrons than to place them in the higher energy set of d orbitals. Two general cases exist:

(a) *the strong-field case*, where Δ is high and hence electrons are paired (low-spin complexes).

(b) *the weak-field case*, where Δ is low, and it is more favourable to have a maximum number of unpaired electrons (high-spin complexes).

The number of unpaired electrons may be determined by the use of a Gouy balance. If there are no unpaired electrons complexes will be diamagnetic, while the presence of unpaired electrons results in paramagnetic behaviour. The measured magnetic moment will indicate the number of unpaired electrons present.

For octahedral complexes only one possible arrangement of electrons exists for d^1, d^2, d^3, d^8, and d^9 configurations, but both high- and low-spin possibilities exist for configuration d^4 to d^7, as shown in Figure 2.3. d-Orbital splitting will therefore explain why change of ligand can produce complexes with different magnetic properties. Thus $[CoF_6]^{3-}$, a weak-field species, is paramagnetic, while $[Co(NH_3)_6]^{3+}$ is diamagnetic, as are practically all cobalt(III) complexes.

A similar treatment may be applied to tetrahedral complexes. In theory d^3-d^6 species should show both high- and low-spin complexes. However, in practice, as Δ_t values are low, only high-spin complexes are formed.

Figure 2.2 shows why square-planar d^8 complexes are diamagnetic, the separation between the d_{z^2} and $d_{x^2y^2}$ orbitals being greater than the pairing energy of the electrons.

The spectrochemical series

The colour of transition metal complexes results from electronic transitions between the two sets of d orbitals. It follows therefore that the $d \rightarrow d$ spectra of complexes will give values of Δ for different ligands. For octahedral complexes of the first transition series, values of Δ_o range from 7500 to 12,500 cm^{-1} for divalent ions, and from 14,000 to 25,000 cm^{-1} for trivalent ions. Values of Δ increase as we move down the Periodic Table.

If all other factors are kept constant the value of Δ will depend only on the

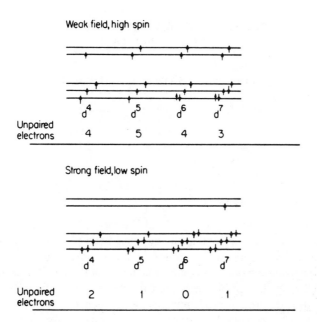

Figure 2.3 Weak- and strong-field arrangements for octahedral complexes.

ligand. If the ligands are listed in order of increasing Δ we have the *Spectrochemical Series*, which then gives a relative assessment of the ability of ligands to cause d-orbital splitting. Some common ligands lie in the following order in this series: $I^- < Br^- < Cl^- < -SCN^- < F^- < OH^- <$ oxalate $\leq H_2O < -NCS^- \leq NH_3 \simeq$ pyridine $<$ ethylenediamine $<$ dipyridyl $< o$-phenanthroline $< NO_2^- < CN^- \simeq CO$. Some ligands of biological interest lie in the sequence: $SH < CO_2^- <$ amide $<$ imidazole. The positions of the ligands at the strong and weak ends of this series reflect the phenomenon of π bonding between ligand and metal, the two cases being considered later. The existence of π bonding is not allowed for in the crystal field postulates. The ligands in the central part of the series are in sequence in terms of their polarizing power, either as an anion or as a dipolar species.

The energies of the bands in the electronic spectrum of a complex will reflect the positions of the ligands in the spectrochemical series. If several different ligands are present, then the measured value of Δ for the complex will represent an averaging out of the individual Δ values. Thus the bands in the spectrum of ML_4Br_2 will lie at lower energies than the corresponding bands in the spectrum of ML_4Cl_2. Again, if we are uncertain of the linkage of the thiocyanate ion in the complex $ML_4(SCN)_2$, then the positions of the bands compared with those of the chloro complex will indicate whether the thiocyanate is nitrogen or sulphur bonded. It may be seen therefore that, if the electronic spectrum of a complex is known, it may be possible to make predictions concerning the nature of the binding groups by a consideration of the measured crystal field splitting.

TABLE 2.2 Crystal field stabilization energies for high-spin complexes.

No. of d electrons	Octahedral	Tetrahedral
1, 6	$\frac{2}{5}\Delta_o$	$\frac{3}{5}\Delta_t$
2, 7	$\frac{4}{5}\Delta_o$	$\frac{6}{5}\Delta_t$
3, 8	$\frac{6}{5}\Delta_o$	$\frac{4}{5}\Delta_t$
4, 9	$\frac{3}{5}\Delta_o$	$\frac{2}{5}\Delta_t$
(0), 5, (10)	0	0

Crystal field stabilization energies

The placing of electrons is the t_{2g} orbitals in octahedral complexes means that they are $\frac{2}{5}\Delta_o$ lower in energy than if they had been placed in the hypothetical degenerate orbitals. Hence a d^1 complex would be more stable than predicted by a simple electrostatic model. This energy is the Crystal Field Stabilization Energy (CFSE). This can be readily calculated for various d configurations from the orbital splitting diagram for the appropriate stereochemistry. Values are given in Table 2.2. Thus, for an octahedral complex, $\frac{2}{5}\Delta_0$ is added for each electron in the t_{2g} level and $\frac{3}{5}\Delta_0$ is substrated for each electron in the e_g level. Clearly for a d^5 high-spin configuration there will be no CFSE, and the values for d^n and d^{n+5} configurations will be the same, except for the pairing energies associated with d^{n+5} configurations.

The magnitude of crystal field stabilization energies can be significant, and is reflected in a number of areas of transition metal chemistry, including redox potentials (CFSE will vary from one oxidation state to another), coordination number, kinetics (highly stabilized complexes will react slowly), and in a range of thermodynamic quantities, including formation constants. Formation of complexes with nitrogen ligands will confer additional stability compared to the aquo

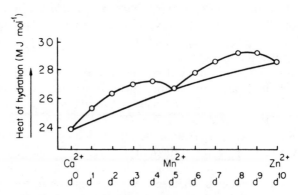

Figure 2.4 Variation in the heat of hydration with d configuration for M^{2+} ions.

complex, provided the ion is subject to CFSE. So for Mn^{2+}, d^5 configuration, complex formation does not result in increased stabilization. A detailed example is given for hydration energies (Equation 2.1). These have been measured for the divalent ions of the first transition series, and are shown in Figure 2.4.

$$M^{2+}_g + 6H_2O \rightleftharpoons [M(H_2O)_6]^{2+}_{aq} \qquad (2.1)$$

The results for the d^0, d^5, and d^{10} systems (no CFSE) fall on a smooth curve. The others lie above this curve by an amount equivalent to the CFSE. This may be calculated and subtracted from the hydration energies, in which case the new values also fall on the curve. Alternatively, those results could have been used to provide independent checks on Δ_o values obtained from electronic spectra.

Ligand field theory

There is good evidence for suggesting that metal orbital–ligand orbital overlap is important. There are two general approaches to the problem of allowing for covalency in the metal–ligand bond. In *Ligand Field Theory* covalency is not formally introduced, but is allowed for by the adjustment of various parameters. For complexes in which the metal is in a normal oxidation state this usually works well as the amount of orbital overlap is small. The alternative approach, that of *molecular orbital theory*, begins with the assumption that overlap of orbitals does occur. By its very nature, however, this theory can allow for differing degrees of overlap and so can cover all possible cases, from the electrostatic picture to that of maximum overlapping of orbitals, together with all intermediate possibilities.

There are three parameters of particular importance in ligand field theory. These relate to various interelectronic interactions and are the spin–orbit coupling constant, λ, and the Racah parameters, B and C, associated with electron repulsion. Fuller reference will be made to these later. The spin–orbit coupling constant is very important in determining the detailed magnetic properties of many transition metal complexes.

The presence of covalency means that the electrons of the ligand will be partly transferred to the metal ion so that the effective positive charge on the metal will be reduced. This will allow the d-electron clouds of the metal to 'expand' outwards from the metal ion. This has been termed the nephelauxetic effect and will vary from ligand to ligand. It is customary to arrange ligands in order of their tendency to cause cloud expansion—this is the nephelauxetic series, such as $F^- < H_2O < NH_3 <$ ethylenediamine $< -NCS^- < Cl^- \sim CN^- < Br^- < I^-$. Repulsion between electrons will be reduced as a consequence of cloud expansion and this is reflected in lower values of the Racah parameters in complexes than in the free ion. It is customary to express values for the complex by B' and C', usually $B'/B \sim 0.7$. Values of B' may be obtained from electronic spectra and provide the basis for the arrangement of ligands in the nephelauxetic series.

Molecular orbital theory

The basic principles are those applied to simpler molecules. Molecular orbitals are constructed by linear combination of atomic orbitals, subject to the appropriate symmetry and energy considerations. Both σ and π molecular orbitals may be formed, but we shall first discuss the bonding in an octahedral complex considering σ interactions only.

The metal can provide nine orbitals, five $3d$, one $4s$, and three $4p$. Of these, the t_{2g} orbitals (d_{xy}, d_{yz}, d_{zx}) will not be suitable for σ bonding as they are not orientated towards the ligand orbitals. The six ligands will each provide an appropriate orbital, which will contain two electrons. We then allow each of the six metal orbitals to overlap with the appropriate symmetry σ orbital, two combinations being possible leading to bonding and antibonding (signified by *) molecular orbitals. This is represented schematically in Figure 2.5. The orbitals are given their appropriate symmetry designations. These are group theoretical in origin and will be used here only as convenient labels to identify the symmetry class to which the metal and ligand orbitals belong. It may be noted, however, the symbols a_{1g}, e_g, and t_{1u} represent sets of singly, doubly, and triply degenerate orbitals.

These molecular orbitals, as they are not constructed from atomic orbitals of equal energy, will resemble one atomic orbital more than the other. The six bonding MO's will have more ligand character than metal character while the converse is true of the antibonding MO's, while obviously the t_{2g} orbitals remain metal orbitals.

We now feed in the available electrons in accord with Hund's rules. Twelve electrons will occupy the six bonding MO's, these may be regarded as essentially ligand electrons. The remaining electrons, equal to the number of d electrons of the metal ion, will then occupy the t_{2g} and e_g^* orbitals. In effect then the situation closely resembles that obtained by crystal field theory, provided we compare the

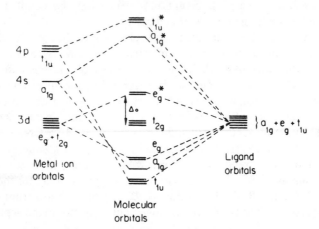

Figure 2.5 Qualitative molecular orbital scheme for an octahedral complex.

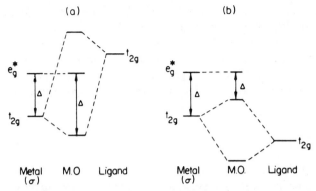

Figure 2.6 Interaction of ligand orbitals and metal t_{2g} orbitals.

e_g orbitals of the metal ion in crystal field theory with the e_g^* MO's of molecular orbital theory. It is then possible qualitatively to account for magnetic and electronic spectral properties in a similar fashion to that of crystal field theory.

π bonding

This possibility must be considered whenever the ligand has suitable $p\pi$ or $d\pi$ atomic orbitals or π molecular orbitals available for overlap with the t_{2g} orbitals of the metal ion.

A number of situations are possible, depending upon whether the ligand π orbitals are filled or empty and whether or not they are higher or lower in energy than the metal t_{2g} orbitals. When ligands contain donor atoms oxygen and fluorine then the only π orbitals available are filled p orbitals, of lower energy than the metal t_{2g} orbitals. Other ligands have higher energy empty π orbitals. The two cases are depicted in Figure 2.6. It may be seen that the effect of (a) is to increase Δ, while in (b) Δ is decreased. Thus π bonding is important in understanding the positions of ligands in the spectrochemical series.

ELECTRONIC SPECTRA

Charge-transfer bands

These intense bands sometimes appear in the visible region of the spectrum but are usually found in the UV region, often tailing off into the visible, however. They are associated with the transfer of an electron from an orbital of one atom to an orbital of another. It is possible for the charge-transfer to be localized on the ligand (e.g. SCN^-) but usually it involves ligand and metal ion. Charge may be transferred from the ligand to the metal in which case electrons in either the σ-bonding orbitals or orbitals of the ligand are excited to the empty t_{2g} or antibonding MO's of the complex. Alternatively, electrons of the σ-bonding orbitals of the complex may be promoted to empty π orbitals of the ligand, in which case metal → ligand charge transfer has occurred.

The energy at which the charge-transfer band appears will depend on both metal ion and ligand, and the relative ease of oxidation or reduction of these species. For a constant metal ion, the charge-transfer band will be diagnostic of the ligand. In the $4d$ and $5d$ transition series, as Δ values are larger, it is sometimes more difficult to sort out charge-transfer and $d \rightarrow d$ bands. In all cases, comparison with the spectrum of a non-transition metal complex may help in the assignment.

$d \rightarrow d$ spectra

The energy difference between the e_g and t_{2g} orbitals of the metal ion is such that the excitation of an electron from a lower to a higher level can be achieved by the absorption of visible light, so accounting for the colour of transition metal complexes. The bands in the visible and near-infrared regions of the spectrum, with the exception of ligand infrared overtones, are therefore the $d \rightarrow d$ bands and are usually weak in intensity. They provide a great deal of information about the structure of the complex.

Crystal field theory can offer a qualitative explanation of the $d \rightarrow d$ transitions in transition metal complexes. A d^1 system, such as Ti(III), should only show one such transition, corresponding to the excitation of the single electron from a t_{2g} to an e_g orbital. This band is seen at around $20,000 \text{cm}^{-1}$ in the spectrum of $[\text{Ti}(\text{H}_2\text{O})_6]^{3+}$ and is a direct measure of Δ_0. The spectrum of a d^9 system, such as Cu(II), may also be simply interpreted. Here, however, it is said that the Cu(II) ion is a one positron ion (or one 'hole'), and that one band should be observed corresponding to the $e_g \rightarrow t_{2g}$ transition of the positron. A number of complicating features in these two cases will be discussed later.

The interpretation of the spectra complexes of other d configurations is less straightforward. It is necessary to consider the energy levels derived from each d configuration for the free ion. For any particular d^n species ($2 < n < 8$) not all the arrangements of the electrons among the d orbitals are of equal energy. This is because the electrons repel each other differently in different orbitals. Thus each d configuration will give rise to a number of energy levels or *terms*, each of which is usually degenerate. The full calculation of the relative energies of these terms is complex. They are described in terms of the Racah parameters B and C, which are in themselves composite quantities.

The terms are characterized by the total spin angular momentum and the total orbital angular momentum quantum numbers S and L respectively. (S and L for the term corresponds to s and l for the individual electron in an atomic orbital.) This involves the use of the Russell–Saunders coupling scheme which assumes that spin and orbital angular momenta do not interact. Terms with L values 0, 1, 2, 3, 4... are designated as S, P, D, F, G etc. and the classification is completed by reference to the spin multiplicity $(2S + 1)$, the number of ways in which the unpaired electrons may be arranged. Thus if $S = 1$ the spin multiplicity is 3, a triplet. If $S = \frac{3}{2}$ and $L = 2$, the term is a ^4D term. The energy of the substates making up the term is given by the vectorial addition of L and S. This is represented by J. Thus if $L = 2$, the four ways of arranging the three unpaired electrons in the ^4D term are $+\frac{1}{2}, +\frac{1}{2}, +\frac{1}{2}; +\frac{1}{2}, +\frac{1}{2}, -\frac{1}{2}; +\frac{1}{2}, -\frac{1}{2}, -\frac{1}{2}$, and $-\frac{1}{2}, -\frac{1}{2}, -\frac{1}{2}$ corresponding to the J values $\frac{7}{2}, \frac{5}{2}, \frac{3}{2}, \frac{1}{2}$. However, in the ^4D term L may take on any value between L and $-L$, so we also have to combine these S values with L values 1, 0, -1, -2, showing that the ^4D term contains 20 'degenerate' states, all in fact having slightly different energies. Similarly a ^3P term will have $3 \times 3 = 9$ substates.

The effect of the crystal field upon the Russell–Saunders terms of the free ion

The crystal field does not split the s or the p atomic orbitals but does split the d and f orbitals into two and three sets respectively. Similarly, the S and P terms of the ion are not split, but the D and F terms are split by the crystal field into two and three components. The Russell–Saunders states will always be affected in the same way, irrespective of the d configuration in which they occur. Table 2.3 gives the effect of the crystal field upon the various Russell–Saunders states. The crystal field states are designated by group theoretical symbols. We recall that the five d orbitals were split into two sets irrespective of whether the crystal field was tetrahedral or octahedral, although the relative energies of the e_g and t_{2g} orbitals were inverted. Similarly, the Russell–Saunders D terms are split into two states by both tetrahedral and octahedral fields and so on, but again the energies of the two substates will be inverted in the two cases. It is usual to plot the Russell–Saunders terms on an energy scale and to plot the splitting against the crystal field energy Δ. This is done in Figure 2.7 for a d^8 ion (Ni^{11}) in an octahedral field. The splittings would be inverted in a tetrahedral field.

We are now able to predict the electronic spectrum of a d^8 octahedral complex. Clearly a number of transitions may occur from the ground state to excited states. However, a quantum mechanical rule says that transitions may only occur between states of the same multiplicity. The ground state in the present case is a triplet state. We may therefore observe three transitions from the $^3A_{2g}$ ground state, to the $^3T_{2g}$, $^3T_{1g}(F)$, and $^3T_{1g}(P)$ excited states. The two different $^3T_{1g}$ states are characterized by (F) and (P), a reminder of the free-ion state from which they originate. The spectrum of a nickel(II) octahedral complex is given in Figure 2.8. It is possible to assign all these bands by comparison with the term-

TABLE 2.3 The splitting of Russell–Saunders states by the crystal field.

Free ion	Crystal field states
S	A_1
P	T_1
D	$E + T_1$
F	$A_2 + T_1 + T_2$
G	$A_1 + E + T_1 + T_2$

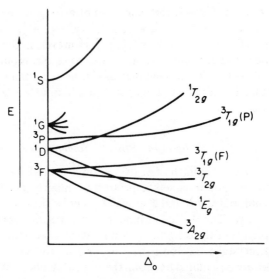

Figure 2.7 The effect of an octahedral crystal field upon the d^8 configuration.

splitting diagram. The lowest energy band, for example, will be the $^3A_{2g} \rightarrow \,^3T_{2g}$ transition, which in the present case is a direct measure of Δ_0.

It is possible to treat all d configurations in this way. The appropriate energy level diagrams are readily available. It should be noted that all crystal field states have the same multiplicity as the free-ion term. Spin-forbidden transitions may occur (i.e. between states of different multiplicity). These will be very weak in intensity and often can be satisfactorily assigned. Table 2.4 lists the Russell–Saunders states of various configurations. It may be seen for the d^5 configuration that only spin-forbidden transitions may occur, as there is no state

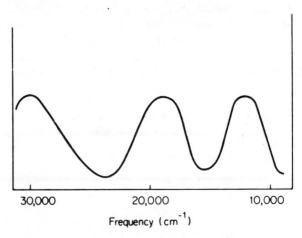

Figure 2.8 The electronic spectrum of $[\text{Ni(en)}_3]^{2+}$.

TABLE 2.4 The Russell–Saunders states of the d configuration.

d^1	d^2	d^3	d^4	d^5	d^6	d^7	d^8	d^9
^2D	^3F	^4F	^5D	^6S	^5D	^4F	^3F	^2D
	^3P	^4P	^3H	^4G	^3H	^4P	^3P	
	^1G	^2H	^3G	^4F	^3G	^2H	^1G	
	^1D	^2G		^4D		^2G	^1D	
	^1S	^2F		^4P		^2F	^1S	

of the same multiplicity as the ground state. This explains why Mn(II) complexes are practically white in colour.

It may be noted that there is a symmetry in Table 2.4. This results from the fact than d^n systems are related to d^{10-n} systems by the electron/positron argument applied to the d^1 and d^9 systems. The splittings of the d^{10-n} states will, however be inverted with respect to those of the d^n states. This means that octahedral d^n species and tetrahedral d^{10-n} species will have similar energy level diagrams. In certain cases transitions may correspond to two electron jumps. The intensity of these bands is usually low, despite being spin allowed.

The weaknesses of the crystal field approach

If the postulates of crystal field theory were fully obeyed, then the intensity of $d \rightarrow d$ bands would be very low indeed. However, in practice the d orbitals are not pure d orbitals, but have some ligand character. As a result, a greater intensity is observed. It should be noted that the intensity of the $d \rightarrow d$ bands is greater for tetrahedral complexes than for octahedral complexes.

If it is wished to account quantitatively for the electronic spectra of complexes, then it is necessary that the energies of the excited states be known relative to the ground state for all values of Δ. This requires a knowledge of the Racah parameters B and C. They may be determined for the free ion from atomic spectra. However, to account satisfactorily for the experimental results, these terms B and C must be treated as parameters, their values being reduced below that of the free ion. This is a consequence of covalency as already described. All this is involved in the ligand field approach to electronic spectra. Values of B and Δ for the complexed ions may be obtained from Tanabe and Sugano diagrams which plot E/B against Δ/B for each state. Values of C may also be obtained.

Some general problems

The spectra of d^1 and d^9 complexes may be understood in simple terms, as already outlined. However, we also have to account for the breadth of these bands and for their lack of symmetry. The band widths in general are proportional to the slopes of the individual excited states in the energy level diagram. The ligand atoms are vibrating and so Δ is continuously changing over a narrow range. However, if the excited state slopes rapidly with change of Δ then a fairly wide range of energies will be associated with that transition.

One result of these broad bands is that it is difficult to note interactions of small energy. Spin–orbit coupling is such a factor. This can cause further splitting of the degenerate crystal field state. However, for the first transition series, the fine structure of a band resulting from this is not likely to be seen. This is more important for $4d$ and $5d$ elements.

A much more important effect in this context is the Jahn–Teller effect. This is associated with the presence of a crystal field of lower symmetry than O_h or T_d. The theorem states 'that any non-linear system which is in a degenerate state will undergo some kind of distortion to remove this degeneracy'. This means therefore that any states still degenerate in the crystal field will split, and implies that spin–orbit coupling has not done this. The Jahn–Teller splitting will result in further stabilization of the system.

The physical basis of the Jahn–Teller effect may be readily seen by considering the effect of unevenly filled e_g and t_{2g} shells on the symmetry of the complex. Let us consider a d^9 system and assume that we have one electron in the d_{z^2} orbital and two electrons in the $d_{x^2-y^2}$ orbital. Repulsion between ligand electrons and metal electrons will then be greater in the $x-y$ plane than on the z axis, so producing a tetragonally distorted complex with short bonds on the z axis. If the opposite arrangement of electrons holds, then the bonds on the z axis will be longer than the bonds in the plane. (This is usually the case.) This explains why the vast majority of copper(II) complexes are tetragonally distorted.

It may readily be seen that for six-coordinate complexes, uneven occupancy of the t_{2g} orbitals is not going to be as important as the uneven occupancy of the e_g orbitals in causing distortion. Table 2.5 lists those configurations for which a Jahn–Teller effect is expected to be observed on this basis. It should be noted that for high-spin complexes, d^n and d^{5+n} configurations should have a similar effect. It is not possible to predict how large the distortion will be or its form, but it may be shown that the effect is of importance in the spectra of octahedral complexes. Obviously, if the ground state is not subject to the Jahn–Teller effect then the excited state will be (unless a two electron jump occurs). If the ground state is split, then a band will appear in the near infrared corresponding to the transition between the two components of the ground state. However, it is unlikely that an effect will be seen in the visible region, as the upper of these two 'ground states' will not be sufficiently populated to allow an observable band to be seen for the

TABLE 2.5 John–Teller distortions of octahedral complexes

	Observed	Predicted
d^1, d^6	Slight distortion	Yes
d^2, d^7	Slight distortion	Yes
d^3, d^8	No distortion	Yes
d^4, d^9	Large distortion	Yes

transition to the excited state. If the Jahn–Teller effect is causing the splitting of the excited state, then transitions to both these may be seen. Clearly, however, the fact that degenerate terms are split by the Jahn–Teller effect, by energies of around a few hundred cm^{-1}, will provide another reason for the width of bands in electronic spectra, and for the presence of shoulders or splitting on bands.

Complexes of lower symmetry

In practice most complexes do not have six or four identical ligands around the metal ion and so only approximate to O_h and T_d symmetry, although their properties may often be understood satisfactorily on such a basis. We have discussed a number of other causes for lowered symmetry. In biological systems we often have to consider metal ion environments which are of very low symmetry. The electronic properties of these metal ions are difficult to explain.

As the symmetry of the complex is decreased beyond cubic, so the number of bands in the electronic spectrum will increase, as the crystal field states are split further, so allowing more transitions. The effect may only show itself in a broadening and splitting of the bands and often may be treated by regarding the low-symmetry component as a perturbation of the main cubic field.

For complexes of non-cubic symmetry, it is necessary to introduce additional-parameters. Clearly the effect of the crystal field cannot now be described by a single parameter. The energies of the five d orbitals will be split in a much more complex way. Thus a tetragonally distorted octahedral complex (of D_{4h} symmetry) requires in-plane and axial crystal fields to be considered. Two parameters are therefore neeeded. Each example of a distorted stereochemistry must be considered independently. Usually this is done in terms of molecular orbital theory.

MAGNETIC PROPERTIES OF TRANSITION METAL ION

In transition metal ions we are specially concerned with the magnetic properties of unpaired electrons and with the information that may be derived from a study of these properties.

Molecules with closed shells of electrons only, have no inherent magnetic properties as there are no degeneracies to be split by the magnetic field. The magnetic field will, however, induce a small opposing moment and so such molecules are repelled by the magnetic field. These molecules are termed *diamagnetic*. An unpaired electron has magnetic properties resulting from both its spin and orbital motion. Molecules with unpaired electrons are therefore attracted into a magnetic field and are said to be *paramagnetic*. The two effects, diamagnetism and paramagnetism, are opposed and in calculating the para-magnetic moment of a transition metal complex due allowance must be made for the diamagnetism of the ligand.

Susceptibilities may be measured by the Gouy and Faraday methods and also by a NMR method (in solution). The advantages of the last two methods lie in the fact that only small amounts of material are required, an important point when considering metalloproteins. The non-heme protein ferredoxin has been studied in the solid state and solution by the Faraday and NMR methods, and this has shown that the protein is diamagnetic in the oxidized form. The Faraday method will allow the study of single crystals and hence the study of anisotropy in magnetic susceptibility. The most often used method is the Gouy method. The advantages here are that the equipment is simple and easy to operate, but comparatively large amounts of material are required. These methods are described in standard texts.

Magnetic moments (expressed in Bohr magnetons) may be calculated readily from susceptibility data. In accurate work a correction should be made for temperature-independent paramagnetism, TIP.

We have noted that there are two sources of the paramagnetic moment, i.e. spin and orbital contributions. If the full orbital contribution is made, then the magnetic moment will be given by the equation:

$$\mu = \sqrt{4S(S+1) + L(L+1)}$$

where L and S are the total orbital and spin angular momentum quantum numbers for the molecule. Clearly, if there is no orbital degeneracy (i.e. an S-state ion, for which $L = 0$) there will be no orbital contribution. The remaining contribution will be the spin-only term. Table 2.6 lists spin-only moments for various numbers of unpaired electrons. For other states, however, the orbital degeneracy could be lost or reduced by the environment of the other atoms. This is termed the quenching of the orbital contribution; quenching may be partial or complete. However, it may be shown that for an octahedral complex an orbital contribution is only to be expected if there is a ground state T_{1g} or T_{2g} term.

An orbital may be occupied by one electron with either $S = +\frac{1}{2}$ or $-\frac{1}{2}$. In the presence of a magnetic field these two arrangements will differ in energy and the more stable arrangement will be more favoured, although the actual occupancy will follow a Boltzmann distribution with respect to temperature. This stabilization of the ground state is the reason for the sample being drawn into the magnetic field. The splittings of spin (and orbital) degeneracy caused by the magnetic field are very small and as a result interactions too small to affect electronic spectra will have considerable effect upon the magnetic moment. In particular they affect the orbital contribution, because they help determine the

TABLE 2.6 Spin-only values.

Unpaired electrons	S	μ
1	$\frac{1}{2}$	1.73
2	1	2.83
3	$1\frac{1}{2}$	3.81
4	2	4.93
5	$2\frac{1}{2}$	5.92

ground state of the complex in the absence of a magnetic field. The presence of low-symmetry crystal fields is one example, but in this case it is difficult to deal quantitatively with the problem. It should be noted, however, that if the ground state contains an odd number of unpaired electrons, then there is a degeneracy (Kramer's degeneracy) which a low-symmetry field cannot remove. This will be split by a magnetic field.

One very important effect is that of *spin–orbit coupling*. Spin–orbit coupling may split states with both spin and orbital degeneracy into substates and can therefore have a marked effect upon the measured magnetic moment. In deriving the Russell–Saunders terms for the free ion, it is assumed that spin and orbital angular momenta will not interact and that their magnitude will determine the energy of a particular arrangement. In practice, however, they couple together to some extent, the result being that certain configurations of a nominally degenerate state are more stable than others.

The extent of spin–orbit coupling is given by the spin–orbit coupling constant λ, the measure of the coupling between resultant spin and orbital angular momenta. There is also a single electron spin–orbital coupling constant. Values for free ions may be obtained from atomic spectra and for complexes by magnetic measurements.

The effect of spin–orbit coupling for d^1–d^4 ions is to reduce the measured magnetic moment (usually down to around the spin-only value) while for d^6–d^9 ions the effect is to increase the measured moment. In Table 2.7 there are listed experimental moments for a number of species for different stereochemistries, together with calculated spin-only values. Orbital contributions will be expected for octahedral Co^{2+}, low-spin Fe^{3+}, Mn^{2+}, Mn^{3+}, and Cr^{2+} and for tetrahedral Ni^{2+}. It is clearly possible to interpret magnetic data in a general way in the light of this information.

In summary we may see that for the formation of an octahedral complex the original degeneracies of the free ion will have been split successively by the crystal field, the tetragonal distortion, spin–orbit coupling, and then by the magnetic field, each resulting in a successive stabilization of the ground state.

TABLE 2.7 Magnetic moments of transition metal complexes

Ion	Spin-only value	Measured moment	
		Octahedral	*Tetrahedral*
Cu^{2+}	1.73	1.7–2.2	
Ni^{2+}	2.83	2.8–3.3	3.3–4.0
Co^{2+} h.s.	3.88	4.5–5.2	4.2–4.8
Co^{2+} l.s.	1.73	1.8	
Fe^{2+} h.s.	4.90	5.1–5.7	5.3–5.5
Fe^{3+} h.s.	5.92	5.7–6.0	
Fe^{3+} l.s.	1.73	2.0–2.5	
Mn^{2+} h.s.	5.92	5.6–6.1	5.9–6.2
Mn^{2+} l.s.	1.73	1.8–2.1	

Temperature-independent paramagnetism (TIP)

This phenomenon is a result of spin–orbit coupling when a ground state (not split by spin–orbit coupling) is coupled by it to a degenerate excited state, well separated in energy. TIP results because the magnetic field, as a result of spin–orbit coupling, pushes electron density into an excited state. The effect is a weak one and, as the thermal population of the excited state is essentially zero, does not vary with temperature.

Magnetically non-dilute species—antiferromagnetism

So far the discussion has been concerned with magnetically dilute systems in which the paramagnetic centres are isolated from each other. However this situation does not always hold, and the paramagnetic centres may interact with each other either through bridging ligands or directly when the distance between the paramagnetic ions is small. In antiferromagnetism this leads to a decreased magnetic moment, and sometimes diamagnetic behaviour, as adjacent metal ions couple their unpaired electrons either directly through covalent bond formation or by superexchange through a ligand. The frequent occurrence of multimetal centres in metalloenzymes and other biological systems means that the topic of antiferromagnetism is particularly important in this context. Thus it is relevant to the understanding of the magnetic properties of the *iron sulphur proteins*, which contain two, four, or eight iron atoms per mole of protein, the dioxygen carriers *hemerythrin* and *hemocyanin*, which contain antiferromagnetically coupled pairs of iron and copper ions respectively, certain *copper oxidases*, and also *oxyhemoglobin*, whose diamagnetism may reflect either the presence of low-spin iron(II) and singlet dioxygen or iron(III) antiferromagnetically coupled to superoxide ion.

While decreased values of magnetic moments indicate the possible presence of antiferromagnetic behaviour, it may be confirmed by magnetic measurements over a range of temperatures. Antiferromagnetic behaviour is shown by the presence of a maximum in the plot of susceptibility against temperature, termed the Neel point. Below this temperature, antiferromagnetic behaviour is observed but as the temperature is raised so the thermal energy overcomes the magnetic interaction and normal paramagnetic behaviour is observed, the Curie–Weiss law being followed. This differs from the normal Curie law in that while a plot of $1/\chi$ versus temperature is linear there is an intercept on the temperature axis. Thus a non-linear plot of $1/\chi$ against temperature with values of χ lower than the spin-only value indicates an antiferromagnetic interaction.

Electron paramagnetic resonance

EPR (or ESR) spectra provide a great deal of insight into iron- and copper-containing proteins and is a most valuable, although complex, technique. The spectra are usually only observed for ions which have an odd number of unpaired electrons in the ground state, i.e. those with a Kramers doublet. This degeneracy is removed by a magnetic field and the resulting transitions between the two components are observed in the EPR spectra. In view of the small energies involved, the spectra will clearly be sensitive to the various factors discussed above which result in splittings equivalent to or larger than those produced by the magnetic field.

From a pictorial point of view, the unpaired spinning electron can be regarded as a magnet which will align itself with a magnetic field. However, absorption of energy could cause it to align itself against the magnetic field. Alternatively, if the system in question is subjected to a microwave beam and the strength of the magnetic field increased, it will be possible to note the particular strength at which the electron changes its alignment. This will be characteristic of the environment of the electron.

The unpaired electrons may interact with the nuclear spin I of the metal, giving $2I + 1$ bands. The separation associated with these is the hyperfine splitting, and will depend upon the stereochemistry of the complex. The presence of covalent interaction with a ligand will result in the appearance of a fine structure of $2I' + 1$ lines, where I' is the nuclear spin of the ligand. The greater the covalency the greater is the ligand hyperfine splitting.

The position of the EPR resonance gives the field at which it occurs. From this the g value (the splitting factor) may be calculated from the equation $h\nu = g\beta H$ where β is the Bohr magneton and H the field. The value of g is a measure of the contribution of the spin and orbital motion to the total angular momentum. For a free electron $g = 2.0023$. For inorganic complexes g values are higher than 2 and will be related to the arrangement of ligands around the metal ion. Inorganic complexes are often anisotropic, having g values dependent upon field direction. Vales are therefore obtained for crystals in terms of $g\parallel$ and $g\perp$. The anisotropy of the g value can give much information.

Organic free radicals will also give rise to EPR signals and these may prove a complicating factor in the interpretation of the EPR spectra of enzyme systems.

The distribution of electrons between two energy states is given by the Maxwell–Boltzmann expression. It is important that EPR measurements be made at low temperature to enhance the occupancy of the ground state at the expense of the excited state, so avoiding line broadening effects.

The value of EPR measurements for metalloproteins is that they will possibly tell us something about the oxidation state of the metal, e.g. for copper; they will confirm the presence of covalency and may help in the characterization of the ligand through the observation of the ligand hyperfine splitting.

ELECTRONIC PROPERTIES OF COMPLEXES OF COPPER AND IRON

These two transition metals figure prominently in later chapters. In order to provide a suitable basis for a discussion of the properties of their complexes, some relevant data are summarized at this point.

Copper

Copper(I) species, being of d^{10} configuration, will only show charge-transfer bands, Cu(I)→L. These have been observed up to 650 nm in the visible region. Copper(II) complexes (d^9) will also show charge-transfer bands, Cu(II)←L. Some complication may arise from the presence of ligand bands. The $d→d$ spectra of copper(II) species are different to interpret in terms of the stereochemistry of the complex. The usual stereochemistry is that of a tetragonally distorted octahedral one. The effect of such a crystal field upon the degeneracy of the d orbitals has been seen in Figure 2.2. As the ligands upon the z axis are withdrawn, so the energy of the d_{z^2} and d_{zx}, d_{yz} orbitals fall while that of the $d_{x^2-y^2}$ and d_{xy} will rise relative to these. Three positron transitions may then occur $(d_{x^2-y^2}→d_{z^2};$ $d_{x^2-y^2}→d_{xy}; d_{x^2-y^2}→d_{xz,yz})$, but when the distortion is slight only one band will be seen as the energy separation between the split bands will be small. A similar state of affairs will hold for a grossly distorted square-planar complex, while only for a medium distortion may three absorptions be seen. However, there may be other stereochemistries of low symmetry of importance in biological molecules and it may well be difficult to interpret their spectra.

Iron

Charge-transfer transitions may be observed for Fe(II) and Fe(III) complexes. For Fe(II), d^6 complexes in an octahedral stereochemistry the ^5D ground state is split into two states. One transition only therefore may be observed. At higher ligand fields it is possible that a singlet state may drop below the ^5D state and provide a new ground state. Fe(II) low-spin complexes will, of course, be diamagnetic. For Fe(III), d^5 complexes with a ^6S ground state, spin-forbidden bands only are possible as there are no other sextuplet states. The $d→d$ bands will therefore be of low intensity. High-spin complexes are usually observed with magnetic moments close to the spin-only value, as there is no orbital angular momentum contribution. Low-spin complexes will have considerable orbital contribution and moments are usually raised above the spin-only value of 1.73 BM to about 2.3 BM. The $d→d$ spectra of Fe(III) in hemoproteins have been a very useful guide to electronic structure and have also thrown light upon the nature of the axial ligand in a number of cases.

KINETICS AND MECHANISM OF REACTION OF TRANSITION METAL COMPLEXES

The reactions of transition metal complexes include

(a) *ligand substitution*, in which one coordinated ligand is replaced by another from solution;
(b) *redox processes* (electron-transfer reactions), in which there is a change in the oxidation state of the metal ion; and
(c) *reactions of coordinated ligands*, which may be affected considerably by the

presence of the metal ion. This topic is of wide fundamental interest and discussion will be reserved until the next chapter.

A useful classification in terms of general reactivity involves the division of complexes into inert and labile. Labile complexes are those whose reactions are over in less than one minute at $25\,^{\circ}C$ for concentrations around $0.1\,mol\,dm^{-3}$, while inert complexes are those whose reactions may be readily followed under these conditions, having half-lives greater than about one minute. Much work has been carried out on the reactions of inert complexes, particularly those of cobalt(III) and chromium(III), but increasing attention is being paid to the reactions of labile complexes. This partly reflects developments in the techniques for following very fast reactions in solution, particularly in the use of stopped-flow systems, relaxation techniques, and magnetic resonance techniques. The use of relaxation techniques allows the study of reactions having first-order rate constants up to $10^{10}\,s^{-1}$.

Ligand substitution in octahedral complexes

Two basic mechanisms are available for this nucleophilic substitution. One is the *Dissociative mechanism* (*D*) and the other is the *Associative mechanism* (*A*). The *D* mechanism involves a five-coordinate transition state while the *A* mechanism involves a seven-coordinate transition state. The alternative mechanisms are shown below for reaction (2.2), in which ligand X^- is replaced by ligand Y^-, and A is a neutral, unidentate ligand. Rate laws for the two pathways are shown in equations (2.3) and (2.4).

$$[CoA_5X]^{2+} + Y^- \rightarrow [CoA_5Y]^{2+} + X^- \tag{2.2}$$

D mechanism.

$$[CoA_5X]^{2+} \xrightarrow{slow} [CoA_5]^{3+} + X^-$$

$$[CoA_5]^{3+} + Y^- \xrightarrow{fast} [CoA_5Y]^{2+}$$

$$\text{Rate} = k_1[CoA_5X^{2+}] \tag{2.3}$$

A mechanism.

$$[CoA_5X]^{2+} + Y \xrightarrow{slow} [CoA_5XY]^+ \xrightarrow{-X} [CoA_5Y]^{2+}$$

$$\text{Rate} = k_2[CoA_5X^{2+}][Y^-] \tag{2.4}$$

An inspection of rate laws (2.3) and (2.4) suggests that the two mechanisms may be readily distinguished from each other on kinetic grounds. However the matter is much more complicated. Thus the solvent is known to participate in some reactions, particularly in aqueous solutions. Reaction (2.2) may occur by formation of an intermediate aquo complex, in which X^- is replaced in an *A* mechanism, followed by fast replacement of H_2O by Y^-.

$$[CoA_5X]^{2+} + H_2O \xrightarrow{slow} [CoA_5H_2O]^{3+} + X^-$$

$$[CoA_5H_2O]^{3+} + Y^- \xrightarrow{fast} [CoA_5Y]^{2+} + H_2O$$

However, as the concentration of the solvent water is effectively unchanged it will not appear in the rate law, which will then be identical to (2.3). This corresponds to a *D* mechanism even though the mechanism is an associative one.

Another complication is that of ion pairing, particularly if the reactions are carried out in non-aqueous solvents. The cationic complex $[CoA_5X]^{2+}$ and the anion Y^- may well exist as the ion pair $[CoA_5X^{2+}]\ldots Y^-$ which may be the species undergoing reaction. In this case the rate law must include a dependence upon the concentration of the ion pair, which will in turn depend upon the concentration of complex *and* entering group, quite independently of the mechanism of the actual act of substitution.

In view of these problems, mechanisms have been assigned indirectly, particularly by observing the effect on the rate of reaction of various modifications to the complex, and then comparing these with the differing requirements of the D and A transition states. For example, a D reaction should be speeded up by the presence of bulky ligands and the A reaction should be slowed, as the formation of the D transition state relieves overcrowding, while the formation of the A transition state emphasizes it. More direct approaches have involved the setting up of competition experiments in which the five-coordinate intermediate of the D mechanism is able to distinguish between different nucleophiles. Such studies as these have confirmed the existence of the D mechanism, but there is no uncomplicated evidence for the A mechanism. It should be stressed that many reactions appear to lie between these two extremes.

One reaction that was thought for some time to involve an associative mechanism is the base hydrolysis reaction.

$$[CoA_5X]^{2+} + OH^- \rightarrow [CoA_5OH]^{2+} + X^-$$
$$\text{Rate} = k_2[CoA_5X^{2+}][OH^-]$$

Much controversy has centred around this reaction, but it is generally accepted now that it takes place *via* a dissociative process, despite the second-order rate law. The rate law arises from the fact that the reactive species is the conjugate base of the complex, the concentration of which depends also upon the concentration of the proton-accepting base, in this case $[OH^-]$.

$$[Co(NH_3)_5Cl]^{2+} + OH^- \rightleftharpoons [Co(NH_3)_4(NH_2)Cl]^+ + H_2O$$
$$[Co(NH_3)_4(NH_2)Cl]^+ \xrightarrow{\text{slow}} [Co(NH_3)_4(NH_2)]^{2+} + Cl^-$$
$$[Co(NH_3)_4(NH_2)]^{2+} + H_2O \xrightarrow{\text{fast}} [Co(NH_3)_5OH]^{2+}$$

A third mechanism, involving electron transfer, has been suggested for this reaction which is of interest as it may have some general relevance to biochemical systems. Hydrolysis of cobalt(III) complexes by OH^- has certain special features. Thus this reaction differs from those of the analogous chromium(III) and rhodium(III) complexes, in that base hydrolysis is very much faster than acid hydrolysis, and that stereochemical change occurs frequently. In addition, ligand substitutions carried out in basic solution usually give equilibrium concentrations of possible products through ligand scrambling, implying the formation of labile cobalt(II) complexes. A number of reactions of cobalt(III) complexes are known to involve redox catalysis. All this suggests that base hydrolysis may also proceed *via* a cobalt(II) intermediate and generation of a radical.

In the following scheme Y represents any base.

$$[L_5Co^{III}X]^{n+} + Y^- \rightleftharpoons \{[L_5Co^{III}X]Y\}^{(n-1)+} \text{ Ion pair}$$
$$\{[L_5Co^{III}X]Y\}^{(n-1)+} \rightleftharpoons \{[L_5Co^{II}X]Y\cdot\}^{(n-1)+} \text{ Ion pair}$$
$$\{[L_5Co^{II}X]Y\cdot\}^{(n-1)+} \rightleftharpoons [L_5Co^{II}Y\cdot]^{n+} + X^-$$
$$\quad\quad\quad \updownarrow \quad\quad\quad\quad\quad\quad \updownarrow$$
$$Co^{II} + 5L + Y\cdot \quad\quad [L_5Co^{III}Y]^{n+}$$

The overall reaction pathway will then depend upon the lability of the cobalt(II) intermediate and the oxidizing power of the radical Y·. For $Y^- = OH^-$, the rate-determining step is the electron transfer from OH^- to Co(III), presumably to an excited state of Co(III) to avoid the generation of a low-spin cobalt(II) state.

This scheme will allow an explanation of the special features of base hydrolysis of cobalt(III) complexes discussed above. Cobalt(III) differs from chromium(III) and rhodium(III) in the ease of reduction of the metal to the bivalent state, so explaining why base hydrolysis is so much faster than acid hydrolysis for cobalt(III). In general, it should be noted that the larger the ligand field strength, the smaller the differences in rates of acid and base hydrolysis for cobalt(III). Stronger ligand fields stabilize cobalt(III) over cobalt(II), (CFSE effect) and hence increase the redox potential for cobalt(II)–cobalt(III). Such an electron-transfer reaction has been suggested for $Co^{3+} + {}^-OOH \rightarrow Co^{2+} + \cdot OOH$.

The formation of complexes. The replacement of coordinated water

The replacement of coordinated water by an entering ligand group is a reaction of fundamental importance. For completeness we shall make brief reference here to the alkali metals and alkaline earth metals in addition to the transition elements. Most of their aquo complexes are six coordinate. The replacement of water and the exchange of water with solvent have been studied by the techniques we have mentioned earlier. The alkali metals form complexes faster than any other ions in the Periodic Table, but the complexes are weak and so the studies have been carried out at high concentration. They clearly show, however, that the rate of substitution varies linearly with the ionic radius of the cation (Table 2.8)

TABLE 2.8 First-order rate constants for complex formation (25°C).

	$10^{-7}k_1(s^{-1})$		$k_1(s^{-1})$		$k_1(s^{-1})$
Li^+	4.7	Be^{2+}	10^2	V^{2+}	30
Na^+	8.8	Mg^{2+}	10^5	Cr^{2+}	10^8
K^+	15	Ca^{2+}	10^8	Mn^{2+}	3×10^7
Rb^+	23	Sr^{2+}	5×10^8	Fe^{2+}	3×10^6
Cs^+	35	Ba^{2+}	9×10^8	Co^{2+}	2×10^6
		Cd^{2+}	5×10^9	Ni^{2+}	2×10^4
		Hg^{2+}	3×10^9	Cu^{2+}	2×10^8
		Pb^{2+}	6×10^8	Zn^{2+}	3×10^7

suggesting that the rate-determining step is the loss of the water molecule. The larger the metal ion (as the charge is constant), the poorer its 'hold' over the water molecule and the faster the rate. The D mechanism is confirmed by only a very slight dependence of rate constant on the nature of the entering ligand. The rate constants for the reactions of the alkaline earth ions show a much wider spread. The reactions of the aquo complexes of Sr^{2+} and Ba^{2+} are very fast. Together with Ca^{2+} they appear to show a correlation between rate constant and ionic radius as observed in Group I. The rate differences between Ca^{2+} and Mg^{2+} are significant in helping to understand the different behaviour of these ions as enzyme activators.

The rates of exchange of coordinated water have been measured by NMR for certain divalent transition metal ions. The results are as follows: $V^{2+} < Ni^{2+} < Co^{2+} < Fe^{2+} < Mn^{2+} < Zn^{2+} < Cr^{2+} < Cu^{2+}$. Crystal field stabilization energy effects are important here. Those ions that are highly stabilized in this way such as Ni(II) will be reluctant to undergo reaction. The transition state will be of different stereochemistry and so there will be a loss of CFSE in the transition state. This will be an added contribution to the activation energy. The greater the CFSE of an ion, the greater will be the activation energy in its reactions. However, as the stereochemistry of the transition state is not known with certainly, it is not possible to make many quantitative predictions, although this has been successfully done for certain reactions. The fast reactions of Cr^{2+} and Cu^{2+} species result from the Jahn–Teller effect.

These reactions have always been thought to involve a dissociative pathway, but recently it has been shown by studies on volumes of activation that the earlier first-row transition metals undergo solvent exchange by an associative mechanism, and that only the later ones ($Fe \rightarrow$) involve dissociative processes.

Substitution in square-planar complexes

Substitution at a four-coordinate planar complex is of biological interest because square-planar Pt(II) complexes are important anti-tumour agents, as discussed in Chapter 10. Substitution at Pt(II) and other square-planar d^8 complexes is nearly always *via* an associative mechanism, reaction rates being dependent upon the nature and concentration of the entering ligand. Five-coordinate intermediates have been isolated in some case. The kinetics are usually of the form:

$$k_{obs} = k_1[\text{Complex}] + k_2[\text{Complex}][Y^-]$$

where Y^- is the entering ligand. This has been interpreted in terms of two reactions taking place simultaneously, one with the solvent (followed by fast attack of Y^-) and the other with Y^-. In general $k_1 < < < k_2$. The magnitude of the two terms reflects the comparative nucleophilicity of Y^- and the solvent. Thus by varying the ligand Y^- and keeping the same solvent, it is possible to build up a nucleophilicity scale of ligands Y^-, given by $n_{Pt}(Y) = \log(k_2/k_1)$. Alternatively, it is possible to convert k_1 to a second-order constant ($k_1^0 = k_1/[S]$) and define a dimensionless scale $n^0_{Pt} = k_2/k_1^0$.

The trans *effect*

The reactions shown below may in theory give a mixture of *cis* and *trans* products. This is determined by the nature of the ligand L.

$$[PtLX_3]^- + Y^- \rightarrow [PtLX_2Y]^- + X^-$$

In general the lability of any group in square-planar complexes is determined by the ligand *trans* to it. This is the *trans* effect, and it is possible to write a sequence of ligands in order of increasing *trans* effect, such as

$$H_2O < OH^- < NH_3 < Cl^- < Br^- < I^- < NO_2^- < CO < CN^-.$$

The effect may be understood in terms of the associative mechanism. The π-acceptor ligands at the strong end of this series aid bond *making* in a *trans* position to themselves by withdrawing electron density from the metal and *trans* ligand, so aiding the nucleophilic substitution at that point. The other ligands operate through a σ effect and assist bond *breaking* for the leaving *trans* ligand.

The *trans* effect is of practical significance, for example in designing synthetic pathways to pure *cis* and *trans* isomers, as illustrated with the following example.

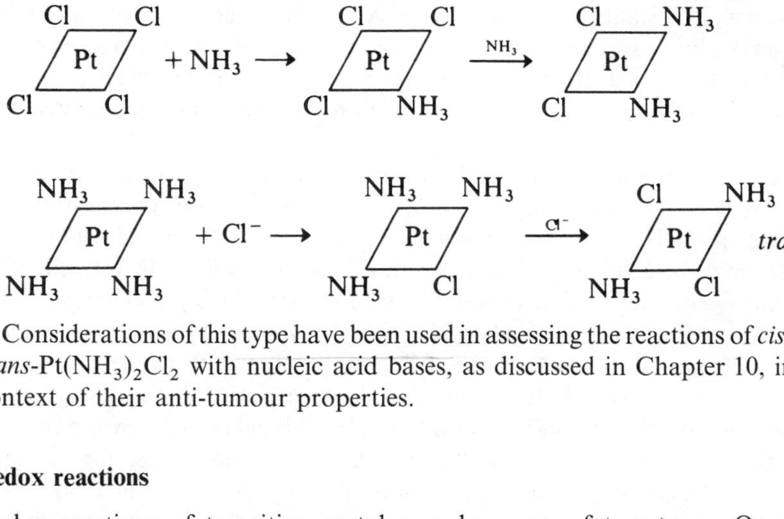

Considerations of this type have been used in assessing the reactions of *cis*- and *trans*-Pt(NH$_3$)$_2$Cl$_2$ with nucleic acid bases, as discussed in Chapter 10, in the context of their anti-tumour properties.

Redox reactions

Redox reactions of transition metal complexes are of two types. One type involves electron transfer with no change in the coordinated ligands, that is, only the oxidation state of the metal ion is changed. The other involves electron transfer coupled with group transfer. An example of the first type would be the reaction between ferrocyanide and ferricyanide, where $[Fe(CN)_6]^{4-}$ loses an electron and $[Fe(CN)_6]^{3-}$ gains one. That this reaction does occur very rapidly is shown by the use of isotopically labelled cyanide. An example of the second type is the oxidation of $[Cr(H_2O)_6]^{2+}$ by $[Co(NH_3)_5NCS]^{2+}$, giving eventually $[Cr(H_2O)_5NCS]^{2+}$. However, the initial product is the *S*-bonded thiocyanate complex, confirming that the transition state in the reaction involves a bridging thiocyanate group, $[(NH_3)_5CoNCSCr(H_2O)_5]^{4+}$.

The actual mechanisms of redox reactions generally follow this same division. *Inner-sphere mechanisms* are those in which the oxidant and reductant are connected by a bridging ligand common to the coordination shell of both metal ions, as in the reaction given immediately above. While group transfer usually occurs in this mechanism, it is not an essential feature of it, as we shall see shortly. The other mechanistic type is that of the *outer-sphere mechanism* in which electron transfer occurs through the intact coordination shell of both metal ions. The ferrocyanide–ferricyanide reaction proceeds by an outer-sphere mechanism. It should be noted that outer-sphere mechanisms are known in which hydrogen ion transfer occurs as well. The following represents the transition state of a typical reaction of this type.

$$(H_2O)_5-Fe^{II}-\overset{\displaystyle H}{\underset{\displaystyle H}{\overset{\displaystyle |}{O}}}\cdots H\cdots\overset{\displaystyle |}{O}-Fe^{III}(H_2O)_5$$

The distinction between inner-sphere and outer-sphere mechanisms is not always easy to establish experimentally. An outer-sphere mechanism may be confirmed if the rate of reaction is greater than the rate of ligand substitution into either coordination shell, as this obviously rules out the bridge mechanism. The rate law should also correspond to a transition state involving all the ligands of both metal ions.

Studies on electron transfer with group-transfer reactions have been extensively carried out by Taube and his group, in particular with chromium(II) as the reducing agent. This is a good choice as chromium(III) is inert and hence group transfer may be confirmed. The reaction of chromium(II) with the following oxidizing agents has been examined: $CrCl^{2+}$, $FeCl^{2+}$, $AuCl_4^-$ and the pentaammine species $[Co(NH_3)_5X]^{2+}$ where $X = NCS^-$, Cl^-, N_3^-, PO_4^-, acetate, oxalate etc. In the appropriate cases the use of ^{36}Cl has confirmed the transfer of Cl^- to the chromium ion, with electron transfer in the other direction. In addition, there is evidence for the *S*-bonded chromium(III) thiocyanate species, while intermediate bridged species have been isolated in other cases, for example $[(NC)_5CoNCFe(CN)_5]^{6-}$ in the oxidation of $[Co(CN)_5]^{3-}$ by $[Fe(CN)_6]^{3-}$.

Group transfer does not always occur in reactions of this type. Thus oxidation of Cr(II) by $[IrCl_6]^{2-}$ gives hexaaquochromium(III) as the major product. This confirms that group transfer and electron transfer occur separately; that bridging serves to bring two metal ions together, but that once electron transfer has taken place the bridge may break either to give group transfer or no group transfer, depending upon the energetics of the overall reaction.

A further point of interest in inner-sphere reactions lies in reactions where organic ligands act as extended bridges connecting the two metal centres, which are then some distance from each other. In some cases the transfer of electrons through the bridge becomes rate determining. This allows an interesting comparison of rates of electron conduction through different ligands. Not unexpectedly this depends upon the polarizability and degree of unsaturation of the ligand. In the Cr(II) reduction of Co(III) complexes of the type $(NH_3)_5CoOC(O)R^{n+}$ evidence has been produced to suggest the organic acid anion is present as a radical anion during the reaction.

Much of the foregoing discussion is relevant to biological redox reactions involving metal ions. Cytochromes interact with each other *via* outer-sphere reactions, while inner-sphere reactions involving simple and extended bridges are known. Reactions involving extended, conjugated bridges are particularly relevant to biological systems.

The rates of redox reactions

An essential feature of biological redox reactions is the fast rate at which they occur. It is important therefore that the factors that control the rates of such reactions in simpler systems be assessed. The rates of both inner-sphere and outer-sphere reactions are dependent upon the nature of the ligands. Electron transfer by the outer-sphere mechanism takes place more rapidly if the ligands are highly polarizable, for example cyanide or bipyridyl. The rates of inner-sphere reactions depend upon the ability of a ligand to form a bridge to the other metal centre. Thus the reaction quoted earlier involving thiocyanate as a bridging ligand is slow because the formation of an *S*-bonded thiocyanate complex of chromium(II) is not favoured. The absence of a ligand able to form a bridge may force a reaction to proceed *via* an outer-sphere reaction even though the ligands are non-polarizable. In such cases the redox reaction will be very slow, for example electron exchange between hexaammine complexes of cobalt(II) and cobalt(III).

One important principle that affects electron-transfer reactions is the Franck–Condon principle which says that there must be no movement of nuclei during the time of electronic transition. This means that the geometry of the two species after reaction must be identical to that existing before the actual electron transfer. The Fe—CN bond length of the ferrocyanide species is slightly longer than that of the ferricyanide molecule. If, therefore, electron transfer took place with reactants in their ground state, both products would be of higher energy than the reactants, as their bond lengths would either be too short or too long. There is in fact no heat change in a reaction of this type. Therefore electron transfer will only occur after the appropriate rearrangements have taken place to give two equivalent molecules. These molecules will then be in vibrationally excited states.

This can provide a useful guide in assessing the relative rates of redox reactions. If substantial changes must take place before the configurations are alike, then the reaction will be slow. Other factors are also important, thus the rate of electron transfer between cobalt(II) and cobalt(III) species is much slower than the one we have just discussed. For cobalt(II) complexes are high spin and cobalt(III) complexes are nearly always low spin. The cobalt(III) complex prepared by electron loss from the cobalt(II) complex will then be a high-spin complex and similarly the cobalt(II) species will be produced as a low-spin complex. That is, the reaction is producing electronically excited products and the activation energy will therefore be appropriately higher.

The redox reactions discussed so far involve two reactants both undergoing a one-electron change. These are complementary reactions. Non-complementary reactions are those where the oxidizer and reducer undergo reactions involving different numbers of electrons, i.e. the stoichiometry is not $1:1$. Non-complementary reactions are usually slower than complementary ones as there is a low probability of termolecular (or higher-order) collisions taking place, or because, alternatively, intermediates in unstable oxidation states have to be formed. Thus the reaction $2Fe^{2+} + Tl^{3+} \rightarrow 2Fe^{3+} + Tl^{+}$ either involves the stepwise formation of Tl^{2+} or a third-order reaction, but will be a slow reaction.

An important biological reaction is the reduction of dioxygen to water, a four-electron process. However this coupled to a one-electron reducing agent, so we have a non-complementary reaction, which we expect to be slow. The slow reduction of dioxygen leads to the possible liberation of harmful species (superoxide and peroxide), so *in vivo* this problem is overcome through the use of a multimetal centre which can rapidly transfer electrons to the dioxygen molecule.

FURTHER READING

A number of excellent general textbooks on Inorganic Chemistry are currently available, and will provide supplementary material on topics covered in this chapter. The following publications cover more specialized topics.

C. S. Phillips and R. J. P. Williams, *Inorganic Chemistry*, Oxford University Press, 1965.

B. N. Figgis, *Introduction to Ligand Fields*, Interscience, New York, 1966.

A. E. Earnshaw, *Introduction to Magnetochemistry*, Academic Press, New York, 1969.

D. Sutton, *Electronic Spectra of Transition Metal Complexes*, McGraw-Hill, London, 1968.

F. J. C. Rossotti and H. S. Rossotti, *The Determination of Stability Constants*, McGraw-Hill, London, 1961.

R. D. Gillard, *J. Chem. Soc. (A)*, **917** (1967) (Redox mechanism for base hydrolysis of Co(III) complexes).

CHAPTER THREE

THE STUDY OF METALLOPROTEINS AND OTHER METAL-CONTAINING BIOLOGICAL MOLECULES

Reference has already been made to a number of possible roles for metal ions in biological systems. These are summarized again for convenience in Table 3.1. In recapitulation, it may be seen that there is a relationship between the role of the metal ion and the binding strength of the protein for the metal. Thus for the alkali metals, whose major role is that of charge carrier, interaction with the protein is slight, whereas the metal ions of the redox metalloenzymes are very firmly bound indeed, usually by macrocyclic molecules of the porphyrin type or by strongly polarizable donors. While in the following discussion much emphasis is laid upon the role of the metal ion, its importance must always be kept in perspective. The catalytic efficiency of the enzyme is an overall effect of the whole enzyme. Thus the role of zinc in certain hydrolytic enzymes is one of a Lewis acid, but the rate enhancement of these enzymatic reactions, compared with non-enzymatic reactions, is such that it is very unlikely that this is the only factor.

In the study of metalloproteins all the factors involved in the study of enzymes in general are important. In this chapter we are concerned particularly with the function of the metal. The first step towards elucidating the function of the metalloprotein must involve the characterization of the metal–protein interaction, in terms of the nature of the binding groups and the strength of the interaction with the metal. This will then lead on towards an understanding of the structure of the active site and the associated electronic properties that are responsible for the catalytic function of the enzyme.

TABLE 3.1 Some roles of metal ions in biological processes.

Metal ion	Role	Metal–protein binding
Na^+, K^+	Charge carriers and control mechanisms; structure stabilization (K^+)	Weak
Mg^{2+}, Ca^{2+}	Structure stabilization Trigger effects (Ca^{2+}) Enzyme activator (Mg^{2+})	Medium
Zn^{2+}	Strong Lewis acid in hydrolytic enzymes	Strong
Fe, Co, Cu, Mo	Redox catalysts	Very strong
Fe, Cu	Dioxygen carriers	Very strong

The identification of metal-binding groups in metalloproteins is a complex problem in view of the large number of potential binding groups present. In certain cases, as noted in Chapter 1, the situation is somewhat simpler, as some of these groups have a much greater affinity for the metal ion than others; the imidazole group of histidine and the —SH group of cysteine are particularly important. However, in other metalloenzyme systems, where the binding is much weaker, the active site will be defined with much greater difficulty. It should also be emphasized that the description of the binding site in terms of the immediate binding groups will not allow the explanation of phenomena resulting from long-range effects.

In this chapter it is intended to survey briefly the methods for measuring metal ion–ligand interaction, to show what broad generalizations can be applied to the behaviour of different metals and ligands, and then to use this information with that provided by other approaches (such as spectroscopy, redox potentials, pH titrations) to indicate how the binding of the metal may be studied. Much emphasis will be laid upon the value of replacing main group metals by either one or a series of transition metals in order to throw further light on this problem. The use and design of model compounds will be discussed together with the information that may be obtained from a study of their differences and similarities when compared with the *in vivo* systems. Mention will also be made of more specialized techniques which will eventually replace many of the approaches described here. Of particular value are the techniques of magnetic resonance. Before discussing these topics, however, reference will be made to structural determinations by X-ray diffraction and EXAFS spectroscopy, the latter being a relatively new technique that appears to be of considerable potential in the study of metalloproteins.

X-ray diffraction

An increasing number of enzymes, including metalloenzymes have been examined by X-ray diffraction methods. These are of particular value when coupled with the results of chemical amino acid sequence determinations. Table 3.2 lists some metalloenzymes and other metal-containing biological molecules for which crystallographic results are available, and includes details of the binding groups for the metal.

It is unlikely that such structural studies will allow a full assessment of metal–protein interaction. Studies on electronic and dynamic aspects of these metalloenzymes will require specific investigation by other techniques.

As noted earlier it is necessary to assume that these structures, determined for the solid state, are relevant to solution kinetics. In the past this proposition has been accepted, with supporting evidence in some cases.[1] However it is possible that enzymes undergo conformational changes in solution and crystallize out in a non-active conformation. Problems of this type have been encountered in the case of carboxypeptidase. Detailed mechanisms for the action of this enzyme

TABLE 3.2 X-ray diffraction studies

	Metal	Ligands	Comment	Foot-note
Oxymyoglobin	Fe	Porphyrin, His, O_2		1
Myoglobin	Fe	Porphyrin, His		2
Metmyoglobin	Fe	Porphyrin, His		3
Oxyhemoglobin	Fe	Porphyrin, His, O_2		4
Hemoglobin	Fe	Porphyrin, His,		5
Erythrocruorin	Fe	Porphyrin, His		6
Azidometmyohemerythrin	Fe⎱ Fe⎰	His-25, His-54, Tyr-109 / His-73, His-101, Tyr-67	Fe—N_3—Fe bridge	7
Aquomethemerythrin	Fe⎱ Fe⎰	His-25, His-54, Tyr-109 / His-73, His-77, His-101	Bridging Asp-106, Glu-58	8
Cytochrome c	Fe	Porphyrin, His-18, Met-80		9
Rubredoxin	Fe	Cys-6, Cys-9, Cys-38, Cys-41	Tetrahedral	10
HiPIP	Fe_4S_4	Cys-43, Cys-46, Cys-63, Cys-77	Fe_4S_4 cube	11
Ferredoxin	⎰Fe_4S_4⎱ ⎱Fe_4S_4⎰	Cys-8, Cys-11, Cys-14, Cys-45 / Cys-18, Cys-35, Cys-38, Cys-41	Two cubes	12
Plastocyanin	Cu	His-37, His-87, Cys-84, Met-92		13
Carboxypeptidase A	Zn	His-69, Glu-72, His-196, H_2O		14
Carbonic anhydrase C	Zn	His-93, His-95, His-113, H_2O		15
Thermolysin	Zn	His-142, His-146, Glu-166, H_2O		16
	Ca(1)	Asp-138, Glu-177, Asp-185, Glu-187, Glu-190, H_2O		
	Ca(2)	Glu-177, Asn-183, Asp-185, Glu-190, H_2O, H_2O		
	Ca(3)	Asp-57, Asp-59, Glu-61, H_2O, H_2O, H_2O		
	Ca(4)	Tyr-193, Thr-194, Thr-194, Ile-197, Asp-200, H_2O		
Trypsin	Ca	Glu-70, Glu-80, Asn-72, Val-75	Bidentate Glu	17
Alcohol dehydrogenase	Zn	Cys-46, Cys-174, His-67, H_2O	Catalytic Zn	18
	Zn	Cys-97, Cys-100, Cys-103, Cys-111		

Table 3.2 *continued*

	Metal	Ligands	Comment	Foot note
Superoxide dismutase	Cu	His-44, His-46, His-61, His-118	Bridging His	19
	Zn	His-61, His-69, His-78, Asp-81		
Leucine aminopeptidase		Preliminary data		20
2Fe-2S protein				21

1. J. C. Kendrew, R. E. Dickerson, B. E. Strandberg, R. G. Hart, D. R. Davies, D. C. Phillips, and V. C. Shore, *Nature*, **185**, 422 (1960).

2. C. L. Nobbs, H. C. Watson, and J. C. Kendrew, *Nature*, **209**, 339 (1966); T. Takano, *J. Mol. Biol.*, **110**, 537 (1977).

3. T. Takano, *J. Mol. Biol.*, **110**, 569 (1977).

4. M. F. Perutz, H. Muirhead, J. M. Cox, L. G. Goaman, F. S. Matthews, E. L. McGandy, and L. E. Webb, *Nature*, **219**, 29 (1968); M. F. Perutz, H. Muirhead, J. M. Cox, and L. G. Goaman, *Nature*, **219**, 131 (1968).

5. W. Bolton, J. M. Cox, and M. F. Perutz, *J. Mol. Biol.*, **33**, 283 (1968).

6. R. Huber, O. Epp, W. Streigemann and H. Formanek, *Eur. J. Biochem.*, **19**, 42 (1971).

7. W. A. Hendrickson, G. L. Klippenstein, and K. B. Ward, *Proc. Natl. Acad. Sci. U.S.A.*, **72**, 2160 (1975).

8. R. E. Stenkamp, L. C. Sieker, and L. M. Jensen, *Proc. Natl. Acad. Sci. U.S.A.*, **73**, 349 (1976).

9. R. Swanson, B. L. Trus, N. Mandel, G. Mandel, O. B. Kallai, and R. E. Dickerson, *J. Biol. Chem.*, **252**, 759 (1977) and following papers.

10. J. R. Herriot, L. C. Sieker, L. H. Jensen, and W. Lovenberg, *J. Mol. Biol.*, **50**, 391 (1970); for EXAFS structure see B. Bunker and E. A. Stern, *Biophys. J.*, **19**, 253 (1977).

11. S. W. Carter, J. Kraut, S. T. Freer, R. A. Alden, L. C. Sieker, E. Adman, and L. H. Jensen, *Proc. Natl. Acad. Sci. U.S.A.*, **69**, 3526 (1972).

12. L. C. Sieker, E. Adman, and L. H. Jensen, *Nature*, **235**, 40 (1972); *J. Biol. Chem.*, **251**, 3801 (1976).

13. P. M. Colman, H. C. Freeman, J. M. Guss, M. Murata, V. A. Norris, J. A. M. Ramshaw, and M. P. Venkatappa, *Nature*, **272**, 319 (1978).

14. W. N. Lipscomb, G. N. Reeke, J. A. Hartsuck, F. A. Quiocho, and P. H. Bethge, *Phil. Tran. Roy. Soc. Lond.*, **B257**, 177 (1970).

15. A. Liljas, K. K. Kannan, P.-C. Bergsten, I Waara, K. Fridborg, B. Strandberg, U. Carlblom, L. Jarup, S. Lovgren, and M. Petef, *Nature New Biol.*, **235**, 131 (1972); K. K. Kannan, B. Nostrand, K. Fridborg, S. Lovgren, A. Ohlsson, and M. Petef, *Proc. Natl. Acad. Sci. U.S.A.*, **72**, 54 (1975).

16. B. W. Matthews, L. H. Weaver, and W. R. Kester, *J. Biol. Chem.*, **249**, 8030 (1974); P. M. Colman, J. N. Jansonius, and B. W. Matthews, *J. Mol. Biol.*, **70**, 701 (1972); B. W. Matthews and L. H. Weaver, *Biochemistry*, **13**, 1719 (1974).

17. W. Bode and P. Schwager, *J. Mol. Biol.*, **98**, 693 (1975).

18. H. Eklund, N. Nordström, E. Seppezauer, G. Söderlund, I. Ohlsson, T. Boiwe, B.-O. Söder berg, O. Tapia, C. Bränden, and A. Akeson, *J. Mol. Biol.*, **102**, 27 (1976).

19. J. S. Richardson, K. A. Thomas, B. H. Rubin, and D. C. Richardson, *Proc. Natl. Acad. Sci. U.S.A.*, **72**, 1349 (1975).

20. F. Jurnak, A. Rich, L. van Loon-Klaasen, H. Bloemendal, A. Taylor, and F. H. Carpenter, *J. Mol. Biol.*, **112**, 149 (1977).

21. K. Ogawa, T. Tsukihara, H. Tahara, Y. Katsube, Y. Matsura, N. Tanaka, M. Kukudo, K. Wada, and H. Matsubara, *J. Biochem.*

54

have been proposed on the basis of the results of X-ray crystallography on the enzyme and its substrate complex. These are described in Chapter 4. It appears[2] however that different conformations of the active site exist in the solid and solution states, and so now it is doubtful whether the structural data are of relevance to the actual mechanism of the enzyme.

Structures of metalloenzymes show some common features.* Thus polar residues are excluded from the interior except when specially required for the function of the enzyme. There is a cleft for substrate binding, together with appropriate binding groups. The interactions with substrate are facilitated by the hydrophobic interior of the protein.

Extended X-ray absorption fine structure (EXAFS) spectroscopy

This relatively new technique[3] has the advantage of being specific to the metal in question, despite its low concentration, and in not requiring crystalline material. However it only gives information on bond lengths and there are certain ambiguities inherent in the interpretation of the experimental data. The validity of the technique has been established by redetermining certain compounds of known structure,[4] and it is now being extended to metalloproteins such as nitrogenase, hemoglobin, rubredoxin, xanthine oxidase, and cytochrome c.

The technique of X-ray absorption spectroscopy gives information on the oxidation state and chemical environment of the element being investigated. The structural measurements arise from the fact that the emitted photoelectron, which may be regarded as a spherical wave radiating from the absorbing atom, will be partially reflected by the electron clouds of the atoms surrounding the absorber. The resulting interference between outgoing and reflected waves is manifested in the fine structure of the X-ray absorption at energies greater than the absorption edge, which may be used to provide very accurate assessments of bond lengths in the first coordination sphere around the metal.

FORMATION CONSTANTS

These provide a quantitative measure of the extent to which a metal will complex with any particular group or ligand. There are now a number of very useful compilations of formation constants including those for a range of metals with many important biological molecules, such as amino acids, peptides, purines, phosphates, nucleic acids etc.[5]

If we have a metal ion M and a monodentate ligand L, then, provided precipitation does not occur, a series of stepwise equilibria will come into existence for which formation constants may be written, the higher the value of K the greater being the concentration of the complex species. 'Stepwise' formation constants are illustrated as follows:

*Structures of electron transfer proteins reviewed by E. T. Adman, *Biochim. Biophys. Acta*, **549**, 107 (1979).

$$\mathrm{M + L \rightleftharpoons ML} \qquad\qquad K_1 = [\mathrm{ML}]/[\mathrm{M}][\mathrm{L}]$$

$$\mathrm{ML + L \rightleftharpoons ML_2} \qquad\quad K_2 = [\mathrm{ML_2}]/[\mathrm{ML}][\mathrm{L}]$$

$$\mathrm{ML_{(N-1)} + L \rightleftharpoons ML_N} \qquad K_N = [\mathrm{ML}_N]/[\mathrm{ML}_{N-1}][\mathrm{L}]$$

An alternative way of expressing these results involves the use of overall formation constants.

$$\mathrm{M + L \rightleftharpoons ML} \qquad\qquad \beta_1 = [\mathrm{ML}]/[\mathrm{M}][\mathrm{L}]$$

$$\mathrm{M + 2L \rightleftharpoons ML_2} \qquad\quad \beta_1 = [\mathrm{ML_2}]/[\mathrm{M}][\mathrm{L}]^2$$

$$\mathrm{M + NL \rightleftharpoons ML_N} \qquad\quad \beta_N = [\mathrm{ML}_N]/[\mathrm{M}][\mathrm{L}]^N$$

It may be shown that the two types of formation constants are interrelated by $\beta_N = K_1 . K_2 . K_3 \ldots K_N$. Strictly speaking, they should be expressed in terms of activities rather than concentrations. While formation constants have been measured over a concentration range in order to allow extrapolation to zero ionic strength, when the activity will be represented by the concentration (thermodynamic formation constants), it is more customary to carry out measurements at constant ionic strength using an added electrolyte. Comparisons between systems at the same ionic strength are then meaningful. These are stoichiometric formation constants.

The measurement and use of formation constants

This subject has been discussed in detail in a number of publications.[6] Most measurements are made at 25 °C. The analytical method must be one that will not disturb the equilibrium. For the formation of a 1:1 complex, provided the initial concentrations of metal and ligand are known, then it is only necessary to know the concentration of one species at equilibrium to allow the calculation of the formation constant. Usually, however, a variety of complex species will exist in stepwise equilibria and a number of formation constants have to be determined, so the situation is obviously much more complex. The calculations themselves are tedious and computing techniques are usually employed.

One widely used method, that of potentiometric titration, may be applied if the ligand may be protonated and the pK_a of its conjugate acid is known. The accuracy of this method falls away rapidly as the pK_a drops towards 2. The method itself involves setting up a competition for the ligand between metal ion and hydrogen ion. Usually a known solution of metal salt and mineral acid is titrated with a standard solution of ligand while the pH is read at appropriate intervals. Alternatively, a mixture of ligand and acid may be titrated with metal salt, or the metal ion and ligand may be titrated with alkali. The difference between total added acid and free acid (pH) is a measure of the amount of protonated ligand. From the pK_a the amount of free ligand may be calculated. The amount of ligand complexed by a known amount of metal ion is then derived from the equation

$$\mathrm{L_{total} = LH^+ + L + LM^{n+}}$$

It is then possible to work out $\bar{n}$, the average number of ligand molecules complexed with each metal atom. A plot of $\bar{n}$ against pA, where A is the concentration of the free ligand, is known as the formation curve. A curve may level off, for example, at $\bar{n} = 2$. This indicates that a 1:2 species is the highest species formed in solution. By interpolating at $\bar{n} = 0.5, \bar{n} = 1.5$ etc. it is possible to obtain rough values of formation constants. Accurate values of overall stability constants are usually obtained by a curve-fitting procedure. Measurements have to be carried out over a range of metal ion concentration to ensure that dimerization is not occurring.

This method is probably the quickest and most widely applicable of all methods (provided that the pK_a of the ligand allows its application) but there are many others. These include a variety of optical methods, the use of EMF measurements, polarography, and solvent extraction.

Job's method may often prove to be useful in determining the stoichiometry of complex formation in solution and also in obtaining the formation constants in favourable cases. This method only applies where the complex absorbs in a region of the spectrum where the ligand and metal ion do not. A number of solutions are prepared in which the concentrations of the two reagents are continuously varied, keeping the total molarity constant. The optical density of each solution is measured at λ_{max} and plotted against solution composition. The absorbance (and hence the concentration of complex) will be a maximum when the two reagents are present in the same stoichiometry as the complex. The stoichiometry of the solution at maximum optical density will therefore correspond to the stoichiometry of the complex. Figure 3.1 illustrates the formation of 1:1 and 1:2 complexes. The method becomes less practical at higher stoichiometries.

Some formation constant data are presented in Table 3.3.

A number of general points must be borne in mind when considering formation constant data.

(a) Successive formation constants will decrease in the absence of any special

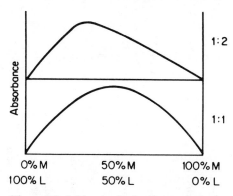

Figure 3.1 Job's method of continuous variation.

TABLE 3.3[a] Logarithms of formation constants in aqueous solution at 25°C.

Ligand	Ionic strength (mol dm^{-3})	Constant	Mg^{2+}	Ca^{2+}	Mn^{2+}	Fe^{2+}	Fe^{3+}	Co^{2+}	Ni^{2+}	Cu^{+}	Cu^{2+}	Zn^{2+}
Glycine	$\to 0$	K_1	3.44	1.38	3.44	4.3	—	5.23	5.77	—	8.62	5.52
	$\to 0$	K_2	3.01	—	—	3.5	—	4.02	4.80	—	6.97	4.44
Cysteine	0.01	K_1	<4	—	4.1	—	—	9.3	—	19.2	—	9.86
	0.01	K_2	—	—	—	—	—	7.6	—	—	—	8.84
	$\to 0$	β_2	—	—	—	11.77	32.1	—	19.3	—	—	—
Serine	Varied	K_1	—	~0.5	—	—	—	—	—	—	—	—
	Varied	β_2	—	—	—	7.0	—	8.0	—	—	14.54	—
Arginine	0.15	K_1	—	—	—	—	—	3.87	—	—	—	—
	0.15	K_2	—	—	—	—	—	3.20	—	—	—	—
	0.15	K_3	—	—	—	—	—	2.08	—	—	—	—
Aspartic acid	0.01	β_2	—	—	—	—	—	—	—	—	13.90	7.80
	0.10	K_1	2.43	1.60	3.74	—	—	5.90	7.12	—	8.57	5.84
	0.10	K_2	—	—	—	—	—	4.28	5.27	—	6.78	4.31
Histidine	0.01	β_2	—	—	7.74	9.3	—	13.86	15.9	—	18.33	12.88
	0.01	K_1	—	—	—	—	—	6.8	8.8	—	—	7.6
Lysine	0.01	β_2	—	—	—	—	—	—	—	—	13.7	—
	0.01	K_1	—	—	2.0	4.5	—	—	—	—	—	—
Glutamic acid	Varied	K_1	1.9	2.05	3.3	4.6	—	5.06	5.90	—	7.85	5.45
	Varied	K_2	—	—	—	—	—	3.40	4.44	—	6.55	4.01
Glutamine	0.16	K_1	—	0.18	—	—	—	—	—	—	—	—
Asparagine	0.01	β_2	~4.0	—	~4.5	6.5	—	8.13	10.6	—	14.9	8.7
Glycylglycine	$\to 0$	K_1	1.06	1.24	2.15	—	—	3.49	—	—	6.04	3.80
	$\to 0$	K_2	—	—	—	—	—	2.39	—	—	5.62	2.77
Glycylserine	0.01	K_1	—	—	—	—	—	—	—	—	—	3.7
Alanylglycine	0.01	K_1	—	0.66	—	—	—	—	—	—	—	3.0

Ligand	Conc.	Constant	1	2	3	4	5	6	7	8	9	10
Glycylalanine	0.01	K_1	—	—	—	—	—	3.15	—	—	—	4.1
	0.01	K_2	—	—	—	—	—	2.58	—	—	—	3.33
Glycylglycylglycine	→0	K_1	—	—	1.41	—	—	2.98	—	—	5.41	2.99
		K_2	—	—	—	—	—	2.61	—	—	5.15	2.58
Imidazole	Varied	K_1	—	0.08	—	—	—	—	2.94	—	4.20	2.19
		K_2	—	—	—	—	—	—	2.41	—	3.47	2.41
		K_3	—	—	—	—	—	—	1.99	—	2.84	—
Proline	0.03	β_2	<4	—	5.5	8.3	—	9.3	—	10.87	—	—
1,10-Phenanthroline	Vaired	β_2	—	—	—	—	—	—	—	—	16.8	10.2
		K_1	—	—	—	—	—	—	—	—	6.30	—
		K_2	—	—	—	—	—	—	—	—	6.15	—
		K_3	—	—	—	—	—	—	—	—	5.50	—
		β_3	—	—	7.35	21.0	—	—	—	—	—	—
2-Methyl-1,10-phen-anthroline		β_3	—	—	—	10.8	14.10	—	—	—	—	—
Ammonia		K_1	0.23	−0.2	—	—	—	2.11	2.79	5.93	4.15	2.37
		K_2	−0.15	−0.6	—	—	—	1.63	2.24	4.93	3.50	2.44
		K_3	−0.42	−0.8	—	—	—	1.05	1.73	—	2.89	2.50
Ethylenediamine	1.0	K_1	0.37	—	2.73	4.28	—	5.93	7.66	—	10.72	5.92
		K_2	—	—	2.06	3.25	—	4.73	6.40	—	9.31	5.15
		K_3	—	—	0.88	1.99	—	3.30	4.55	—	—	1.86
1,3-Diaminopropane	0.15	K_1	—	—	—	—	—	—	6.98	—	9.77	—
		K_2	—	—	—	—	—	—	4.93	—	7.17	—
Ethanolamine	→0	β_2	—	—	—	—	—	—	—	—	6.68	—
2-Mercaptoethylamine	1.0	K_1	—	—	—	—	—	9.38	10.96	—	—	8.07
N,N-Dimethyl-ethylenediamine	0.1	K_1	—	—	—	—	—	—	—	—	9.23	—
		K_2	—	—	—	—	—	—	—	—	6.73	—
N,N'-Dimethyl-ethylenediamine	0.1	K_1	—	—	—	—	—	—	—	—	9.69	—
		K_2	—	—	—	—	—	—	—	—	6.65	—

TABLE 3.3 (continued)

Ligand	Ionic strength (mol dm⁻³)	Constant	Mg^{2+}	Ca^{2+}	Mn^{2+}	Fe^{2+}	Fe^{3+}	Co^{2+}	Ni^{2+}	Cu^{+}	Cu^{2+}	Zn^{2+}
2-Methyl thiourea	Varied	K_1	—	—	—	—	—	—	3.23	—	5.42	—
	—	K_2	—	—	—	—	—	—	2.79	—	5.11	—
Acetic acid	Varied	K_1	0.51	0.41	—	—	—	—	0.67	—	2.16	1.03
Oxalic acid	—	β_2	2.55	3.0	3.82	4.7	—	4.7	—	—	8.5	7.02
Salicylic acid	→0	K_1	—	0.36	—	—	15.2	—	—	—	10.6	—
Malic acid	—	K_1	1.55	2.66	—	—	—	—	—	—	—	3.32
Uramil-N,N-diacetic acid	→0	K_1	3.1	5.2	4.0	—	—	3.2	3.3	—	—	3.2
Nitrilo triacetic acid	0.1	K_1	7.0	8.17	7.44	8.84	15.87	10.6	—	—	12.68	—
		K_2	—	3.43	—	—	8.45	3.9	—	—	—	—
EDTA	0.1	K_1	9.12	11.0	14.04	13.9	25.1	16.2	18.56	—	18.79	—
Hydroxide	—	K_1	2.6	1.1	2.8	3.2	—	3.6	3.8	—	6.5	4.7
Azide ion	0	K_1	—	—	—	—	5.06	—	—	—	2.56	—

ᵃTaken from 'Stability Constants', Special Publication of the Chemical Society No. 6, 1957. Further data are contained in Chapter 9. A few constants were not measured at 25°C.

effect. This is a statistical phenomenon reflecting the decreasing number of available coordination sites as more ligands complex.

(b) Steric repulsions between ligands may result in a lowered stability.

(c) It is customary to speak of a 'chelate effect', in which chelate complexes are supposedly more stable than analogous complexes with unidentate ligands, e.g. $[Ni(en)_3]^{2+}$ and $[Ni(NH_3)_6]^{2+}$, en $=$ $NH_2CH_2CH_2NH_2$. While statistical and thermodynamic considerations may be used to demonstrate the existence of the chelate effect, numerical comparisons in support of it are usually based upon a comparison of formation constants of different dimensions. It is more appropriate to express concentrations as mole fractions (compared to the solvent) as these are dimensionless. A comparison of formation constants calculated on this basis shows the chelate effect to be much less important in a quantitative sense than it is usually suggested to be.

Five-membered chelate rings are more stable than six-membered rings, and increasing ring size leads to increased instability.

A comparison of formation constant data for a series of ligands with one metal ion will indicate the preferred ligand for that metal, and calculation will show the distribution of the metal ion amongst its different complexes. Similarly it is possible to say which metal ion out of a number available will complex to a particular ligand. Competition by metals for a particular ligand will be complicated by the presence of other ligands of higher affinity for one metal than others, as this effectively reduces the concentration of that metal ion and so allows the remaining metals to compete more effectively. This situation can be treated by using conditional formation constants that allow for the presence of such a ligand. An example will be discussed in Chapter 10, namely competition between Li^+ and Mg^{2+} for binding sites in the presence of ATP.

FACTORS AFFECTING THE STABILITY OF METAL COMPLEXES

It is clearly of great importance to be able to assess the factors that influence complex stability. If we are given certain metals and a range of donor groups, it should then be possible to make sensible predictions regarding the mode of metal binding. A consideration of the available results on complex stabilities allows us to make certain broad generalizations and also to observe more refined trends superimposed upon these.

Class (a) and class (b) metals

Chatt and Ahrland have divided metal ions into these two types. Class (a) metals include the alkali and alkaline earth metals, zinc, and the early members of the first transition series (up to Cr). These form more stable complexes with ligands having donor atoms from the first short period (N, O, F), than with analogous ligands having donor atoms from the second short period (P, S, Cl). Class (b) behaviour does not extend to many metals of biological interest but the later transition elements (Mn on) are borderline in behaviour between class (a) and class (b). The reverse stability situation is true. For class (b) metals, the stability of the complex will be greater when the donor atom is from the second short period. More generally the stability will decrease with donor atom in the sequence $S > I > Br > Cl > N$. For class (a) metals there will be a useful correlation between the base strength of the ligand and complex formation, that is regarding H^+ and M^{n+} as being equivalent. For class (b) elements, simple consideration of this type plays a less important role and other factors will be more important; thus they will form π bonds with ligands having empty π orbitals such as CO and CN^-. Crystal field effects will also be important as have already been discussed.

'Hard' and 'soft' acids and bases (HSAB)

This classification is a very useful one. 'Hard' metal ions or acids are like the proton, small and not easily polarized ($\equiv$class (a)), while 'soft' acids are large and easily polarized. Ligands with highly electronegative donor atoms are hard bases while polarizable ligands are soft. A general rule is that stable complexes are those formed between hard acids and hard bases, and soft acids and soft bases. Table 3.4 classifies some acids and bases. While we will reserve our summary of conclusions until later, it may be seen that the information contained in Table 3.4 will allow a number of phenomena to be appreciated. Sulphur donors are soft bases while nitrogen donors are hard. It is not surprising therefore that the thiocyanate ion is nitrogen bonded to the elements of the first transition series but sulphur bonded to those of the second and third.

An examination of Table 3.4 also shows that the metal ions of biological significance are hard or borderline hard. Furthermore, fundamental cellular constituents and potential binding groups are hard elements. In other words, living systems are 'hard'. In contrast the common toxic heavy metals found in metal poisoning or environmental pollution are soft. In general, soft metals are toxic; for example, As, Pb, Hg, Tl, and Cd compounds are well known to be poisonous. The principles of HSAB theory allow a rationalization of biological phenomena such as heavy metal poisoning and, while not allowing of mechanistic interpretations do provide clues to target molecules, and the loci of action.[7]

The Irving–Williams series of stability

It is often valuable to prepare a series of model compounds with varying transition metal ions and to compare their formation constants with those of the apoenzyme of the system in study with the same series of metals. Normally in model compounds, the formation constants of complexes of the bivalent ions of the first transition series follow the Irving–Williams sequence of stability (Mn(II) < Fe(II) < Co(II) < Ni(II) < Cu(II) > Zn(II)). The complexing power of a metal ion will vary with the charge/radius ratio of that ion, i.e. polarizing

TABLE 3.4 'Hard' and 'soft' acids and bases.

Acids			Bases		
Hard	*Intermediate*	*Soft*	*Hard*	*Intermediate*	*Soft*
H^+	Zn^{2+} Sn^{2+}	Cu^+ Pd^{2+}	H_2O RNH_2	Pyridine	$-SCN^-$ R_3P
Li^+	Cu^{2+} Pb^{2+}	Ag^+ $Pt^{2+,4+}$	ROH $-NCS^-$	Br^-	$-CN^-$ R_3As
Na^+	Ni^{2+}	Au^+ Cd^{2+}	R_2O Cl^-	N_3^-	RSH H^-
K^+	Fe^{2+} Group V	Tl^+	OH^- PO_4^{3-}	NO_2^-	R_2S $S_2O_3^{2-}$
Group IIA/IIIA			OR^- SO_4^{2-}		RS^-
			NH_3		

power. For the above ions this ratio varies in the sequence Mn(II) < Fe(II) < Co(II) < Zn(II) < Ni(II) > Cu(II). The departure of the results from this sequence reflects the varying CFSE contributions of each ion. However, the situation with regard to the metalloenzymes may not always be straightforward, stability constant data may or may not follow the same pattern of behaviour as model compounds and this must be accounted for in the mechanistic scheme. This may be of use in testing a suggested mechanism.

Some conclusions

The general trends discussed above and the detailed formation constant data allow useful conclusions to be drawn. It is possible to predict which ligands will combine with a wide range of metal ions and those which will be much more selective in their action. Much of this information provides the underlying reasons for the choice of certain analytical reagents for certain operations. The hard–soft concept seems to be particularly useful in rationalizing the data. It is also possible to understand some data in terms of steric hindrance in the ligand and hence a selectivity for metal ions based upon the preferred stereochemistry of that metal. Thus o-phenanthroline is a good ligand for both Fe^{2+} and Cu^{+}, which favour octahedral and tetrahedral stereochemistries respectively. The presence of methyl groups adjacent to the nitrogen donor atoms in a phenanthroline makes it a selective reagent for copper in the presence of iron. The methyl groups prevent the ligand satisfying the stereochemical requirements of Fe^{2+}.

Considering the data from the point of view of the metal ion, it may be seen that interesting parallels and differences between certain biologically important metal ions emerge. This data is important when we come to consider metal-for-metal replacement, as metal ions must bind to the same extent and to the same binding groups in order for successful replacement to be feasible. There is a good

TABLE 3.5 Preferred ligand binding groups for metal ions.

Metal	Ligand groups
K^{+}	Singly charged oxygen donors or neutral oxygen ligands
Mg^{2+}	Carboxylate, phosphate, nitrogen donors
Ca^{2+}	$\equiv Mg^{2+}$ but less affinity for nitrogen donors, phosphate, and other multidentate anions
Mn^{2+}	Similar to Mg^{2+}
Fe^{2+}	—SH, —NH$_2$ > carboxylates
Fe^{3+}	Carboxylate, tyrosine, —NH$_2$, porphyrin (four 'hard' nitrogen donors)
Co^{3+}	Similar to Fe^{3+}
Cu^{+}	—SH (cysteine)
Cu^{2+}	Amines $\gg$ carboxylates
Zn^{2+}	Imidazole, cysteine
Mo^{2+}	—SH
Cd^{2+}	—SH

parallel between Mg^{2+} and Mn^{2+}, and manganese will replace magnesium in many biological systems. However, Mg^{2+} and Ca^{2+} are rather more different. The ligand preferences for certain metal ions are listed in Table 3.5.

METAL ION INTERACTION WITH AMINO ACIDS, PEPTIDES, AND PROTEINS

A discussion of metal ion interaction with some model ligands is given later in this chapter. In view of the importance of metal–protein interactions we will discuss separately model metal complexes with amino acids, peptides, and proteins[8] and consider the implications of these results in terms of metal ion interactions in metalloproteins.

Amino acids and peptides

Each of the naturally occurring amino acids may form a stable five-membered chelate ring with a metal ion. If there are no complicating donor side chains, then the donor groups will be the amino and carboxylate groups. Under other pH conditions this may not be the case; thus at lower pHs the amino acid may coordinate as a neutral ligand. When the carboxylate group is not part of such a chelate ring, then often four-membered rings are formed in which both oxygen atoms are bound to the metal. Alternatively the carboxyl group may bridge two metal centres.

Simple peptides (again other than those with side chains) will combine less strongly with metal ions than do the amino acids, as is indicated by formation constant data. In order to avoid the formation of large rings it is not likely that the terminal groups alone would be utilized. Much work, including X-ray studies, has been carried out on metal complexes of peptides.[9] These show the importance of planar peptide links in metal–peptide complexes. It appears that non-deprotonated N-peptide atoms are not used as metal-binding sites. These would involve a tetrahedral environment. When a metal is bonded to three donor groups of a single peptide molecule (the central one of which is the deprotonated nitrogen atom), then the three donor atoms will be in the same coordination plane. A number of peptide complexes are known in which a tautomeric form of the peptide is stabilized by coordination so allowing the coordination of the

$$\left(\begin{array}{c} -C=N- \\ | \\ OH \end{array} \right)$$

nitrogen atom in a planar environment.

The presence of side chains will markedly affect the behaviour of amino acids and peptides. Histidine and cysteine are both examples where strong affinities for metals may be explained by the side chains (imidazole and SH), although these groups may remain uncoordinated in certain cases. The common nitrogen donors found in proteins lie in the spectrochemical series: N(imidazole) < N(peptide) < $-NH_2$.

Cobalt(II) peptide complexes appear to involve amino nitrogen, side chain and oxygen donors and no deprotonated peptide nitrogen atoms. These latter donors are important in cobalt(III) complexes. This is because the higher ligand field of the deprotonated N(peptide) link will allow the CFSE advantage of cobalt(III) over cobalt(II) to be utilized, so giving stable cobalt(III) complexes. The stability of a number of cobalt(II) and cobalt(III) complexes with amino acids may be understood on this basis.

Metal–protein complexes

A number of additional points must be borne in mind when comparing complexes of amino acids and peptides with those of proteins. It is reasonable that the measured first formation constant for a metal–protein system should relate to that of the simple model approximating to that of the binding group in the protein. Second and third formation constants etc., are not so meaningful as coordination of these groups may be determined by what groups are brought near by the coordination of the first group. Despite this, however, in general the behaviour of model metal–protein complexes is like that of peptide complexes, that is they are normal coordination complexes. This is not necessarily true of naturally occurring metalloproteins.

The presence of side chains (carboxyl, sulphur, imidazole groups) may well dominate coordination by proteins, with formal peptide links playing little part. Charges on the protein will be important as also will be the basic stereochemistry of the protein molecule, as this may result in certain groups being presented in a favourable position for coordination and other groups in unfavourable positions. This too will depend upon the favoured coordination number and stereochemistry of the metal ion and upon the size of the metal ion. The binding of heavy metals to proteins has been reviewed.[10]

A statistical factor will also be operative. This will depend upon the number of different donor groups available in the protein in addition to the reactivity of those groups for the metal ions. If there are many more less-reactive groups than than reactive ones, then this will result in a greater possibility of their coordination despite their lower reactivity.

In natural systems there will also be competition for the donor site with H^+. Protonation of a donor group prevents its coordination. Metal binding will be favoured by a large acid dissociation constant for the protonated ligand, ligand H^+. This may well determine which of two ligands will be coordinated to a particular metal ion at any pH. We will illustrate this with reference to ammonia and imidazole. These have a similar affinity for Zn^{2+} as is shown by the following formation constant data: $\log K_1$ for $Zn:NH_3 = 2.37$, $\log K_1$ for $Zn:imidazole = 2.58$. However, the pK_a values for NH_4^+ and imidazole H^+ are 9.26 and 7.03 respectively. At pH 7, therefore, less than 1% of ammonia is present as NH_3, while over 50% of imidazole is unprotonated. Thus on a statistical basis, metal–imidazole interaction is favoured by a factor of 10^3. As the pH increases over 7, coordination of ammonia relative to imidazole will become more effective.

THE STUDY OF METALLOPROTEINS AND OTHER METAL-CONTAINING SYSTEMS

The characterization of metal-binding sites

The environment of the metal in metal–enzyme systems may include ligand groups from one or both of the protein and substrate, i.e. M–E–S, E–M–S, or E–S–M. The simplest function of the metal ion, that of bringing the reacting groups into the correct orientation, must involve coordination of both protein and substrate. Such behaviour may usually be understood on the basis of formation constants and the stereochemistry of model compounds. This is often the case for systems involving Mg^{2+}, Mn^{2+}, and Ca^{2+}.

We will briefly describe the application of the methods we have just considered to the problem of elucidating metal-binding sites, prior to considering other methods for examining metalloproteins in more detail. It should be emphasized, however, that these methods cannot lead to definite conclusions regarding the full details of metal-binding sites. The results of X-ray structural studies in every case studied so far have not been in accord with the suggestions put forward on the basis of chemical evidence, even though features of schemes have been confirmed. Data on metalloproteins have been reviewed by Vallee and Wacker.[11]

Metal-for-metal replacement. Metal ions as probes[12]

Certain of the techniques described in the following sections may be applied to a wider range of metalloenzymes by metal-for-metal replacement in the metalloenzyme. Thus a main-group metal ion may be replaced by a transition element or by a metal that has useful probe properties, for example one that allows the use of Mössbauer or NMR spectroscopy. The probe must be chosen with care, as ideally it should occupy exactly the same site as the native metal ion, i.e. the principle of isomorphous replacement must hold. If the biological activity of the enzyme is maintained it may be assumed that the new metal ion has successfully replaced the native metal ion.

The ionic radius of the metal ion is an important consideration, while the probe metal should also have similar requirements to the native metal in terms of the preferred stereochemistry of the metal-binding site and the nature of the binding ligands. Sometimes a compromise choice is effective. Thus Ni^{2+} should be a good probe for Mg^{2+} in terms of ionic radii (Mg^{2+}, 0.65 Å; Ni^{2+}, 0.69 Å) while they prefer the same stereochemistry. However the similarity in chemistry is not very good, and so Ni^{2+} does not replace Mg^{2+} effectively.[13] On the other hand, Mn^{2+} which has an ionic radius of 0.80 Å, is an effective probe for Mg^{2+} as their chemistry is very similar. Table 3.6 lists some probes that have been used

TABLE 3.6 Metal-for-metal substitutions

Native cation	Ionic radius (Å)	Probe with ionic radius and properties	Chemistry
K^+	1.33	Tl^+ (1.40, NMR, fluorescence)	Moderate
		Cs^+ (1.69, NMR)	Good
Mg^{2+}	0.65	Mn^{2+} (0.80, EPR, paramagnetic)	Good
		Ni^{2+} (0.69, d–d spectra, paramagnetic)	Poor
Ca^{2+}	0.99	Mn^{2+} (0.80)	Poor
		Eu^{2+} (1.12, Mössbauer, paramagnetic)	Redox reactions!
		Lanthanide^{3+} (1.15–0.93, range of probe properties)	Moderate
Zn^{2+}	0.69	Co^{2+} (0.72, d–d spectra, paramagnetic, resonance Raman)	Excellent
Fe^{3+}	0.64	Gd^{3+} (1.02, paramagnetic)	Moderate
		Tb^{3+} (1.00), Eu^{3+} (1.12),	
		Ho^{3+} (0.97, fluorescence)	
		Ga^{3+} (0.62, NMR)	

in the study of biological systems, and also assesses the extent of similarity in chemistry between probe and native metal ions. The lanthanide elements are particularly useful in certain cases as this offers the possibility of having a series of probes differing only slightly in ionic radii. It should be noted that Ln^{3+} has been used to replace Ca^{2+} despite the difference in charge. Probes for the IA and IIA cations will be discussed in Chapter 9.

If the criteria for isomorphous replacement are relaxed then a wide range of substitutions may be carried out. The divalent transition metals $Mn(II)$–$Zn(II)$ offer a series of probes that vary in size and geometry and provide access to a wide range of spectroscopic techniques. The catalytic efficiency of these cations may also be correlated with their changing properties (geometry and chemistry) and so give insight into kinetic factors. The protein transferrin transports iron. The iron centre allows the use of EPR, Mössbauer, absorption spectra, and resonance Raman techniques, while the substitution of iron by Cu, Zn, Al, Ga, Tb, Co, and Cr has been achieved. The information obtained from such substitutions will be discussed in Chapter 8.

It is also possible to use probe metal ions and complexes independent of any metal substitution. In this case the probe will either bind to particular sites or remain free in the solution under study. Substitution-inert metal ions have received much attention. Complexes of Co(III) and Cr(III) are notable examples. Thus Cr(III)–nucleotide complexes such as Cr(ATP) have been used to probe the function of Mg^{2+} in a number of kinases, which transfer a phosphoryl group from nucleotides to a variety of substrates.[14] The other important area involves various NMR probes, as developed by Williams and his coworkers.[12,15]

Chemical testing

In a number of cases, the chemical testing of metalloenzyme and apoenzyme for the presence of specific groups has been of value. The underlying principle here is that a binding group for a metal in a metalloenzyme is bound so firmly that it does not react with the analytical reagents. Thus the —SH content of proteins may be found by titration with a variety of reagents. Titration with Ag^+ shows that hemocyanin, the oxygen-carrying copper protein, does not have any free —SH groups, while the results for the copper-free apoenzyme show the presence of free —SH. Hence it is suggested that copper is bound by —SH in the enzyme. On the other hand, there are examples where this chemical testing method has given incorrect results, possibly because a group is inaccessible in the metallo-enzyme and accessible in the apoenzyme as a result of changes in the protein structure.

pH methods

Another approach involving a comparison of metalloprotein and apoenzyme is that of hydrogen ion titration. A large number of groups are involved in protonation equilibria but some attempt has been made to determine pK_as of metal-binding groups by a comparison of titration curves in the presence and absence of the metal. However, it is known that pK_a values of protein groups may be different from those of the group in isolation and so while the pK_a of the binding group may be calculated, this does not necessarily allow the identification of the binding group.

Thermodynamic measurements

Formation constants may be measured for reversible metal–protein interaction by potentiometric methods and compared with those of model compounds. It is certainly possible to predict the favoured groups out of those available on the protein but the statistical factor may cause complication. Thus the data in Table 3.3 show that Cu(II) has a much stronger tendency to coordinate with nitrogen donors than oxygen donors. We would expect, therefore, that Cu(II) would be bound to imidazole rather than to a carboxylate group. But carboxylate groups are usually more plentiful, so in practice we may find Cu(II) combined with both groups, even though formation constant data would not suggest this. Hydrogen ion competition is often a useful guide alongside this particular approach.

Comparisons between the reactivity of different metals for the same site are useful but are only valid when 'isomorphous replacement' occurs. Irving and Williams have shown[16] that the slope of the plot of formation constant against atomic number of the metal is greater for nitrogen and sulphur ligands than for oxygen, reflecting the different CFSE contributions. This approach has been used in several cases, including one, carboxypeptidase, where it led to the incorrect suggestion that RS⁻ was a binding group. Again, Lindskog and Nyman[17] have measured formation constants for a number of metal carbonic anhydrase complexes and by a suitable comparison with data for the corresponding ethylenediamine complexes were able to suggest that the probable binding of zinc in carbonic anhydrase was through three nitrogen bases and did not involve RS⁻.

The redox potentials of transition metal ion couples will provide information on the nature of the ligand groups as has already been discussed. The redox potentials of metalloenzymes are often rather different from those observed for model compounds. The copper 'blue' proteins provide an example of this behaviour. This is associated with the irregular stereochemistry of the metal-binding site, as an unusual geometry will usually destabilize the higher oxidation state of an element more than the lower.

Absorption spectra

The band pattern and band intensity in the $d-d$ spectra of transition metal complexes will give information regarding the stereochemistry of the site. For metalloenzymes this has led to the conclusion that the geometry is usually rather different from that observed for complexes of the metal. Thus for cobalt(II) carbonic anhydrase (i.e. Zn^{2+} replaced by Co^{2+}) the visible absorption spectrum indicates the presence of a distorted tetrahedral site, a proposition that is supported by X-ray studies. In addition the spectrum is pH dependent in a way that may be correlated with enzymatic activity. By contrast, cobalt(II) complexes of the metal-activated enzymes yeast and rabbit muscle enolases, pyruvate kinase, and β-methyl aspartase do not show this distorted spectrum. The role of the metal ion here will be more difficult to elucidate.

The band positions in the $d-d$ spectra, together with a knowledge of the spectrochemical series, will allow the postulation of reasonable binding groups, while it may also be possible in the appropriate cases to note something of the nature of the spin state of the metal ion and hence, perhaps note indications of bond length changes (alongside Mössbauer and EPR spectroscopy and magnetic susceptibility data where possible). An example of the use of the position and intensity of bands lies in the work of Brill, Martin, and Williams[18] on the visible spectrum of the copper protein erythrocuperin (superoxide dismutase). It is suggested that the cupric ion is surrounded by at least four nitrogen donors while the high extinction coefficient (for $d-d$ bands) is attributed to a degree of asymmetry such as is observed for complexes of the ligand bis(salicylaldehyde-)ethylenediimine.

Charge-transfer bands will also help define the ligand, knowing the metal ion. An often quoted example of the use of charge-transfer bands in model systems involves the Fe(III) catalysis of the decarboxylation of dimethyloxaloacetic acid. A purple colour appears during the reaction which is characteristic of a charge-transfer band involving a ferric enolate group. This shows that keto-enol tautomerism is involved. The scheme was given in Chapter 1.

Vallee[19] has used charge-transfer bands in the study of metallothionein. This contains zinc and cadmium and has an intense band at 250 nm, similar to that observed for cadmium mercaptide complexes. Again, cadmium can replace zinc in liver alcohol dehydrogenase giving a band at 250 nm. On the other hand, if cadmium in metallothionein is replaced completely by zinc, then a charge-transfer band at 215 nm is observed. The zinc–mercaptoethanol complex has a band at 215 nm. This is good evidence for metal binding by —SH in these proteins.

Optical rotatory dispersion and circular dichroism. The Cotton effect

Optical rotatory dispersion (ORD) is concerned with the variation of optical activity (i.e. the angle of rotation α) with the wavelength of the incident light. The ORD curve is a plot of α against wavelength. If the medium through which the light passes exhibits circular dichroism then it will absorb right and left circularly polarized light unequally. A CD spectrum involves a plot against wavelength of the difference in extinction coefficient for left- and right-hand circularly polarized light $(E_1 - E_r)$. The two phenomena are collectively known as the Cotton effect.

The Cotton effect is depicted in Figure 3.2. In the region of an optically active absorption band the ORD curve becomes more negative or positive. It reaches a maximum near the absorption band, reverses its direction to a minimum and then gradually increases again. A Cotton effect is positive when the peak is at the longer wavelength and negative when the trough is at the longer wavelength.

The measurement of circular dichroism is concerned only with light absorption and so is of particular relevance to optically active chromophores, such as chelate complexes. Much has been done on correlating the configuration of optically active complexes with their Cotton effect and also in applying this

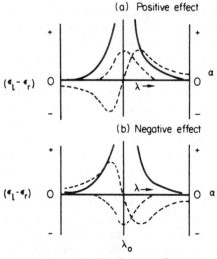

Figure 3.2 The Cotton effect.

technique to the assignment of electronic spectra. Cotton effects have been observed for a number of metalloproteins, indicating that the metal-binding sites are asymmetric.

ORD allows the study of general protein structure and the configuration of particular areas such as the enzyme active site. The Cotton effects of proteins basically arise from the asymmetry of the polypeptide skeleton and will be sensitive to the secondary and tertiary structure of the protein. These have been termed *Intrinsic Cotton effects*. Superimposed upon this will be *side-chain Cotton effects*, resulting from amino acid residues in asymmetric environment, and *Extrinsic Cotton effects*, resulting from the asymmetric interaction of the protein or peptide with other groups such as coenzymes, inhibitors, metal ions, or substrates. The Cotton effect will therefore be a reflection of the asymmetry of the metal-binding site, either as a direct result of the conformations of the ligands or as a result of their interaction with metals. Alternatively, the binding of an additional (not necessarily optically active) group to the metal may produce the asymmetry.

Vallee and Riordan[20] have used the extrinsic Cotton effect to study conformational changes during the action of carboxypeptidase. The coupling of *p*-azobenzenearsonate with tyrosyl residues of carboxypeptidase A leads to the presence of an extrinsic Cotton effect. On addition of substrate glycyl-L-tyrosine the band at 530 nm disappears while lower wavelength bands are also affected. This shows that binding of substrate to the enzyme alters the conformation of azotyrosine.

Resonance Raman spectroscopy

Vibrational spectroscopy offers insight into the geometry and bonding arrangements of localized groups of atoms in molecules. The technique of infrared spectroscopy is clearly only of very limited use in biological systems in view of the aqueous media utilized. Laser Raman spectroscopy may be used in aqueous solutions but complications arise from the complexity of biological molecules. These problems are avoided in resonance Raman spectra where only the vibrations associated with a chromophoric group are observed.

In resonance Raman spectroscopy, the exciting line is tuned into an electronic absorption band. There is a coupling of electronic and vibrational transitions and some of the Raman bands which are related to the electronic transition that is responsible for the absorption will be greatly enhanced. If a biological sample has a suitable chromophore then only the vibrations in the vicinity of the site will be enhanced. The remainder of the vibrational modes will not be enhanced and so cannot complicate the spectra. Biological chromophores associated with this effect include transition metal ions and heme groups, and these are usually the sites of biological function. Thus resonance Raman spectroscopy is a highly selective and sensitive technique for exploring biological chromophores.[21]

Heme systems have been much studied. In this case the Raman spectra are dominated by the porphyrin vibrational modes, which are enhanced by resonance with the allowed electronic $\pi \rightarrow \pi^*$ transitions of the porphyrin in the visible and near-UV region. Vibrations involving the iron atom are not strongly enhanced. Certain of the porphyrin frequencies show substantial shifts as the Fe(III) changes to Fe(II) (without change in spin state) and as the spin state changes without change in oxidation state. The latter phenomenon is attributed to doming of the heme group. Correlations have also been made with the nature of the axial ligand.

The resonance Raman technique has also been used to study interactions between hemes in biological membranes, while resonance Raman spectra of whole mitochondria have been reported and assigned to various cytochrome components.[22]

Magnetic resonance techniques

Nuclear magnetic resonance and electron paramagnetic resonance spectroscopy have proved of immense value in the study of biologically important molecules. Many illustrations of the use of these techniques may be found in recent reviews and symposia reports.[23-26]

EPR. This technique was discussed in Chapter 2. It will give information regarding the redox state of the metal (although this may be ill-defined), covalency of the metal–ligand bond, the nature of the ligands, and the stereochemistry around the metal ion. The spectrum itself will be defined in terms of *g* values, hyperfine structure constants, and the area under the absorption curve. For metalloproteins EPR measurements have to be made on frozen solutions, rather than a crystal, and there is, therefore, an averaging out of all orientations of the molecules compared with the fixed orientation of a single crystal. The interpretation of such EPR data is difficult and often rather empirical conclusions have been reached for metalloproteins, mainly by comparison with the spectra of model compounds. Programmes are now available for the computer simulation of spectra from frozen aqueous solutions.[27]

An example of the use of EPR in detecting the binding groups for a metal is the study of a single crystal of copper-substituted insulin. It has been suggested,[28] on the basis of the ligand hyperfine structure, that the spectra are typical of copper interacting with two magnetically equivalent nitrogen donor atoms in a complex of trigonal symmetry. Associated titration studies involving a comparison between experimental and computed titration curves lead to the suggestion that the cupric ion is bound to identical nitrogen donors, either imidazole or the amino groups of *N*-terminal phenylalanines.

EPR studies have also been carried out on enzymes which contain a probe paramagnetic ion. Cohn and her colleagues have pioneered the use of Mn(II) in this connection and have been able to make substantial contributions to the understanding of the behaviour of a number of Mg(II)-catalysed enzymes, usually associated with phosphate groups. An alternative approach[29] is to use a 'spin label' which is a stable free radical such as a nitroxide which will bind to a particular site on the enzyme. It is not always possible specifically to label an amino acid. EPR has been used to detect conformational changes induced by metal ions and substrates in creatine kinase. The study of enzymes containing Mn(II) and a spin label has been of particular value, as the interaction between these two centres may be detected in the EPR spectrum. Thus in creatine kinase the geometry at the active site has been clarified in that it has been possible to estimate that the distance between the Mn(II) ion and the spin label in the Mn−ADP complex is 8.0 ± 0.2 Å. These will be discussed further in Chapter 4.

Reference was made in Chapter 1 to the enzyme glutamine synthetase. This is made up of twelve identical subunits and catalyses the biosynthesis of glutamine from glutamate, ammonia, and ATP. Each subunit binds two divalent cations for catalysis to occur. Binding at the first site (of high affinity for cation) produces conformational changes in the protein, while the second metal ion site is the metal−nucleotide site. These two sites for Mn(II) binding have been investigated by EPR,[30] while Cr(III) nucleotides have been used as paramagnetic substrate analogues.[31] The latter data has allowed estimates of the distances between Mn(II) bound at the high-affinity site and Cr(III) nucleotide bound at the second metal ion site (7.1 Å). In the presence of substrate this distance became 5.2 Å showing that substrate binding causes the two sites to move closer together through conformational changes in the protein. This same conclusion has been reached[30] from the observation that binding of nucleotide to the second Mn(II) site renders this Mn(II) signal unobservable and causes a reduction in signal from the high-affinity Mn(II) site. These results imply that the two sites have moved together to allow interaction and also that the Mn(II) bound at the first site may also be involved in substrate binding and catalysis in addition to inducing conformational change.

NMR. Proton magnetic resonance techniques have been extensively applied to the study of the detailed structure of proteins and nucleic acids, particularly since high-resolution spectrometers have been available. The use of computer averaging has helped improve the signal/noise ratio so that low concentrations can now be studied. Conformational changes in the metalloproteins have been monitored by PMR. Thus when the sixth ligand in myoglobin is fluoride or cyanide the complex is high spin or low spin respectively. This changeover in spin state is associated with a change in metal−ligand bond length of only 0.1 Å, yet the PMR spectrum of myoglobin is considerably affected, reflecting the conformational changes that have taken place. A range of other nuclei in addition to ^1H have now been utilized.

Metallothionein has been much studied by NMR techniques. This is a cysteine-rich protein (some 35% of amino acid residues) of molecular weight 6000−7000. It contains between 6 and 10% metal, usually cadmium and/or zinc. The protein has been suggested to be involved in metal detoxification. PMR studies have shown that the structure of the apoprotein is such that each residue is exposed to solvent, but the metallo forms have well-defined tertiary structure.[32] ^{113}Cd NMR studies show that there are seven magnetically non-equivalent CdS_4 sites in Cd−metallothionein.[33]

Binding of metals to other macromolecules has been studied by the shifting of resonances in ^{1}H and ^{13}C NMR spectra on addition of metal. Thus the binding of Zn(II) to the antibiotic bleomycin[34] has been linked to pyrimidine and imidazole groups in the molecule. A much more specific approach[35] has utilized the binding of ^{199}Hg(II) to bleomycin and the observation of spin–spin coupling in the NMR spectrum. Such studies have shown that unlike Zn(II), Hg(II) is not bound to pyrimidine residues. Confirmation that a histidyl group is a ligand is obtained from splitting of the C-2(H) and C-4(H) resonances of the histidyl group due to ^{199}Hg–^{1}H coupling.

One very important application of NMR lies in the study of the effect of a paramagnetic ion upon the spectrum of a metal-containing system. Metal-for-metal substitution has allowed this approach to be applied quite widely. One example[36] involves the preparation of an active manganese carboxypeptidase complex and the demonstration that one molecule of water is bound by the metal ion and that this is displaced by the inhibitor β-phenylpropionate.

Cohn[24] has looked in particular at the use of Mn(II) as a paramagnetic probe. She has shown that the effect of the paramagnetic ion upon the relaxation rate of the NMR signal of water protons is very sensitive to the type of metal complex formed, and that it can, in appropriate cases, provide much useful information in terms of the role of the metal ion and the nature of the metal–enzyme and metal–enzyme–substrate complexes, including their binding constants. She has applied this technique in particular to enzymes which catalyse phosphoryl transfer reactions of the type $ATP^{4-} + XH \rightleftharpoons ADP^{3-} + XPO_3^{2-} + H^+$.

The study of NMR relaxation rates (I/T_1) of substrate nuclei in the presence of paramagnetic probes has allowed the estimation of distances from the probe to specific atoms. Some fourteen such distances have been compared[25] with the corresponding crystallographic distances, and appear to have been calculated with precisions of $\pm 10\%$. Conformational changes have been studied by this technique. Paramagnetic sources have included Mn(II), Cr(ATP), and also spin-labelled compounds such as analogues of NAD. Spin labels are usually less effective in causing nuclear relaxation.

Mössbauer spectroscopy

The elements Fe, K, I are all Mössbauer nucleides of potential biological interest but ^{57}Fe is the most easily used and at present the only one used extensively. Mössbauer spectra are presented in terms of three parameters; the isomer shift, the quadruple splitting, and the magnetic hyperfine splitting. Isomer shifts reflect the electronegativities of the nearest-neighbour atoms and hence the ionic character of bonds. Quadruple splittings reflect the distribution of electric charge in atoms, while the hyperfine splitting often enables a distinction to be made between the different possible spin states. Low temperature is essential for observing the hyperfine splitting while the variation of the other parameters with temperature provides useful information. Most iron-containing systems, however, involve too dilute an iron concentration for the hyperfine splitting to be

TABLE 3.7 Some results of the application of chemical and instrumental methods (other than X-rays) to the determination of metal-binding groups in metalloproteins.

Metalloprotein	Binding groups and experimental Method	Comments
Carboxy peptidase (Zn)	RS⁻ and NH_2: chemical tests on apoenzyme and enzyme for the —SH group; a comparison of stability constants for a range of metal carboxy-peptidases with those for complexes of mercaptoethylamine	X-ray studies have shown that —SH are *not* binding groups
Carbonic anhydrase (Zn)	Three nitrogen bases: a comparison of stability constant data for some metal–carbonic anhydrases and metal–ethylenediamine complexes; EPR data on the Cu enzyme and model complexes; chemical evidence against —SH binding	X-ray studies indicate binding *via* three imidazole groups
Ferredoxin (Fe)	Inorganic sulphide and —SH: requirement for inorganic sulphide for reformation of the enzyme from the apoenzyme; chemical studies on enzyme and apoenzyme for binding *via* —SH	X-ray studies show the iron–sulphur cubes
Alkaline phosphatase (Zn)	Three imidazoles of histidine: photo-oxidation of native and apoenzymes shows that three histidyl residues are protected in the native enzyme and are not oxidized, the treated apoenzyme has a reduced zinc-binding capacity	
Plastocyanin (Cu)	Stereochemistry suggested to be tetrahedrally distorted from square-planar by EPR, CD and MCD. Chemical and instrumental evidence for binding of Cu by Cys and two His. Tyr and peptide N (deprotonated) also suggested to be ligands.	X-ray studies show irregular geometry, and Cys, two His and Met residues to be ligands
Transferrin and conalbumin (Fe)	Tyrosine and histidine residues: pH studies indicate that the metal binding groups have pK_a values about 11.2 and 7; also EPR studies on copper-substituted proteins (see above)	
Cytochrome *c* (Fe)	The sixth ligand has been suggested to be histidine, lysine, and methionine by different workers.	X-ray studies show it to be methionine

observed. The iron–sulphur proteins provide one exception here. The low natural abundance of ^{57}Fe may necessitate ^{57}Fe enrichment of the iron-containing protein by appropriate methods. Mössbauer studies have been carried out on hemoglobin and myoglobin and on iron–sulphur proteins. Thus the oxidized spinach iron–sulphur protein (containing two iron atoms per 13,000 molecular weight) has been shown by studies at 4.6 K to involve identical atom environments with iron as Fe(III). After reduction with excess dithionite the Mössbauer spectrum indicated that one half of the iron had been reduced to Fe(II). The quadruple splitting confirmed that this was high-spin Fe(II). Chemical studies had already shown that ferredoxin was a single-electron transporter.

EPR and Mössbauer are useful complementary tools. EPR is very sensitive to paramagnetic species, and allows the characterization of intermediate species at low concentration. Mössbauer spectroscopy is sensitive to all the iron atoms present and so will not emphasize paramagnetic forms.

Table 3.7 lists some conclusions that have been drawn concerning the nature of the metal-binding groups in metalloproteins on the basis of the results of applying the techniques described in the preceding sections.

The stereochemistry and oxidation states of metals in metalloenzymes

It has already been emphasized that the stereochemical requirements of the protein molecule may result in an abnormal site symmetry for the metal, and that, for redox metalloenzymes, this is often associated with unusual redox potentials and difficulties in defining the oxidation state of the metal. When in a subsequent section we come to talk about model compounds, we will then be placed in a difficulty inasmuch as it is not possible to obtain models that duplicate these factors. Vallee and Williams[37] have put forward the general suggestion that the active metal site of the enzyme is in a geometry approaching that of the transition state of the appropriate reaction and as such is uniquely fitted for catalytic action. They have termed this an entatic state. In later chapters some reference will be made to this. Table 3.8 lists a number of metalloproteins for which irregular physical properties have been observed. Some have already been discussed in the text.

Studies with inhibitors

A study of the role of inhibitors in enzymatic reactions has contributed greatly to an understanding of enzyme mechanisms. In this section the inhibition of metalloenzymes by added chelating agents will be discussed. One method of showing the presence of a metal in an enzyme is to demonstrate inhibition by added chelating and other coordinating agents. Those widely used include EDTA, nitrilotriacetic acid, Tiron (4,5-dihydroxy-m-benzene sulphonic acid), ethylenediamine, dimethylglyoxime, mercaptoethanol, dipyridyl, dithizone, and the inorganic ions azide and cyanide. It is assumed that any effect is due solely to

TABLE 3.8 Irregular binding sites for metals in metalloenzymes

Protein	Metal	Stereochemistry/coordination number	Techniques
(a) Native			
Hemoglobin	Fe^{2+}	5-Coordinate	X-rays
Myoglobin	Fe^{2+}	5-Coordinate	X-rays
Hemoglobin	Fe^{3+}	Iron displaced out of	
Myoglobin	Fe^{3+}	porphyrin plane. Strong axial field	X-rays/EPR
Cytochromes	Fe^{3+}	Rhombic	EPR
Ferredoxins	Fe^{3+}	EPR g value = 1.94 not understood	EPR
Rubredoxin	Fe^{3+}	Asymmetric	EPR/CD
Conalbumin Transferrin	Fe^{3+}	Very low symmetry	EPR/CD
Vitamin B_{12} coenzymes	Co^{3+}	5-Coordinate	d–d/NMR
Copper blue proteins	Cu^{2+}	Low symmetry	d–d/EPRCD
Oxyhemocyanin	Cu^{2+}	Distorted orthorhombic	d–d
(b) Substituted			
Carbonic anhydrase	Co^{2+}	Distorted, 5-coordinate?	d–d
Carboxypeptidase	Co^{2+}	Irregular	d–d
Alkaline phosphatase (two atoms Co^{2+})	Co^{2+}	Distorted	d–d
Phosphoglucomutase	Co^{2+}	Irregular, 5-coordinate?	d–d

coordination to the metal ion and that there is no other interaction between chelating agent and protein, substrate or coenzyme, i.e. that the inhibitors will preferentially replace the protein groups binding the metal. This problem really involves a comparison of formation constants, but there are severe complications in this. Thus the metal-binding groups may not be necessarily known. Even if they are known, then the reactivity of the metal ion towards the inhibitor may not be that expected as the basis of the study of model compounds. Structural features in the protein may prevent the operation of the chelating agent, while the strength of any metal–inhibitor binding will reflect the stereochemistry and strength of the binding of the metal to the protein. Quantitative comparison is therefore difficult, but despite this much of value can come from a study of metalloenzyme–inhibitor interaction.

The inhibiting action may either involve complete removal of the metal leaving the apoenzyme, which may or may not be reactivated by addition of the original metal ion, or the replacement of some of the protein groups only, so giving a mixed enzyme–inhibitor metal complex. Prevention of inhibition by having free metal ion present, with which the chelate may react, is usually taken as confirmation that the chelating properties of the inhibitor are responsible for the inhibitory effect. If a mixed complex is formed then the inhibition should be reversed by adding excess metal ion to compete for the inhibitor. This reversal of inhibition can be used to demonstrate the presence of the mixed complex.

The measurement of the stoichiometry of metalloenzyme–inhibitor complex is also of value here. Job's method may be applied, using either a maximum of light absorption or of ORD to monitor the system. If a 3:1 complex is formed with a bidentate inhibitor, then clearly the metal ion has been removed completely from the protein. The observation of the formation of a 1:1 complex would suggest the presence of the mixed complex.

Inhibition by chelating agents will throw light on mechanism in a number of ways. Kinetic studies on inhibition can be of value, while in general the effect of inhibition on the physical properties of the system allows the correlation between inhibition and the change in those properties. In Chapter 4 we will consider the way in which the addition of inhibitors to cobalt carbonic anhydrase at different pH values causes changes in the d–d spectrum of the cobalt which may be correlated with inhibitor function and hence the mechanism of action of carbonic anhydrase.

KINETIC STUDIES ON
MODEL SYSTEMS AND METALLOENZYMES

In this section we are concerned particularly with studies on model systems for metalloenzyme reactions and also with studies involving the metal in metallo-enzymes. Again, the theme of the value of *comparing* metal ions is stressed. For a quantitative comparison it is necessary to have kinetic data as well as formation constant data.

In many model systems, the catalytic effect of various metal ions follows the Irving–Williams series, that is, it is the same as the sequence of formation constants. This is of interest mechanistically. Most metal aquo ions form complexes in aqueous solution by fast diffusion of a ligand molecule to the aquo complex forming an outer-sphere complex, while the rate-determining step involves the replacement of the aquo group by the ligand (by the appropriate mechanism). The rate of replacement[38] of water in octahedral complexes is $Ca^{2+} > Cu^{2+} > Zn^{2+} > Mn^{2+} > Fe^{2+} > Co^{2+} > Mg^{2+} > Ni^{2+}$, which is of course different from the Irving–Williams series. It follows therefore that in model systems the rate-determining step is not the formation of the complex. Rather, as the catalytic order follows the Irving–Williams order, it may mean that the rate-determining step is one involving a rearrangement of the metal–substrate complex in which the metal ion assists in orientating the substrate for reaction. The rate, from metal ion to metal ion, will then depend upon the extent to which complex formation has taken place, in which case there should be a direct relationship between the catalytic rate constant, k, and the formation constant K for the different metals. If the formation constant doubles, so will the rate constant.

$$M + S \overset{\text{fast}}{\rightleftharpoons} MS \qquad K$$

$$MS \overset{\text{slow}}{\longrightarrow} \text{products} \quad k$$

Alternatively, the activity of the metal may be one of a Lewis acid. The catalytic effect should then depend as before on the extent of complex formation, together with the differing Lewis acidity of the various metal ions. As Lewis acidity also follows the Irving–Williams series, the overall effect follows this series, but now there will not be the direct correlation between rate constant and formation constant observed for the first case. The rate constant will increase more rapidly than the formation constant. It may be seen therefore that variation of the metal ion and a comparison of rate constant and stability constant may allow some insight into the role of the metal. An example[39] of such Lewis acid catalysis lies in the metal ion catalysed decarboxylation of acetone dicarboxylic acid, where a comparison of rate constant and formation constant over a range of metal ions clearly shows the Lewis acid requirement for the metal ion to catalyse the removal of CO_2.

An increasing number of studies on the kinetics of complex formation in model systems have been reported. These are very fast reactions and are usually studied by the temperature-jump relaxation technique. An early example[40] is the study of the kinetics of complex formation between magnesium(II) aquo ions and 8-hydroxyquinoline ('oxine'). In order to attempt to represent enzyme–metal–substrate complexes more closely in the matter of the presence of ligands other than water in the coordination shell, these workers studied oxine substitution into magnesium ATP, magnesium polytriphosphate, and magnesium uramil-N,N-diacetate complexes in addition to hexaaquomagnesium(II). This meant that a statistical correction had to be introduced to allow for the binding of different numbers of water molecules to the metal ion in the different cases. It appears, however, that the first ligand has comparatively little influence on the rate at which the metal aquo complex reacts with a second ligand.

On the basis of the calculated rate constants for the different stages of the reaction, an attempt was made to rationalize the differing catalytic powers of Ca^{2+} and Mg^{2+}. While not differing very much in terms of formation constant, their catalytic powers can be quite different. Thus Ca^{2+} will inhibit the reactions of some Mg^{2+}-activated enzymes and *vice versa*. It appears that the rate constants for the formation of a complex are about 1000 times larger for Ca^{2+} than for Mg^{2+}. As the formation constants are fairly similar, this must mean that calcium complexes dissociate about 1000 times more rapidly than magnesium complexes, as $K = k_f/k_b$. In other words, the extent of complex formation stoichiometrically may be the same, but the lifetime of the complexes is much shorter for Ca^{2+}.

The relative catalytic properties of the two ions may then be understood by a scheme in which an internal rearrangement of the complex E–M–P (P = product) may or may not be necessary before the final product is liberated by complex dissociation. If the rearrangement *must* occur (e.g. a conformational change of the protein) then the necessary changes must take place in a time shorter than the lifetime of the complex. For calcium, whose complexes have a shorter lifetime, this cannot occur, but a comparison of reaction rates suggests that it may be possible for magnesium. For such a reaction Ca^{2+} will be an inhibitor.

On the other hand, if there is no need for the rearrangement, the product will be produced more quickly for Ca^{2+}, as the rate of dissociation is faster. Ca^{2+} will therefore be a more effective catalyst of such reactions.

It should be noted that it is possible to explain the differing catalytic efficiencies of Mg^{2+} and Ca^{2+} purely in terms of their preferences for different binding sites (i.e. binding groups and site symmetry). This seems a more reasonable explanation, and will be discussed further in Chapter 9.

A second study[41] with oxine has involved measuring the kinetics of its complex formation with molybdate. This has been used as a model system for enzyme–substrate binding in xanthine oxidase. This is a 290,000 molecular weight protein which catalyses the oxidation of xanthine to uric acid by molecular oxygen. EPR studies have shown that electron transfer from substrate to oxygen proceeds *via* molybdenum, FAD (flavin adenine dinucleotide) and iron. A Mo(V) signal is observed within 10 ms reaction time between substrate and enzyme, so indicating that molybdenum forms part of the active site of the enzyme. The formation of a complex between molybdate and oxine will serve as a model system to indicate the feasibility of this suggestion.

The results in the pH region 7.9–8.9 show that a 1:1 complex is formed, to an extent dependent upon the pH, reflecting protonation both of the molybdate and oxinate anions. The enzymatic reaction has been studied, and it appears that the overall forward rate constant for the oxine–molybdate reaction is about 4–5 orders of magnitude lower than that[42] estimated for the enzyme–substrate binding reaction. However, there is EPR evidence that xanthine oxidase involves a proton in association with the molybdenum, so it may be more meaningful to compare with oxine substitution into $[MoO_3OH]^-$, data for which were calculated.[41] The rate constants are in fact fairly close to the estimated one for the enzyme–substrate binding reaction and are therefore compatible with the suggestion that molybdenum is bound in the suggested way in xanthine oxidase (Figure 3.3).

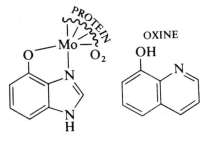

Figure 3.3 Suggested binding of xanthine to molybdenum in xanthine oxidase.

To turn from model compounds to metalloenzymes, kinetic studies of particular interest here include metal-for-metal exchange rates. These are very much slower than those for model compounds, the rate-determining step probably being the relaxation of the protein.

The binding of zinc by the apoenzyme of carbonic anhydrase has been studied.[43] The recombination is accelerated by the dissociation of protons from groups with apparent pK_a values of 5.4 and 7.2. Again the rate constants are orders of magnitude smaller than for the formation of zinc complexes where the rate-determining step is dissociation of water from the first coordination sphere. The kinetic parameters also differ in the two cases.

THE USE OF MODEL COMPOUNDS

The complexity of biological macromolecules, and the associated problems in their study, means that it is often worthwhile to examine the behaviour of simpler molecules that contain what appear to be the essential features of the original molecule, uncomplicated by other features of that molecule. The model compounds may also have other useful properties such as solubility, not possessed by the original species. Provided that the comparison is a real one, then the conclusions reached for model compounds may be extrapolated to the parent molecule and some light thrown upon its reactions and functions. On the other hand, differences between the behaviour of the models and the molecule may also have some significance, as will be demonstrated for certain hydrolytic enzymes.

Despite the fact that good chemical models are available for most enzymatic reactions, it should still be emphasized that a comparison of rates, in general, shows that the reactions undergone by the models are very much slower than the enzymatic reactions. It is possible to attempt to correct for certain catalytic features in the enzyme but even so, the enzymatic reactions may be up to 10^{11} times faster than the model reactions. This serves to re-emphasize the cautionary note expressed in the introduction to this chapter.

A great deal of information is available on amino acids and simple peptides as models for proteins. The conclusions discussed earlier in this chapter on the binding groups of metals in proteins are based upon this type of information. Here, however, there are problems which we have not yet discussed in detail. The reactivity of amino acid side chains (such as pK_a) may be changed on incorporation into the protein, due to the hydrogen bonding or other interactions in the protein. Often the catalytic function of an enzyme may well depend upon the protein environment generating an abnormal physical property for some group or grouping. Such a feature obviously cannot be incorporated into the design of model compounds. Again, other protein groups may be rendered inaccessible by the protein structure, and so the reactivity for that group implied by model compounds may not mean anything in that particular example. This explains why chemical tests for the presence of certain residues in the protein may not function. —SH groups are well known for demonstrating lowered reactivity on incorporation into the protein.

We have already discussed the unusual stereochemistry of the metal ions in many metalloenzymes. This cannot be readily represented by model compounds, so their value is limited. The order of catalysis of hydrolysis in model compounds by various metal ions is not always followed in enzyme systems for the same metals. Again the relationship to the models must be examined closely. Finally, when considering models for redox enzymes, it must be emphasized that the electronic state of the model compound must be known before it can be compared with the metalloenzyme.

The use of model compounds has allowed the build up of much information regarding the reactivity of coordinated ligands. The presence of a metal ion often has a very marked effect upon the reactivity of the group. The Lewis acid behaviour of metal ions will lower the basic properties of coordinated groups and so enhance attack by nucleophiles or cut down the possibility of electrophilic attack. The use of model compounds also illustrates the role of metal ions as templates and their possible role in biosynthesis of macrocyclic molecules. There are many examples in model compounds of the condensation of amines and carbonyl groups to give Schiff bases being aided by metal ion templates. Other condensations include[44] the formation of square-planar nickel(II) chelates in which α-diketones condensed with β-mercapto amines.

Without the metal ion, the yield is very poor, as the amine and mercapto groups act as competitive nucleophiles. The chelate may then be closed by reacting with an appropriate difunctional molecule.

Applications of this type of behaviour may well be important in the biosynthesis of certain natural products and have in fact been used in the synthesis[45] of corrin complexes, as noted in Chapter 1.

Models for vitamin B$_{12}$ coenzymes

The corrin and porphyrin ring systems are of great biological importance. They are shown in Figure 3.4.

The porphyrins[46] are derived from porphin, varying according to the nature of the substituents. They are all intensely coloured and highly conjugated. The group itself is tetradentate with four pyrrole-like nitrogens surrounding a central site for metals. On coordination the two NH protons are lost, as is the NH proton in corrin. To illustrate the importance of the tetrapyrrole group it may be noted that iron complexes of the porphyrins are the hemes, and that depending upon

84

Corrin   Porphin

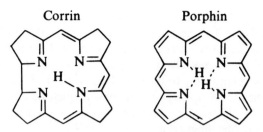

Figure 3.4 Corrin and porphin ring structures.

the nature of the protein in the hemoprotein and the substituent on the porphyrin, we have (i) the hemoglobins and myoglobins in which the iron(II) hemoprotein will combine reversibly with oxygen without oxidation to the iron(III) state; (ii) peroxidases, which catalyse the hydrogen peroxide oxidation of substrates, and (iii) catalases, which catalyse the decomposition of hydrogen peroxide. On the other hand, if magnesium lies at the centre of the hydroporphin nucleus, we have the chlorophylls, while if cobalt is the central atom of the corrin ring system we have vitamin B_{12} and the series of cobamide coenzymes, depending in both cases upon the nature of the substituents.

From the chemical point of view the 'organometallic chemistry' of cobalt in vitamin B_{12} and its derivatives is unusual and interesting. This may be parallelled quite remarkably in many cases by some quite simple model compounds.[47,48]

The corrinoid ring systems provide a square-planar set of donor atoms, while the cobalt usually has axial ligands above and below the plane, although sometimes one only. If one of the axial ligands is the benzimidazole of the associated nucleotide, the complex is a cobalamin; if this is replaced by a water molecule, then we have the cobinamides. Cobalt may be formally regarded as cobalt(III). Vitamin B_{12} is cyanocobalamin (i.e. the sixth group is cyanide, a result of the isolation procedure), B_{12a} aquocobalamin, B_{12c} nitrocobalamin. The B_{12a} form may be reduced first to B_{12r} and then to B_{12s}. B_{12r} is a low-spin cobalt(II) complex, while B_{12s} is apparently a cobalt(I) species.

The 5-deoxyadenosyl cobalamin has the adenosyl group carbon bonded to the cobalt. This was the first naturally occurring Co—C bonded structure to be recognized. Previously only a small number of highly unstable compounds with cobalt–carbon bonds had been known. Alkylcobalt systems are now well known in models based upon the vitamin B_{12} structure. For the cobinamides in particular the alkyl species exist partly as five-coordinate structures, with the water molecule displaced. Structures are given in Figure 3.5.

Porphyrins have been used as models for corrin ring systems. They will add groups readily in the axial positions. This is not typical of all porphyrins. Thus Cu(II) and Ni(II) porphyrins have a low affinity for extra ligands, while Mg(II), Zn(II), and Cd(II) porphyrins add on one ligand giving a square-pyramidal five-coordinate structure. Fe(II), Mn(II), and Co(II) porphyrins all add on two moles of ligand. Cobalt porphyrins[46] can be converted into organic derivatives, but they cannot be reduced to the cobalt(I) state in aqueous solution. The reduction

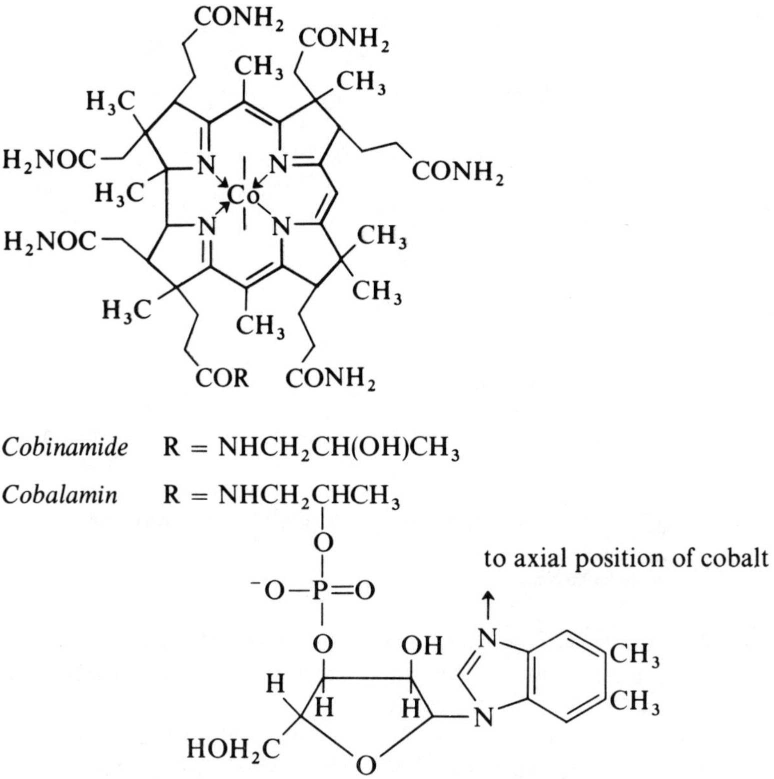

Figure 3.5 The structure of cobalamin and cobinamide.

is usually carried out with a Grignard reagent in non-aqueous solution. However, bis(dimethylglyoxime)cobalt complexes (the cobaloximes) (Figure 3.6) show very many of the reactions of the cobalt atoms in corrins. They too add on axial groups and form stable organic derivatives readily, and also they can be reduced to cobalt(I) species. These, though, are very sensitive to oxygen and only exist in alkaline pH. In acid solution, like vitamin$_{12s}$ they decompose liberating hydrogen and giving cobaloxime. The comparison between cobaloximes and B_{12s} has contributed to an understanding of the latter. It appears that the close similarly between cobalamins and cobaloximes is due to the presence of an in-plane ligand of similar strength and is independent of the axial ligands. This is supported by physical methods,[49] while the similarity is also justified by theoretical methods.[50]

In order to illustrate the comparison with the model compounds, it should be noted now that the reactions catalysed by the cobamide coenzymes include

(a) methyl group transfer in which the cobalt(III) presumably is the methyl-accepting group,
(b) reduction reactions, and
(c) rearrangement reactions.

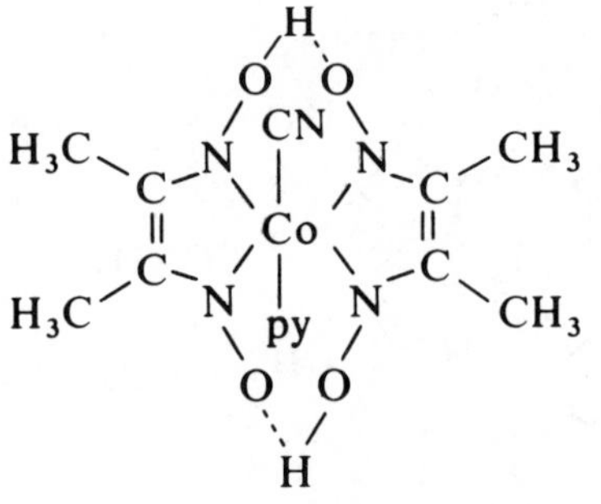

py = pyridine

Figure 3.6 Cyanopyridine cobaloxime.

The first type of behaviour is that shown in the biosynthesis of methionine:

$$\underset{\text{(methylcobalamin)}}{\overset{\displaystyle \overset{\text{CH}_3}{\underset{|}{\text{Co}}}}{}} + \text{homocysteine} \rightarrow (\text{Co}^\text{I})^- + \text{methionine}$$

Methylcobaloximes will similarly methylate homocysteine, although the re-action is not reversible. Demethylation is possible, however, provided it is first converted to the S-adenosyl derivative.

Both vitamin B_{12} coenzymes and cobaloximes will also catalyse reduction reactions involving the synthesis of N-methyl groups from formaldehyde and amines in the presence of a reducing agent such as molecular hydrogen.

$$\text{HCHO} + \text{NH}_2\text{C}_6\text{H}_5 \rightleftharpoons \text{C}_6\text{H}_5\text{NHCH}_2\text{OH} \rightarrow \text{C}_6\text{H}_5\text{NHCH}_3 + \text{H}_2\text{O}$$

Intermediates have been isolated of the type

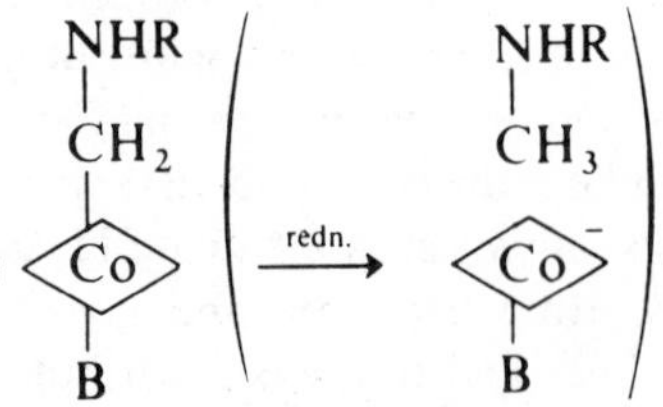

Other aspects of model compounds for vitamin B_{12} will be discussed later; the work of Schrauzer and his group has, however, shown the value of a comparison of alkycobalamins and alkylcobaloximes.

GENERAL CONCLUSIONS

The methods outlined in the preceding sections indicate the general approaches that may be followed in the study of the binding and function of metals in biological macromolecules. A number of examples have been given for each method. In the following chapter we will examine some hydrolytic enzymes in detail to show how these methods may be integrated together. We have chosen to study, in more detail, the role of zinc in carboxypeptidase and carbonic anhydrase in view of the experimental information, particularly X-ray results, available. We will also comment in more general terms on other hydrolytic enzymes.

REFERENCES

1. A. A. Timchenko, O. B. Ptitsyn, D. A. Dolgikh, and B. A. Fedorov, *FEBS Letters*, **88**, 105, 109 (1978).
2. G. M. Alter, D. L. Leussing, H. Neurath, and B. L. Vallee, *Biochemistry*, **16**, 3663 (1977); C. A. Spilberg, J. L. Bethune, and B. L. Vallee, *Biochemistry*, **16**, 1142 (1977).
3. S. P. Cramer and K. O. Hodgson, *Progr. Inorg. Chem*, **25**, 1 (1979).
4. S. P. Cramer, K. O. Hodgson, E. I. Stiefel, and W. E. Newton, *J. Amer. Chem. Soc.*, **100**, 2748 (1978).
5. *Stability Constants*, Special Publication No. 25, The Chemical Society, London, 1971.
6. e.g. F. J. C. Rossotti and H. Rossotti, *The Determination of Stability Constants*, McGraw-Hill, 1961.
7. T. -L. Ho, H. C. Ho, and L. D. Hamilton, *Chem. Biol. Interactions*, **23**, 65 (1978).
8. C. A. McAuliffe, in the Chemical Society Specialist Periodical Report on *Inorganic Biochemistry* (ed. H. A. O. Hill), **1**, 1 (1979).
9. H. C. Freeman, *Adv. Protein Chem.*, **1968**, 342.
10. T. L. Blundell and J. A. Jenkins, *Chem. Soc. Rev.*, **6**, 139 (1977).
11. B. L. Vallee and W. E. C. Wacker, in *The Proteins* (Ed. H. Neurath), (2nd edn.), Vol 5, Academic Press, 1970.
12. J. I. Legg, *Coord. Chem. Revs.*, **25**, 103 (1978); R. J. P. Williams, *Chem. in Britain*, **14**, 25 (1978).
13. E. J. Peck and W. J. Ray, *J. Biol. Chem.*, **244**, 3748 (1969).
14. K. D. Danenberg and W. W. Cleland, *Biochemistry*, **14**, 28 (1975).
15. e.g. C. F. G. C. Geraldes and R. J. P. Williams, *Eur. J. Biochem.*, **85**, 463 (1978) and following papers.
16. H. M. N. H. Irving and R. J. P. Williams, *J. Chem. Soc.*, **1953**, 3192.
17. S. Lindskog and P. O. Nyman, *Biochim. Biophys. Acta*, **85**, 461 (1964).
18. A. S. Brill, B. R. Martin, and R. J. P. Williams, in *Electronic Aspects of Biochemistry* (Ed. B. Pullman), Academic Press, 1964, p. 554.
19. R. Druyan and B. L. Vallee, *Fed. Proc.*, **21**, 247 (1962).
20. B. L. Vallee and J. F. Riordan, *Brookhaven Symp. Biol.*, **21**, 91 (1968).
21. T. G. Spiro, *Acc. Chem. Res.*, **7**, 339 (1974); D. P. Strommen and K. Nakomoto, *J. Chem. Ed.*, **54**, 474 (1977); P. R. Carey and H. Schneider, *Acc. Chem. Res.*, **11**, 122 (1978).
22. F. Adar and M. Erecińska, *Biochemistry*, **17**, 5484 (1978).
23. A. Ehrenberg, B. G. Malmstrom, and T. Vanngard (Eds.), *Magnetic Resonance in Biological Systems*, Pergamon Press, 1967.
24. M. Cohn, *Q. Rev. Biophys.*, **3**, 61 (1970); M. Cohn and J. Reuben, *Acc. Chem. Res.*, **4**, 214 (1971).
25. A. S. Mildvan, *Acc. Chem. Res.*, **10**, 246 (1977).
26. P. I. Quilley and G. A. Webb, *Coord. Chem. Revs.*, **12**, 407 (1974).
27. D. J. Lowe, *Biochem. J.*, **171**, 649 (1978).
28. A. S. Brill and J. H. Venable, *The Biochemistry of Copper* (Eds. J. Peisach, P. Aisen, and W. E. Blumberg), Academic Press, 1966, p. 67.

29. H. M. McConnel and B. G. McFarland, *Quart. Rev. Biophys.*, **3**, 91 (1970).
30. G. E. Hofmann and W. S. Glaunsinger, *J. Biochem. (Tokyo)*, **83**, 1769, 1779 (1978).
31. M. S. Balakrishnan and J. J. Villafranca, *Biochemistry*, **17**, 3531 (1978).
32. A. Galdes, H. A. O. Hill, I. Bremmer and B. W. Young, *Biochem. Biophys. Res. Comm.*, **85**, 217 (1978); A. Galdes, M. Vasak, H. A. O. Hill, and J. H. R. Kagi, *FEBS Letters*, **92**, 17 (1978).
33. P. J. Sadler, A. Bakka, and P. J. Beynon, *FEBS Letters*, **94**, 315 (1978).
34. T. C. Dabrowiak, F. T. Greenaway, and R. Grulich, *Biochemistry*, **17**, 4090 (1978); A. E. G. Cass, A. Galdes, H. A. O. Hill, and C. E. McClelland, *FEBS Letters*, **89**, 187 (1978).
35. A. E. G. Cass, A. Galdes, H. A. O. Hill, C. E. McClelland, and C. B. Storm, *FEBS Letters*, **94**, 311 (1978).
36. R. G. Shulman, G. Navon, B. J. Nyluda, D. C. Douglass, and T. Yamane, *Proc. Natl. Acad. Sci. U.S.A.*, **56**, 39 (1966).
37. B. L. Vallee and R. J. P. Williams, *Proc. Natl. Acad. Sci. U.S.A.*, **59**, 498 (1968); *Chem. in Britain*, **4**, 397 (1968).
38. M. Eigen, *VII Int. Conf. Coord. Chem.*, Butterworths, 1963, p. 97.
39. J. E. Prue, *J. Chem. Soc.*, **1952**, 2331.
40. D. N. Hague and M. Eigen, *Trans. Far. Soc.*, **1966**, 1236.
41. P. F. Knowles and H. Deibler, *Trans. Far. Soc.*, **1968**, 977.
42. H. Gutfreund and J. M. Sturtevant, *Biochem. J.*, **73**, 1 (1959).
43. R. W. Henkens and J. M. Sturtevant, *J. Amer. Chem. Soc.*, **90**, 2669 (1968).
44. M. C. Thompson and D. H. Busch, *J. Amer. Chem. Soc.*, **84**, 1762 (1962).
45. A. Eschenmoser, R. Scheffold, E. Bertele, M. Pesaro, and H. Gschwend, *Proc. Roy. Soc.*, **A288**, 306 (1965).
46. J. E. Falk, *Porphyrins and Metalloporphyrins*, Elsevier, 1964.
47. G. N. Schrauzer, *Acc. Chem. Res.*, **1**, 97 (1968).
48. J. M. Pratt, *Inorganic Chemistry of Vitamin B_{12}*, Academic Press, 1972.
49. G. N. Scrauzer and L. P. Lee, *J. Amer. Chem. Soc.*, **90**, 6541 (1968); G. N. Schrauzer, R. J. Windgassen, and J. Kohnle, *Ber.*, **98**, 3324 (1965).
50. G. N. Schrauzer, L. P. Lee, and J. W. Sibert, *J. Amer. Chem. Soc.*, **92**, 2997 (1970).

Washington

Publication Notice No. 59

Available Date April 22, 1981
Publication Date April 22, 1981

Second Edition
Paper Only

Author & Title: M. N. <u>Hughes</u>: The Inorganic <u>Chemistry</u> of Biological Processes

LC# 80-40499

Product Line	L	**Compositor**	Wiley Ltd.
Title Code	471-27815-7	**Printer**	Wiley Ltd.
List Price	$27.95	**Binder**	Wiley Ltd.
Subject Code	C100/BC00	**Pages**	338
Import Code		**Trim Size**	9 X 6
Type Book		**1st. Yr. Sales Estimate:**	800
Weight ____lbs. ____oz.	57	**Remarks:**	(950 Paper)
Restricted Market Code			
Bindery Code			
Editor Code			
W.I.E. **Yes** **No**			

cc: 1972, 1981 by John Wiley & Sons, Ltd.

Market Rights: World excluding Australia, New Zealand, Papua New Guinea, United Kingdom, Continental Europe (including Turkey), Central Africa and French speaking Africa

Linda Sybert

D. P. No. 1

CHAPTER FOUR

METALLO- AND METAL-ACTIVATED ENZYMES IN HYDROLYSIS AND GROUP-TRANSFER REACTIONS

Some metalloenzymes and metal-activated enzymes involved in hydrolysis and group-transfer reactions are listed in Table 4.1. The metals involved in these enzymes are Mg^{2+}, Ca^{2+}, and Zn^{2+}, whose role is essentially that of Lewis acids.

The main classes of hydrolytic enzymes may be summarized in terms of the nature of the substrate being hydrolysed. Peptides will be hydrolysed from the carboxy terminus by carboxypeptidases and from the amino terminus by amino peptidases. These will have varying degrees of specificity towards particular amino acid residues. Thus leucine aminopeptidases are particularly active towards leucine residues of L configuration. Zinc is necessary for the activity of many of these enzymes, although a second metal, usually calcium, is sometimes required as in thermolysin. A whole range of microbial neutral proteases are known which appear to be metal ion dependent in that they are inhibited by added chelating agents. *Bacillus subtilis* proteases are dependent upon zinc and

TABLE 4.1 Some hydrolytic enzymes.

Enzyme	Reaction catalysed	Metal ion
Carboxypeptidase	Hydrolysis of *C*-terminal peptide residues	Zn^{2+}
Leucine aminopeptidases	Hydrolysis of leucine *N*-terminal peptide residues	Zn^{2+}
Dipeptidase	Hydrolysis of dipeptides	Zn^{2+}
Neutral protease	Hydrolysis of peptides	Zn^{2+}, Ca^{2+}
Collagenase	Hydrolysis of collagen	Zn^{2+}
Phospholipase C	Hydrolysis of phospholipids	Zn^{2+}
β-Lactamase II	Hydrolysis of β-lactam ring	Zn^{2+}
Thermolysin	Hydrolysis of peptides	Zn^{2+}, Ca^{2+}
Alkaline phosphatase	Hydrolysis of phosphate esters	Zn^{2+}
Carbonic anhydrase	Hydration of CO_2	Zn^{2+}
α-Amylase	Hydrolysis of glucosides	Ca^{2+}, Zn^{2+}
Phospholipase A_2	Hydrolysis of phospholipids	Ca^{2+}
Inorganic pyrophosphatase	Hydrolysis of pyrophosphate to orthophosphate	Mg^{2+}
ATPase	Hydrolysis of ATP to ADP and inorganic phosphate	Mg^{2+}
Na^+–K^+–ATPase	Hydrolysis of phosphate with transport of cations	Na^+, K^+
Mg^{2+}–Ca^{2+}–ATPase	Hydrolysis of phosphate with transport of cations	Mg^{2+}, Ca^{2+}
Various phosphatases e.g. fructose diphosphatase	Hydrolysis of phosphate	Mg^{2+}, (Zn^{2+})

calcium while *B. thermoproteolyticus* thermolysin, *B. megaterium* peptidase, and *Streptomyces naraensis* neutral protease are all zinc dependent.

A wide range of enzymes catalyse the hydrolysis of organic and inorganic esters. Thus zinc is found in alkaline phosphatase and phospholipase C which catalyse the hydrolysis of phosphate esters and phospholipids respectively. Magnesium ions are necessary for the activity of many enzymes involved in phosphate hydrolysis and transfer.

In this chapter we will first consider some model studies on metal ion catalysis of hydrolysis and then look in considerable detail at two well-studied zinc metalloenzymes, carboxypeptidase and carbonic anhydrase. The latter enzyme catalyses the hydration of carbon dioxide. It is better to begin by looking at those metalloenzymes for which X-ray diffraction data of high resolution are available, as this eliminates some of the doubts associated with deciding the nature of the metal binding groups. We will, however, look at the chemical and other evidence that has been collected on this matter in order to see how successful those methods have been in determining the nature of the zinc ligands in these enzymes. We will examine the conclusions obtained from the use of models and consider also how the techniques discussed in Chapter 3 have contributed to our understanding of the mode of action of these enzymes. Finally, we will survey other hydrolytic enzymes in more general terms.

MODELS FOR HYDROLYTIC METALLOENZYMES

Much information is available on the hydrolysis of simple peptides and esters. It has been shown that the hydrolysis of esters and peptides is subject to both acid and base catalysis. Particular attention has been paid to the role of metal ions in the hydrolysis of small peptides and correlations drawn between metal ion–peptide stability constants and rate constant over a range of metal ions. The order of reactivity of metal ions in such model reactions is not necessarily parallelled by the order of metal ion activity in a series of metalloenzyme-catalysed reactions, but even so some useful points may be noted. The role of the metal ion as a Lewis acid has been clearly demonstrated.

Bender and Turnquest[1] have studied the copper(II)-catalysed hydrolysis of esters of α-amino acids. On the basis of oxygen-18 exchange data and kinetic studies they have suggested a mechanism involving the formation of a complex in which the ester is chelated *via* the amino and carbonyl groups. Polarization of the carbonyl group then results in enhanced susceptibility of the carbon to nucleophilic attack by a base. Nucleophilic attack is also assisted by the departure of the bonding around the carbonyl group from planarity.

$$\begin{array}{c} X \\ \diagdown \\ \diagup \\ Y \end{array} C{=}O + M^{2+} \rightarrow \begin{array}{c} X \\ \diagdown \\ \diagup \\ Y \end{array} \overset{\delta+}{C}{-}\overset{\delta-}{O}{\rightarrow}M^{2+} \xrightarrow[\text{base}]{H^+-OH^-} XH + YCOOH + M^{2+}$$

X = RNH or EtO

Results such as these obtained with labile complexes are difficult to interpret and other mechanisms cannot be excluded. Of greater significance[2] is work involving inert cobalt(III) complexes, in particular bisethylenediamine complexes with coordinated glycine esters, cis-[Co(en)$_2$(NH$_2$CH$_2$COOR)Cl]Cl$_2$, where R = CH$_3$, C$_2$H$_5$, and iso-C$_3$H$_7$. In these complexes the ester is coordinated via the amino group, with the ester group free. They are stable in aqueous solution for several hours with respect to hydrolysis. In this case the carbonyl group is not coordinated. This is consistent with the mechanism outlined above.

In acid solution Hg(II) reacts with the complex, with hydrolysis of the ester and formation of a glycinatobis(ethylenediamine)cobalt(III) complex. Spectrophotometric studies show that the coordinated halide is removed by mercury(II) (a well-known catalyst of aquation) in the first step and that hydrolysis of the ester occurs in a subsequent step. Infrared spectroscopy suggest that the intermediate is the chelated ester complex. The mercury(II) ion induced aquation of the chloro complex results in the formation of a five-coordinate species, the sixth position then being filled by the ester carbonyl group. In the absence of

$$\left[(en)_2Co \underset{NH_2}{\overset{O=C-OR}{\diagup\hspace{-0.3em}\diagdown\hspace{-0.3em}CH_2}} \right]^{3+}$$

other nucleophiles water attacks the positive centre of the polarized carbonyl group. The hydrolysis reaction is subject to general base catalysis.

More recently[3] further oxygen-18-tracer studies have been carried out on the base hydrolysis of glycine esters coordinated to cobalt(III), in cis-[Co(en)$_2$X(glyOR)]$^{2+}$, (X = Cl, Br) and β_2-[Co(trien)Cl(glyOC$_2$H$_5$)]$^{2+}$. Evidence has been produced for two competing processes, intermolecular

$$cis[(Co(en)_2X(glyOR)]^{2+} + OH^- \underset{fast}{\rightleftharpoons} cis[Co(en)(en-H^+)X(glyOR)]^+ \\ + H_2O$$

hydrolysis of the chelated ester and intramolecular attack of coordinated OH^- at the carbonyl group of the monodentate ester. Both reaction pathways arise from competition for the five-coordinate deprotonated intermediate formed by loss of the halide ion *via* a dissociative conjugate-base mechanism.

However, these reactions could not be observed directly as they are fast compared to hydrolysis of coordinated halide. Therefore a study[4] of the Co(III)-promoted hydrolysis of glycine amides has been of particular value, as base hydrolysis of chelated glycine amides is much slower than that of esters. This allows the hydrolysis of the chelated amide to be observed following loss of Br^- in *cis*-$[Co(en)_2Br(glyNR_1R_2)]^{2+}$ and hence the possibility of confirming intramolecular hydrolysis by coordinated OH^-. Evidence for the existence of an intermediate of reduced coordination number, which may be competed for by species in solution, was provided by showing that azide ion could be incorporated into the complex during hydrolysis of *cis*-$[Co(en)_2Br(glyNH_2)]^{2+}$ giving $[Co(en)_2N_3(glyNH_2)]^{2+}$. No evidence could be found for the formation of *cis*-$[Co(en)_2N_3(glyO)]^+$ where glyO is the nitrogen-bound monodentate glycinate anion, so implying that hydrolysis of the monodentate amide does not occur before or during loss of Br^-. Oxygen-18 studies indicate that the two possible products resulting from competition for the sixth coordination position (i.e. by the carbonyl oxygen of the amide or the OH^- group) are formed about equally. Oxygen-18 studies also show that reaction *via* the chelating amide pathway involves attack of solvent OH^- without opening the ring, and the second pathway to occur *via* intramolecular attack of coordinated OH^- at the carbonyl carbon of the monodentate amide.

It is clear that cobalt(III) is very effective in inducing hydrolysis of amino acid esters and amides. The cobalt(III)-catalysed intramolecular hydrolysis reaction in *cis*-$[Co(en)_2OH(glyNH_2)]^{2+}$ is about 10^7 times faster than the uncatalysed reaction. This mechanism appears to be more efficient than that resulting from direct metal ion polarization of the carbonyl group. For it to occur it is necessary that the coordinated water molecule is acidic enough to exist in the hydroxo form.

Other support for the effectiveness of metal-bound hydroxide in the hydrolyses of esters and amides comes from work on labile metal ions, which catalyse the hydrolyses of esters more efficiently than does OH^-. The pH dependence of the observed rate constant for certain of these reactions suggest that the catalytically active form of the metal complex of the ester contains $M—OH$.[5] Model systems of this kind (i.e. other than Co(III)) often fail to show very significant enhancement of hydrolysis rates. However a millionfold rate enhancement of amide hydrolysis has been observed for certain Cu(II) and Zn(II) complexes, where the cation is forced to lie perpendicular to the plane of the amide group. In these examples the amide, a lactam, acts as a terdentate ligand and the coordination sphere is completed by a water molecule. In the Cu(II) complex the pK_a of the aquo group is 7.6, and the hydrolysis to the amino acid in the Cu(II) and Zn(II) complexes was faster than that of the free ligand by factors of 1.6×10^6 and 1.9×10^5 respectively. The reaction is suggested to occur *via* nucleophilic addition of the metal-bound hydroxide to the carbonyl group. The dependence of such spectacular rate enhancements on subtle geometric factors serves to re-emphasize the difficulties implicit in designing good model systems.[5 a]

These observations are also relevant to carbonic anhydrase, as one possible mechanism involves the conversion of CO_2 to bicarbonate ion by the action of a zinc-bound hydroxide group. Furthermore, the action of carboxypeptidase is usually discussed in terms of two possible mechanisms, 'zinc–carbonyl' and 'zinc–hydroxide'.

Model systems for carbonic anhydrase have usually involved the study of the reactivity of hydroxo complexes of metals towards CO_2 and other substrates for carbonic anhydrase, such as esters. Thus $[Co(NH_3)_5(OH)]^{2+}$ acts[6] as a nucleophile towards the acetyl group of 4-nitrophenyl acetate. In aqueous solution however it is less reactive than the complex $[Co(NH_3)_5(imid)]^{2+}$ (imid = imidazolate ion). It appears that the imidazolate group attacks the carbonyl group of the ester, giving an acetyl imidazole species which is then hydrolysed by water. In view of this and earlier work[7] implicating an imidazole group in carbonic anhydrase activity for the human B enzyme only, it is suggested that a Zn—Imid scheme may hold for ester hydrolysis. A corresponding mechanism for hydration of CO_2 was rejected in view of the properties of the resulting carbamate intermediate.

The reactions of CO_2 as models for carbonic anhydrase have been discussed in detail by Dennard and Williams[8]. Here, the reactions of CO_2 may be understood in terms of the charge arrangement in the molecule, i.e. $\overset{\delta-}{O}=\overset{\delta+}{C}=\overset{\delta-}{O}$, with the CO_2 molecule adding its carbon to a negative centre. Thus the reaction with a Grignard reagent (4.1) may be understood in terms of the polar Mg—R link polarizing the carbon dioxide molecule. Such a reaction involves combined acid–base attack.

$$CO_2 + R—Mg—X \rightarrow RCO_2^-Mg^+ - X \qquad (4.1)$$

A number of decarboxylation reactions[9] have been examined. These are the back reactions corresponding to the overall reaction (4.2) and therefore involve the same transition state.

$$CO_2 + RCOCH_3 \rightarrow RCOCH_2COO^- + H^+ \qquad (4.2)$$

Here too the catalytic role of metal ions as Lewis acids has been clearly demonstrated.

Dennard and Williams[10] have studied the rate of reaction between carbon dioxide and water over a wide pH range, together with the catalytic effect of a number of anions. In contrast to the results of earlier workers they found that there did not appear to be any relationship between the pK_a of the conjugate acid of the anion and its catalytic efficiency. The effect of added anions was measured over a range of pH by using buffers. If catalysis did occur it was then possible to confirm that this was only through the anion by comparing the experimental variation of rate constant with pH and a theoretical curve constructed on the basis of the pK_a of the acid, assuming that the anion was the catalytic species. Good catalysts were species such as arsenite, sulphite, hypochlorite, hypobromite, while carbonate, bicarbonate, nitrite, nitrate, sulphide, sulphate, phosphate are typical of non-active anions. It is suggested that the requirements for catalytic activity are: (a) that the bases should be oxyanions of non-metals in their lower oxidation states, having at least one lone pair, thus HPO_3^{2-} is non-catalytic having no lone pair of electrons; (b) if the oxyanion involves a high oxidation state of a non-metal then there must be no equivalent oxygen atom in the anion to that from which the proton has been removed. This condition prevents the possibility of charge distribution away from the oxygen function involved in catalysis.

Certain mechanisms for base catalysis of CO_2 hydration such as scheme (A) may be eliminated as they would require a correlation between activity and pK.

Dennard and Williams suggest that scheme (B) is a feasible alternative, for here depending upon the stability of the first intermediate there may or may not be a dependance of k upon pK.

Another mechanism could involve catalytic function through X rather than the oxygen atom, in accord with the requirement (a) for catalytic activity. This is illustrated in scheme (C), one involving concerted acid–base attack upon CO_2.

In concluding the section on model compounds it should be noted that infrared studies on the CO_2 molecule bound in the hydrophobic slit at the active site of carbonic anhydrase have indicated that CO_2 is not coordinated to the metal but that the bicarbonate ion is bound to zinc *via* its negatively charged oxygen atom. This will be discussed in more detail when the enzyme itself is considered, but it is clear that the zinc ion itself in carbonic anhydrase can still be put forward as the Lewis acid partner in *concerted* acid–base attack, provided that hydroxide is the other partner, as this will generate bound bicarbonate ion. It will not involve specifying CO_2 coordinated to zinc.

CARBONIC ANHYDRASE[11]

This was the first zinc metalloenzyme to be recognized as such. It occurs in animals, plants, and some microorganisms, although certain carbonic anhydrases in plants do not contain the metal. It catalyses the reversible hydration of carbon dioxide, and so is essential for respiration. It also catalyses the hydrolysis of certain esters and aldehydes.

The enzyme contains one atom of zinc per mole and has a molecular weight of 30,000, with some 260 amino acid residues. Bovine and human carbonic anhydrases have been most studied. The enzyme occurs in three forms, A, B, and C, form C being 30 times more active than form B.

X-ray studies

The dimensions of the enzyme are approximately $40 \times 45 \times 55$ Å. The zinc lies near the bottom of a 12 Å-deep cleft, and is bound by three imidazole groups of histidyl residues, while the fourth position is probably filled by a water molecule.[12] The overall environment around the zinc is distorted tetrahedral, as suggested by the electronic spectra of the cobalt enzyme.

The crystal structure[13] of the human carbonic anhydrase B–imidazole complex suggests that the imidazole is bound to the zinc without displacing the aquo group, so giving a five-coordinate zinc species. This result will be discussed later and is of some significance, as the imidazole is unique in being a competitive inhibitor for the carbonic anhydrase catalysed hydration of CO_2. Structures of sulphonamide complexes of carbonic anhydrase have also been determined.[12,14]

Metal—for zinc substitution

Metallocarbonic anhydrases have been prepared containing Mn(II), Fe(II), Co(II), Ni(II), Cu(II), Cd(II), Hg(II), Pb(II), Be^{2+}, and the alkaline earth metals. The cobalt enzyme is active, with activity up to 100% depending upon the conditions. The Ni(II), Fe(II), and Mn(II) enzymes are slightly active.

Studies on the cobalt carbonic anhydrase have contributed a great deal to our understanding of this enzyme, as discussed in other sections. ^{13}C NMR relaxation studies[15] on the Co(II) human carbonic anhydrase B with ^{13}C-enriched HCO_3^- and CO_2 have shown that the distance of these species from the Co(II) centre is independent of pH and about 3.6 ± 0.2 Å. These results indicate therefore that the CO_2 and HCO_3^- are bound close to the metal and that the CO_2 could be coordinated to the metal. This view is not supported by infrared studies discussed below, or by ^{13}C NMR studies on Cu(II) carbonic anhydrase.[91]

Infrared studies on carbonic anhydrase

Riepe and Wang[16] have measured the infrared spectrum of the CO_2 molecule bound in the cavity at the active site of bovine carbonic anhydrase. The spectra were measured on 33% aqueous solutions of the enzyme at pH 5.5 under equilibrium CO_2 pressure against a reference cell containing the same enzyme solution without CO_2. The characteristic band of CO_2 observed at 2343 cm^{-1} for an aqueous solution, appears at 2341 cm^{-1} in the enzyme. That is, the spectrum is that of a normal CO_2 molecule.

Studies on the infrared spectra of carbonic anhydrase solutions equilibrated with CO_2 and N_2O mixtures show that both compete for the active site suggesting that the CO_2 is not coordinated to the zinc. This is in accord with the fact that the $d–d$ spectrum of the cobalt carbonic anhydrase is not affected by the binding of CO_2. Similar infrared studies on the enzyme in the presence of azide ion show that the carbonic anhydrase bound azide is in fact coordinated to zinc, as is reflected in the presence of an azide band at 2094 cm^{-1} compared with 2046 cm^{-1} for azide in inert protein solution. Bands at both 2094 and 2046 cm^{-1} are observed showing the presence of free and coordinated azide ion.

These workers have also studied the effect of azide ion on the enzyme/CO_2 system. Here the band at 2341 cm^{-1} due to bound CO_2 decreases as the 2094 cm^{-1} band due to coordinated azide increases. They deduced therefore that the hydrophobic cavity was right alongside the zinc ion, so that coordinated azide ion protrudes into the cavity, displacing CO_2. It was found also that bicarbonate ion displaced both CO_2 from the cavity and azide ion from the metal. This means that the bicarbonate ion must be coordinated to the zinc through its negative oxygen with the remainder of the molecule in the

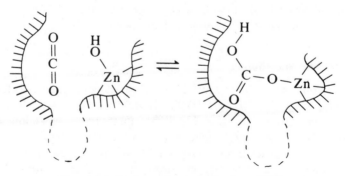

Figure 4.1 A scheme for the hydration of CO_2.

hydrophobic cavity. The implications of this are important. In the dehydration reaction proton transfer will accompany the C—O bond breaking, leaving CO_2 in the cavity and OH^- coordinated to zinc. In the hydration reaction, therefore, we can say at this point that the OH^- on the zinc must attack the bound CO_2 converting it to HCO_3^- (Figure 4.1).

Metal-binding groups in carbonic anhydrase

X-ray studies have indicated that zinc is bound by three imidazole groups of histidine residues. It is clearly shown that the cysteine group is too far away to be involved in metal binding. Chemical evidence was available for the implication of imidazole in the active site, but the situation regarding —SH groups was not clear cut. Thus titrations with Ag^+ had given conflicting results for bovine and human carbonic anhydrase, suggesting in the latter case only that —SH was a binding group, even though the d–d spectra of both cobalt-substituted enzymes were similar.

Stability constant measurements for metal-substituted carbonic anhydrases and a comparison of log K and $3/2$ log K_{en} suggest that zinc is bound by three nitrogen bases. The EPR spectrum of the copper(II) enzyme is similar to that of $Cu(dipyridyl)_2^{2+}$, while Δ_t the crystal field splitting, of the cyanide complex with cobalt(II) carbonic anhydrase in which cobalt is in a tetrahedral environment, was found to be 5300 cm^{-1}, a value similar to that for $Co(benzamidazole)_4^{2+}$. In this particular case, therefore, chemical and physical evidence had given a satisfactory conclusion.

pH studies

These[17,18] have been of value in understanding the activity of carbonic anhydrase. Much has been done on the cobalt(II) enzyme. The d–d spectrum is pH dependent, suggesting that there are two forms of the enzyme, related by a protonation equilibrium. The variation of the d–d spectrum with pH suggests that the group involved in this has a pK_a of 7.1. Figure 4.2 gives the d–d spectrum over a range of pH values.

This equilibrium between a protonated and non-protonated site is associated with the catalytic function of the enzyme. It appears that the non-protonated form is involved in hydration. Kernohan has shown, on the basis of pH–rate studies, that the pK_a of the group appears to be 7.1, in good agreement with the spectrophotometrically determined value.

Two alternative explanations have been offered for the pH effect on the cobalt carbonic anhydrase d–d spectrum. One suggestion is that the proton is released from a protein group which then coordinates to the metal ion. Support for this idea comes from a consideration of the d–d spectrum of the alkaline form of the cobalt enzyme. While the acid form is that of a tetrahedral complex, the spectrum of the alkaline form is different from that of either tetrahedral or octahedral complexes, but may be interpreted in terms of a five-coordinate complex. The

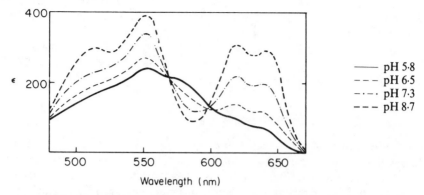

Figure 4.2 The effect of pH on the d–d spectrum of cobalt carbonic anhydrase (Reproduced by permission of The American Society of Biological Chemists, Inc, from S. Lindskog, *J. Biol. Chem.*, **238**, 945 (1963)).

second suggestion is that the ionization is that of a coordinated water molecule, generating a coordinated hydroxide group. This is in accord with several kinetic studies which implicate bound hydroxide, i.e. E—Zn—OH. This latter explanation has been criticized on the grounds that proton release from a coordinated water molecule is unlikely to occur at the low pH implicated by enzyme studies. The catalytic group referred to above has a pK of about 7. However Woolley[19] has demonstrated that metal-bound water in certain five-coordinate complexes of Zn(II) and Co(II) has pK_a values below 9. Thus the ligand shown below **(I)**

gives a five-coordinate Zn(II) complex, with one aquo group which has a pK_a of 8.69. Ni(II) and Cu(II) complexes are diaquo six-coordinate species with p$K_a > 11$. The Co(II) and Zn(II) complexes of **(I)** catalyse the hydration of CO_2 and CH_3CHO.

Inhibitors

Carbonic anhydrase is inhibited by several classes of compounds. Sulphonamides (X—SO_2—NH_2) are efficient and selective inhibitors, while a number of simple anions such as azide, cyanide, iodide, cyanate, and hydrosulphide are also well-known inhibitors. Imidazole is unique in that it is the only compound known to act as a competitive inhibitor for the hydration of CO_2.[20] In general one mole of each inhibitor is added to each mole of enzyme.

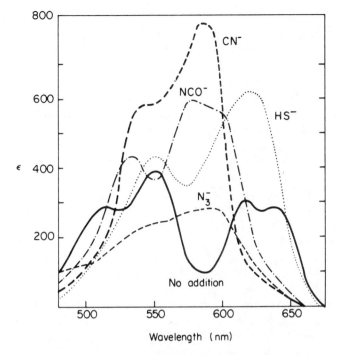

Figure 4.3 d–d spectra of inhibitor complexes with cobalt carbonic anhydrase (Reproduced by permission of The American Society of Biological Chemists, Inc. from S. Lindskog, *J. Biol. Chem.*, **238**, 945 (1963)).

The d–d spectrum of cobalt carbonic anhydrase is affected by the addition of inhibitors, suggesting they are bound to the metal: in fact the shift in band position with change of inhibitor is in accord with the position of the ligand in the spectrochemical series. The d–d spectra of the CoCA complex with the inhibitors cyanide, cyanate, acetazolamide (a sulphonamide), and hydrosulphide show that these complexes have an essentially tetrahedral structure (Figure 4.3). This suggests that the inhibitor displaces the zinc-bound water molecule. In accord with this view, the release of one mole of H^+ on addition of zinc to the apoenzyme between pH 6 and 9 does not occur in the presence of inhibitors, suggesting that the inhibitor competes with this ligand group for the metal.

More recently it has been shown by EXAFS measurements[21] that in the iodide complex with carbonic anhydrase, the iodide is bound directly to the zinc, with a zinc–iodide distance of 2.65 ± 0.06 Å. Earlier X-ray diffraction data had suggested zinc–iodide distances of ~ 3.7 Å. Evidence[22] for direct binding of Cl^- to zinc in carbonic anhydrase was obtained from ^{35}Cl NMR line broadening effects. A similar result was obtained[23] with ^{81}Br NMR to show direct binding of Br^- to zinc in bovine carbonic anhydrase.

Sulphonamide complexes of carbonic anhydrase have been studied by X-ray crystallography. This confirms that the inhibitor is bound directly to the zinc, *via* the nitrogen or an oxygen atom. A second oxygen atom of the sulphonamide group forms an additional long bond to the metal, making it effectively five coordinate. It appears therefore that both sulphonamide[24] and anionic inhibitors bind to the zinc, replacing the aquo molecule. It should be noted however that there is some evidence[25] based on the electronic spectra of Co(III) carbonic anhydrase to suggest that halide, acetate, and benzoate complexes exist in equilibrium between four- and five-coordinate species. A longstanding suggestion postulates that the spectra of the complexes with inhibitors (Figure 4.3) show a trend towards that of the alkaline form of the enzyme which is thought by some to be five coordinate, as a result of the availability of a deprotonated ligand group.

The inhibitor imidazole forms a unique five-coordinate complex[13] with carbonic anhydrase in which the aquo group is not displaced. This is the only competitive inhibitor of the hydration reaction. The position of the imidazole group approximates to that occupied by the second donor atom of the sulphonamide group in the structure described above. This has led to the suggestion that CO_2 binds weakly to this fifth coordination site of the metal in the enzyme.

The kinetics of reaction of aromatic sulphonamides with human carbonic anhydrase B and C have been investigated[26] by stopped-flow and fluorescence techniques. The forward and back rate constants k_1 and k_{-1} in the reaction $E + S \rightleftharpoons ES$ have been measured over the pH range 5.0 to 10.8 for a number of inhibitors. There is a 240-fold variation in k_1, while there is evidence for an effect on k_1, produced by an *ortho* substituent in the aromatic ring.

It is suggested that the inhibitor–enzyme complex is formed *via* the rapid pre-equilibrium formation of at least one less stable intermediate, and that the first intermediate could involve the hydrophobic interaction of the enzyme with the

$$E + S \rightleftharpoons (ES)_1 \rightleftharpoons (ES)_2$$

aromatic ring of the inhibitor. This then undergoes rearrangement to the more stable product involving metal ion coordination. The presence of an interaction between the aromatic nucleus of the inhibitor and the hydrophobic portion of the active site cleft has been demonstrated[27] for bovine carbonic anhydrase by the spin-label technique using the sulphonamide spin label 4-[(p-sulphonamido)-benzoyloxy]-2,2,6,6-tetramethylpiperidine-1-oxyl). The presence of substituents in the aromatic ring could affect the relative position of the ring and the protein surface, and so the stability of the first intermediates.

The measured overall association rate constant is pH dependent, but the dissociation is pH independent. The results, which are essentially similar for cobalt and zinc enzymes, suggest that combination occurs between a neutral sulphonamide species and the alkaline form of carbonic anhydrase, a view confirmed recently by the use of Raman spectroscopy.[24]

Chemical modification (carboxymethylation) of a single histidine residue in human carbonic anhydrase markedly affects the sulphonamide binding kinetics, the pH-independent dissociation rate is increased 45 times, while the pH dependence of association is shifted to higher pH. This same chemical modification results in a change in the pH dependence of the equilibrium between the coordination forms of the enzyme (i.e. $d-d$ spectra of CoCA) which is in good agreement with the effect on association rate–pH profile just mentioned.

The pH–rate profile for the inhibitor–enzyme association is bell shaped, with pK's for the two functions of (for the inhibitor p-nitrobenzenesulphonamide) 6.60 and 9.30. The value of the second pK varies with the nature of the inhibitor and corresponds closely with the ionization constant of the sulphonamide group in the inhibitors. The lower pK does not correspond to any sulphonamide ionization, but does correspond to the pK derived from the spectral changes involved in the pH dependence of the cobalt carbonic anhydrase C (6.60), suggesting that the pK derived from the inhibitor studies is associated with the same function involved in the pH dependence of the enzyme, as is confirmed by the carboxymethylation results. There is a similar correlation between the association rate pK of zinc carbonic anhydrase B and the spectrophotometric pK of the cobalt enzyme B. Carboxymethylation can be prevented by the presence of sulphonamides or certain anions.

The spectrum of the Co(II) enzyme–sulphonamide complex is pH insensitive, consistent with the pH independence of the dissociation rate. It seems therefore that the pH-sensitive ligand has been replaced from the metal by the sulphonamide, again suggesting that this is the aquo group.

Sulphonamide inhibitors have been used to probe structural features in carbonic anhydrase. Thus the folding[28] of the bovine enzyme to give the native structure has been studied in the active site region by the use of an azosulphonamide inhibitor, and stages in the folding process characterized prior to the formation of the active site. The depth of the cleft in the enzyme has been explored[29] by the use of a range of spin-labelled sulphonamide inhibitors.

Mechanism

The detailed mechanism of action of carbonic anhydrase is still uncertain. The identity of the group with a p$K_a \sim 7$ that is involved in catalysis is a key issue. Two possibilities have been suggested so far for this: either the ionization of a zinc-bound water molecule, or the ionization of some protein group. There is some evidence for the involvement of an imidazole group in the activity of the enzyme and this appears to be the most likely candidate for an appropriate protein group. Such an imidazole group is located at the entrance of the active site and on deprotonation it may coordinate to the zinc, increasing the coordination number to five. Several mechanisms[30] may be written for each of these acidic groups. An alternative third acidic group is one of the three histidine ligands binding Zn^{2+}.

Several lines of argument seem to support a zinc–hydroxide type of mechanism, *Ab initio* molecular orbital calculations[31] lead to the conclusion that Zn(II) can bind CO_2, and that binding of water to the metal enhances its ionization more than ionization of imidazole. The imidazolate model is also incompatible with NMR studies on the N—H resonances in carbonic anhydrase, inasmuch as this data does not show the rapid exchange of the N—H proton with the solvent that should be a feature of the imidazolate model.[32]

Attempts have been made to measure the proton exchange of the aquo group in cobalt(II) carbonic anhydrase by NMR and hence establish a pK_a value for this group by studying the effect of pH on this phenomenon. However it has been found[33] for human and bovine B cobalt carbonic anhydrase that the NMR pK_a value is 8.6, while the pK_a from visible spectra has a value of 7.3. This result clearly could be important in assessing the mechanism of carbonic anhydrase.

It is probable on balance that some type of zinc–hydroxide[34] mechanism is operative, but there is a further unresolved matter, namely whether or not the CO_2 is bound to the metal. At present contradictory data cannot be resolved directly, but there seems to be increasing support for binding of CO_2 to the Zn(II). Some alternative mechanisms are given below. Hydrogen bonding of the aquo group to Thr-199 and Glu-106 is probably important[13] in controlling the pK_a of the aquo group, but will not be shown in the schemes. Schemes 1 and 2 assume the acidic group is an aquo group, while Schemes 3 and 4 assume it to be imidazole. Scheme 4 involves five-coordinate zinc(II).

1. Nucleophilic attack of Zn—OH on CO_2

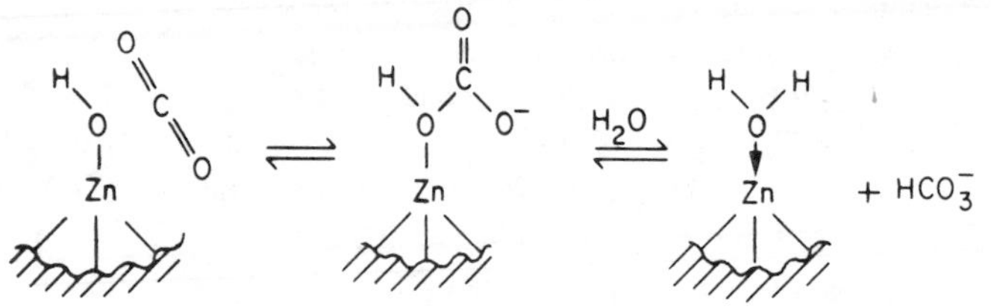

2. General base-assisted attack of H_2O on CO_2

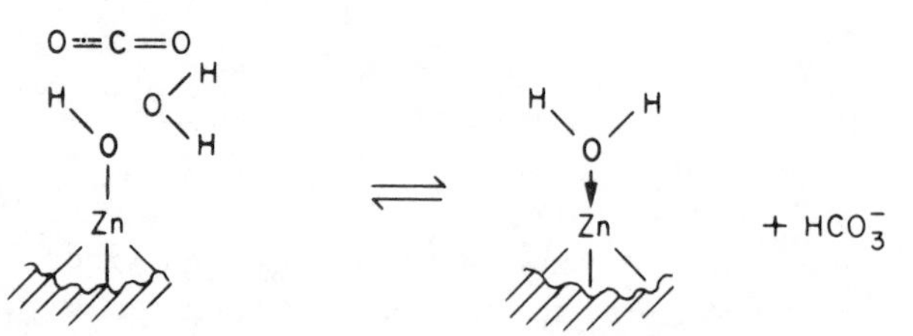

3. General base-assisted attack of $Zn—OH_2$ on CO_2

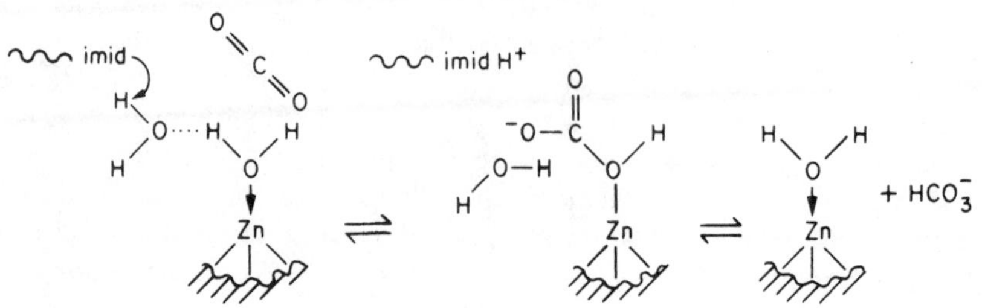

4. General base-assisted attack of H_2O on CO_2

CARBOXYPEPTIDASE

The carboxypeptidases are zinc-containing enzymes that are released from their inactive precursors or zymogens (procarboxypeptidase) in the pancreatic juice of animals for the digestion of proteins. Carboxypeptidase A exists in several forms, depending upon the size of the fragment lost from the zymogen. These forms are listed in Table 4.2, and have similar properties. The commonly met form of carboxypeptidase A is CPAα, which has a molecular weight of 34,600 and 307 amino acid residues. The enzyme contains one mole of Zn^{2+}.

The activity of the enzyme is directed specifically towards the peptide bond at the carboxy terminus of peptides and proteins, for which the side chain of the C-terminal residue is aromatic or branched aliphatic of L configuration. It appears that enzyme activity is affected by at least the first five terminal residues in the substrate. The enzyme also possesses esterase activity towards esters, and this aspect of the enzyme has also been much studied.

$$-CONHCH(R^1)CONHCH(R)CO_2^- \xrightarrow{\text{CPA}} -CONHCH(R^1)CO_2^- + \overset{+}{N}H_3CH(R)CO_2^-$$

The zinc may be removed from carboxypeptidase to give an inactive apoenzyme. Activity is restored on readdition of zinc, but a range of other metals

TABLE 4.2 Forms of carboxpeptidase.

Enzyme	No. of residues	N-Terminus
CPAα	307	Alanine
CPAβ	305	Serine
CPAγ	300	Asparagine
CPAδ	300	Asparagine

have also been successfully incorporated into the apoenzyme. Cobalt(II) carboxypeptidase has greater peptidase activity than the zinc enzyme while other metallocarboxypeptidases are also catalytically active. As will be demonstrated in a later section, the study of these metallocarboxypeptidases has been a particularly rewarding one.

X-ray studies

These show[35] the enzyme molecule to be ellipsoidal in shape, with dimensions $50 \times 42 \times 38$ Å. The zinc is bound inside a cleft in the enzyme molecule by three protein groups, two imidazoles of histidine residues, His-69 and His-196, and one glutamic acid residue (Glu-72). The remaining coordination position is filled by a water molecule, and the overall stereochemistry around the Zn^{2+} is distorted tetrahedral. This water molecule is probably displaced by the substrate on formation of the enzyme–substrate complex.

The X-ray studies show that only certain protein residues are near enough to the substrate to be involved in enzyme activity. These include Glu-270, Tyr-248, and Arg-145. These have detailed roles in the mechanisms for carboxypeptidase A which have been put forward on the basis of crystallographic evidence. Before discussing these mechanisms, it should be recalled that there is no guarantee that information obtained from crystallographic studies in the solid state also holds for solutions of the enzyme. Indeed considerable evidence has been accumulated for carboxypeptidase to suggest that in this case the assumption does not hold. Thus it appears from recent resonance Raman studies[36] on arsanilazotyrosine-248 carboxypeptidase in solution that a number of interconvertible species are present, which probably represent different conformations of the enzyme. In such a situation there is always the possibility that the form which actually crystallizes out may not be the most efficient catalytic species present in the solution. Furthermore, kinetic studies on solutions and crystals of carboxypeptidase show considerable differences in reactivity between these two forms, with reduction in catalytic efficiency in the crystalline enzyme of up to 1000-fold compared with the solution.[37]

Carboxypeptidase forms an extremely stable enzyme–substrate complex with the dipeptide glycyl-tyrosine, the crystals of which are isomorphous with those of the enzyme. A comparison of the electron density map of the enzyme with that of the complex allows the clarification of the binding of the substrate. The C-terminal side chain of the substrate fits into the pocket or cleft of the enzyme. There appears to be no specific binding group involved here in accord with the lack of high specificity for the nature of the side chain. The terminal carboxylate group interacts with the positively charged guanidinium group of arginine-145, while the carbonyl oxygen of the peptide link which is to be split is probably bound to the zinc, with resulting displacement of the coordinated water group.

The enzyme appears to undergo dramatic conformational changes when Gly–Tyr is bound at the active site (Figure 4.4). Interaction between the

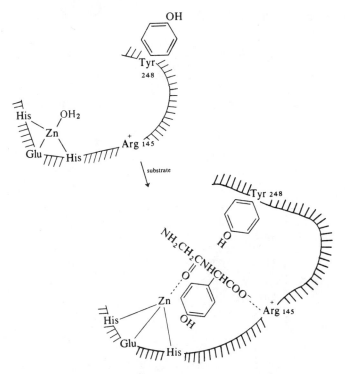

Figure 4.4 Conformational changes induced in carboxypeptidase A by substrate binding.

carboxylate group and the arginine residue results in this residue moving towards the substrate. This movement is magnified through the protein structure so that the phenolic group of a tyrosine residue (Tyr-248) moves about 12 Å with a twisting of the carbon–carbon chain so that the OH group lies in the vicinity of the substrate's peptide link.

General features of this mechanism are certainly correct. Tyr-248 probably serves to donate a proton to the NH group of the hydrolysable bond in the peptide. The Glu-270 residue is close to the carbonyl carbon of this group and could act by promoting general base catalysis of attack of a water molecule or by direct nucleophilic attack at this carbon atom. The pH–rate profile of the enzyme has been interpreted in terms of the ionization of this glutamic acid residue. The role of the Zn^{2+} has not been resolved with certainty. It may be involved in catalysis or it may serve to hold the constituents of the active site in a geometry suitable for binding of substrate.

Figure 4.5 illustrates two possible binding schemes which are consistent with the X-ray data. In the 'zinc–carbonyl' mechanism the substrate is bound to the metal *via* the carbonyl group. The zinc thus serves to orientate the carbonyl group of the substrate in a position suitable for nucleophilic attack, and also

106

(a) *Zinc-carbonyl mechanism*

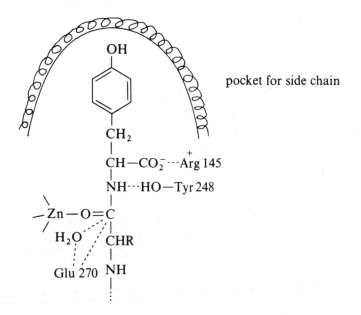

pocket for side chain

(b) *Zinc-hydroxide mechanism*

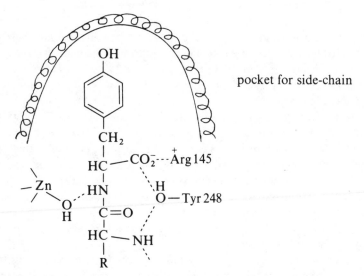

pocket for side-chain

Figure 4.5 The binding of a peptide substrate to carboxypeptidase.

polarizes the carbonyl group making the carbon atom more susceptible to nucleophilic attack, as noted for the model systems. As noted above an unresolved question here is whether or not attack by Glu-270 is direct or *via* a water molecule. In the former case an acyl intermediate would be formed. No evidence has been obtained so far for the existence of such an intermediate, but it would, of course, be readily hydrolysed. In contrast, support for an anhydride intermediate has been found in the hydrolysis of an ester substrate by this enzyme.[38,92]

In the alternative mechanism (the zinc–hydroxide mechanism) the susceptible carbonyl bond is directed away from the zinc, and the zinc still binds the water molecule (or hydroxide group, depending upon the pH). This hydroxide group then attacks the carbon atom of the peptide bond.

The X-ray work has naturally only been possible on a substrate that is not cleaved by the enzyme, and this leads to queries about the binding of reactive true substrates. Nevertheless it is noteworthy that these two mechanisms have been established for studies with model compounds. They will be discussed further in a later section.

Chemical studies on the nature of the zinc-binding groups

A comparison[39] of stability constants for a series of metallocarboxypeptidases with those of bidentate metal complexes suggested that the metal ion was bound by nitrogen and sulphur ligands. The change in relative stability for the metalloenzymes is very similar to the change in stability from metal ion to metal ion when the ligand is mercaptoethylamine but is different from the behaviour with other ligands. In addition, the apoenzyme gave[40] the normal tests for the presence of an —SH group and an RNH_2 group but the tests were negative on the zinc enzyme, suggesting that the metal ion was bound to these groups so modifying their reactivity towards the reagents used to test for their presence. The reagents for the thiol group were Ag^+, *p*-mercuribenzoate, and ferricyanide. The results appeared convincing in that the zinc content of the enzyme and the Ag^+-titratable groups always gave a constant molar total over a range of zinc concentrations.

Despite this evidence, it is quite clear from X-ray studies that a thiol group is not implicated in zinc binding in carboxypeptidase. The chemical evidence may be explained if it is assumed that conformational changes occur in the protein on the binding of the zinc by the apoenzyme and that the —SH group, once accessible to the reagents, becomes inaccessible in the metalloenzyme as a result of this conformational change.

Metal for zinc substitution

The zinc in carboxypeptidase has been replaced by Mn^{2+}, Fe^{2+}, Co^{2+}, Ni^{2+}, Cu^{2+}, Rh^{3+}, Cd^{2+}, Hg^{2+}, and Pb^{2+}. The results described in the last section

TABLE 4.3 Hydrolysis of Bz-(Gly)$_2$-L-Phe and its ester analogue by various metallocarboxypeptidases at 25°C, pH 7.5.

Metal	Bz-(Gly)$_2$-L-Phe		Bz-(Gly)$_2$-L-OPhe	
	k_{cat}(min^{-1})	$10^3 K_m$(mol dm^{-3})	$10^{-4} k_{cat}$(min^{-1})	$10^4 K_m$(mol dm^{-3})
Co	6000	0.66	3.9	3.03
Zn	1200	1.00	3.0	3.33
Mn	230	0.36	3.6	15.2
Cd	41	0.77	3.4	83.3

have illustrated one use of a series of such metal-substituted enzymes. Of greater significance are kinetic studies on these metallocarboxypeptidases which have given much insight into the role of the metal ion in catalysis, for example through a comparison of their peptidase and esterase activities. Table 4.3 presents[41] results for the hydrolysis of a peptide substrate and an analogous ester substrate catalysed by Zn(II)-, Co(II)-, Mn(II)-, and Cd(II)-metallocarboxypeptidases. The data show clearly that in peptidase activity the binding of substrate (given by K_m values) is only slightly dependent upon the metal but that the rate constant for catalysis is very dependent upon the metal. The converse is true for esterase activity. This shows that the primary role of the metal in peptide hydrolysis is associated with catalysis rather than with peptide binding, and that the opposite situation holds for the hydrolysis of esters.

In common with other zinc metalloenzymes, the zinc in carboxypeptidase is replaced effectively by cobalt(II), while the resulting d–d spectrum of the cobalt enzyme suggests a distorted tetrahedral environment around the cobalt atom. This is confirmed by circular dichroism, magnetic circular dichroism, and low-temperature EPR spectra of Co(II)-CPA.

NMR studies[42] on manganese carboxypeptidase have shown that water or hydroxide is bound to the metal, and that this is displaced on formation of the enzyme–inhibitor complex with β-phenylpropionate.

A feature of interest in metallocarboxypeptidases is the way in which metals which prefer different coordination geometries still give active CPA derivatives. This must reflect some flexibility in the binding site for metal ions. The structure of Hg^{2+}-CPA has been reported.[35] In this case the metal ion is displaced by about 1.2 Å from the zinc site. This results from the need for larger metal–ligand bonds and the minimizing of van der Waals repulsion. The overall effect is to move the susceptible peptide bond away from Glu-270 in Hg-CPA, so peptidase activity is lost.

Chemical modification of protein groups

Vallee and his coworkers have carried out a series of elegant studies on the effect of the modification of amino acid side chains on the peptidase and esterase activity of carboxypeptidase. The implication of tyrosyl residues in the catalytic function has been shown by modification[43] through acetylation, nitration, iodination, and coupling with 5-diazo-IH tetrazole or p-azobenzenearsonate.[45] In all cases enzymatic activity was altered, usually peptidase activity being reduced while esterase activity increased.

When acetylation is carried out in the presence of the inhibitor, β-phenyl-propionate, two tyrosine residues are protected, suggesting that these two residues are of particular importance. The X-ray studies have shown that a second tyrosine (Tyr-198), in addition to Tyr-248, is in the general area of the active site. A great deal of work has been carried out on the modification of these two groups and has been discussed in detail.[44] We will only give one further example. The coupling of the tyrosyl group with p-azobenzenearsonate[45] introduces an asymmetric centre into the protein. On addition of either inhibitor or glycyl-L-tyrosine, the circular dichroism spectrum is changed, so indicating that the conformation of the azotyrosyl residue is altered on substrate/inhibitor binding.

In his original mechanism, (1963), Vallee suggested that the C-terminal carboxyl group specificity requirement of the enzyme could be understood in terms of an interaction of this group with a positively charged group, such as a lysyl or arginyl residue. No evidence could be obtained by chemical modification for the implication of lysyl groups, but subsequent to the earlier X-ray studies implicating Arg-145, Vallee and Riordan were able to modify arginyl residues in carboxypeptidase by the use of a 150-fold excess of diacetyl and to demonstrate that peptidase activity was lost. Esterase activity was increased, so indicating that the arginyl residue was not essential for the binding of the ester. It has also been shown chemically that the modification of a histidyl residue resulted in decreased peptidase activity.

Studies on inhibitor and substrate binding

The binding of substrates and inhibitors to the apoenzyme and the various enzymatically inactive metal-substituted enzymes has been well studied and has contributed a great deal to the understanding of the mode of action of carboxypeptidase.

The apoenzyme will bind peptide substrates but will not bind esters. It is reasonable to believe therefore that the interaction between apoenzyme and peptide must be through the peptide nitrogen group via hydrogen bonding to a base, presumably the tyrosyl group already implicated by the work of Vallee. The binding of ester substrates by, for example, the copper enzyme, suggests that in this case the interaction is through metal ion and carboxylate group.

The inhibitor β-phenylpropionate is also only bound by the metalloenzyme; this too probably involves carboxylate–metal ion interaction. It has already been mentioned that, in the presence of this inhibitor, chemical modification of tyrosyl groups does not affect enzyme activity, although it does have a marked effect in the absence of the inhibitor. It appeared, therefore, that the inhibitor interacts also with the tyrosine, as was confirmed by X-ray studies.

Kinetic studies

The kinetic behaviour of carboxypeptidase with a range of substrates and inhibitors has been examined and shown to be very complex. These difficulties are often carried through from one metal to another, suggesting that they do not depend upon the metal ion. In fact they appear to be associated with dipeptide and analogous ester substrates or products having aromatic N-acyl groups.

A study of the rates of metal exchange in the presence of substrates shows that the metal is held more firmly in the presence of substrates, particularly by LL dipeptides but also by DL dipeptides but not by LD dipeptides. This clearly shows the importance of the L configuration of the terminal peptide group in increased metal binding.

The kinetics of peptidase and esterase activity will not be discussed in detail but some mention will be made of pH dependence. The pH–rate profile is different for peptidase and esterase activity. That for peptidase activity is bell shaped having inflexions at pH 6.7 and pH 8.5. If it is assumed that this behaviour reflects protonation equilibria of enzyme groups (e.g. the increase in activity as the acidity is decreased being due to the increasing ionization of the carboxyl residue of Glu-270, while the final decrease in activity at higher pHs being due to the deprotonation of the tyrosyl residue 248), then it may be shown[46] that the two residues have pK_a values of 6.9 and 7.9, values that would not be expected for Glu-270 and Tyr-248. However, in view of the comments in Chapter 3 on the possibility of abnormal pK_a values for groups associated with the active site, it may be seen that these two pK_a values are rather difficult to interpret anyway. It should be noted in this context that the phenolic group Tyr-248, after conformational change, is near the positive residue Arg-145 and this could result in an increased acidity for Tyr-248.

The pH–rate profile for esterase activity, apart from a class of short esters, is different from that for peptidase activity. A rise in rate from pH 5.5 to 7.0 is observed, followed by a plateau to pH 9.0 and a further rise to a maximum at pH 10.5. This suggests that the peptides and esters are hydrolysed by different mechanisms, but at present it is not possible to write mechanisms for ester hydrolysis with any confidence. It would obviously be of great interest to carry out X-ray studies on carboxypeptidase–ester substrate complexes to see what information this would give on this problem.

Model-building studies

Lipscomb and coworkers[35] have reported the results of model-building experiments and have established two binding modes for substrates which may lead to inhibition. In one case the C-terminal side chains fits into the hydrophobic pocket; the aromatic N-acyl group has an interaction with Tyr-198; the second peptide carbonyl is bound to Arg-71, with the C-terminal carboxyl group bound to the zinc atom. In the second case the acyl group is in the pocket, the terminal carboxyl group is bound to Arg-71 and the C-terminal side chain is near Tyr-198.

Short substrates would bind less well in these positions in accord with their failure to show substrate inhibition. These ideas offer an explanation for certain of the kinetic anomalies in the reaction of CPA with dipeptide and ester substrates having aromatic N-acyl groups or large aliphatic groups, although, of course, other possibilities must be considered.

Lipscomb has also carried out model-building experiments with ester substrates, assuming that the aromatic side chain is placed in the pocket and that the carbonyl group is again coordinated to the zinc. These suggest that binding of esters is essentially similar to peptides in some features but that it differs in that the carboxylate group may be bound to the zinc ion, in agreement with suggestions already made. However, it appears unlikely that carboxylate–Arg-145 interaction is lost completely as the conformational change produced by this is necessary to bring Tyr-248 to the active site. A class of short ester substrates exhibits pH–rate dependencies similar to those for peptides. It is reasonable that these should also bind like peptides.

Model-building studies have also revealed that the 'zinc–hydroxide' mechanism for carboxypeptidase activity involves serious steric interference in the binding stage. However, it is not possible to reject the mechanism purely on these grounds because of uncertainties in the positions of some of the relevant atoms in the X-ray structure determination.

The mechanism of peptide hydrolysis

The important conclusion from model compounds is the demonstration that metal-bound hydroxide ion is a very effective nucleophile. While there must be some uncertainty about the validity of transferring this conclusion to enzyme systems, it does present support for the zinc–hydroxide mechanism. The extent to which zinc-bound hydroxide would be formed from the aquo complex is discussed in detail for carbonic anhydrase and the conclusions are relevant to this enzyme. Arguments against the zinc–hydroxide mechanism include the model-building studies, which reveal steric interference in the binding stage, and the NMR studies which appear to suggest that the aquo (or hydroxo) group is displaced on the binding of the inhibitor β-phenylpropionate to manganese carboxypeptidase. (See Ref. 93.)

Figure 4.5 shows only the basic features of enzyme–substrate interaction. Other aspects have also to be accommodated. The requirement for L configuration in the terminal residue is easily understood, as only this form will undergo these specific interactions, as demonstrated by studies on inhibition by D forms of amino acids. The activity of carboxypeptidase is affected by up to five residues, the second residue being particularly important. Lipscomb has shown that maximum interaction of the aromatic groups of the substrate and the enzyme aromatic groups Tyr-198 and Phe-279 occurs when the phenolic OH group of Tyr-248 is also hydrogen bonded to the second NH group of the substrate. This then allows the CO group of the third peptide link to be placed near residue Arg-71, so producing further stabilizing interaction. This involvement of Tyr-198 may account for earlier results implicating two tyrosyl residues in catalytic activity.

Molecular orbital techniques have been used to study the interaction and hydrolysis of substrates by carboxypeptidase.[47,48] Direct nucleophilic attack of Glu-270 in the zinc–carbonyl pathway has been compared with glutamate-promoted attack of a water molecule, while the zinc–hydroxide mechanism has also been evaluated.[47] It is noteworthy that the active site environment in carboxypeptidase is similar to that found in thermolysin, as discussed later. A zinc–carbonyl mechanism can be postulated for thermolysin in which Glu-143 acts as a general base and promotes the attack of a water molecule on the carbonyl carbon of the scissile peptide bond in exactly the same way as Glu-270 is postulated to do in carboxypeptidase. In thermolysin the possibility of direct nucleophilic attack on the carbonyl cation by Glu-143 is excluded by stereochemical factors.

FURTHER EXAMPLES OF ZINC METALLOENZYMES IN HYDROLYSIS

Leucine aminopeptidase

This exopeptidase, which is widely distributed, catalyses the hydrolysis of peptides from the N-terminus. Peptides with N-terminal leucine are the best substrates, but the enzyme also hydrolyses L isomers of all amino acids except proline. The best studied example of this enzyme is prepared from bovine lens tissue. This has a molecular weight of 326,000 and is made up of six identical subunits.[49] The enzyme contains two Zn atoms per subunit, which can be reversibly removed with loss of activity. The zinc has been replaced by Mn(II), Co(II), Cd(II), and Mg(II), with maintenance of activity in the case of Mn(II) and Co(II).[50] A preliminary X-ray study of the structure of bovine lens aminopeptidase has been reported.[51]

Thermolysin

This is a zinc endopeptidase isolated from *Bacillus thermoproteolyticus*. It has a molecular weight of 34,600 and contains one zinc ion which is necessary for catalytic activity. The bacillus is able to function at high temperature for short time periods, and the enzyme is heat stable. This heat stability is due to the presence of four calcium ions.[52] The structure of thermolysin and several enzyme–inhibitor complexes has been much studied by Matthews and his coworkers. The four calcium sites have been characterized,[53,54] two of them Ca(1) and Ca(2) are only 3.8 Å apart and form a so-called double site, while the remaining sites are some distance away. Details of the ligand groups are given in Table 3.2. These sites have been probed by the use of trivalent lanthanide ions.[54] The binding of the calcium ions at the double site is shown to be completely cooperative.[55]

The apoenzyme shows a thermal transition at about 48 °C, while the native enzyme is heat stable up to about 80 °C. The Zn^{2+}-free thermolysin retains its thermal stability completely, showing that Zn^{2+} does not appear to contribute to the heat stability of the enzyme. Fluorescence spectra have shown[56] that the Ca^{2+} ions prevent the unfolding of the protein, although it is not fully understood why the stabilizing effect of Ca^{2+} is so pronounced in thermolysin compared to other enzymes.

The environment around the zinc(II) in thermolysin has been shown to be similar to that of carboxypeptidase, although the overall folding of the two enzymes differs in a number of ways. In thermolysin the zinc ligands are His-142, His-146, Glu-166, and water. Peptide substrates bind to the enzyme with the carbonyl oxygen of the scissile peptide displacing the bound water molecule. Hydrolysis occurs by a zinc–carbonyl mechanism, in which Glu-133 of thermolysin promotes the attack of a water molecule on the carbonyl carbon of the substrate (as with Glu-270 in carboxypeptidase). The structure of the complex of thermolysin with the inhibitor β-phenylpropionyl-L-phenylalamine shows the carbonyl group to be 3.5 Å from a water molecule, which is hydrogen bonded to the carboxylate group of Glu-143. Thermolysin does not have a tyrosine residue at the active site to parallel the role of Tyr-248 in carboxypeptidase, but the imidazole of His-231 is able to act as a proton donor.[57] Stereochemical effects prevent the direct nucleophilic attack of Glu-133 on the carbonyl carbon atom.

Alkaline phosphatase

This enzyme catalyses the hydrolysis of orthophosphate monoesters, with maximum activity at pH 8 or above. It also shows phosphotransferase activity. These reactions are thought to involve an enzyme–phosphate intermediate formed through the phosphorylation of the OH group of Ser-99 in the enzyme with concommitant production of the hydrolysed substrate. The phospho group may then react with water to give inorganic phosphate or with an acceptor to give a new phosphoester product. These alternatives correspond to the hydrolysis and phosphotransferase reactions of alkaline phosphatase.

The enzyme from *E. coli* has been most extensively studied. It contains four moles of zinc per mole of enzyme (molecular weight 80,000) and splits into two subunits on treatment with acid. Two of the four zinc ions are necessary for catalytic activity[58] and the remaining two for the maintenance of structure.[59] Magnesium is also essential for activity,[58-60] and it appears that up to two moles of Mg^{2+} may bind per mole of enzyme, giving a maximum of six metal sites. The use of differential scanning calorimetry shows that binding of Mg^{2+} further stabilizes the enzyme, but this has no effect when zinc is absent.[61] Tb^{3+} is able to replace Mg^{2+} in stabilizing the protein structure, and provides a probe for conformational changes.[62]

The native enzyme may be regenerated by the addition of Zn^{2+} to the apoenzyme, but ^{31}P NMR suggests that the reconstituted enzyme with four

Zn^{2+} is distinguishable from the native enzyme.[63] The enzyme can be reconstituted with six moles of Co(II), suggesting it is bound at both Zn(II) and Mg(II) sites.[64] The binding of Zn(II), Mn(II), Co(II), and Cd(II) to the *E. coli* apoenzyme has a different effect on the UV spectrum of the protein than do Ni(II) and Hg(II), which do not induce the binding of phosphate. Apparently only two moles of Mn(II) and Cd(II) are necessary for maximum effect.[65]

Ligands for zinc

Studies on the photooxidation of native alkaline phosphatase and the apoenzyme in the presence of methylene blue suggest that zinc in the native enzyme is able to protect histidyl groups against photooxidation. Amino acid analysis of the treated apoenzyme suggests it has lost three histidyl groups, which are not lost in the native enzyme. A comparison of native and apoenzymes suggest that some six additional tyrosyl residues are analysable in the apoenzyme. These too may be ligand groups to zinc, but these results could also reflect conformational changes resulting from zinc binding.

Mechanism

Phosphorous-31 NMR has been much used to explore the effect of various stoichiometric ratios of metal to enzyme on the binding of phosphate and the interaction between metal sites. Replacement of Zn(II) by Cd(II) has allowed the application of ^{113}Cd NMR.[66] Furthermore, the dephosphorylation of the Ser-99 phosphointermediate is much slower in the case of Cd(II) alkaline phosphatase than in the native enzyme, and so its presence has been demonstrated clearly by ^{31}P NMR. The addition of two moles of Cd(II) per dimer still only results in the phosphorylation of one serine residue, even when further phosphate is added. This is in accord with studies on the binding of phosphate, which have shown that the dimeric enzyme binds one phosphate readily, and only binds a second phosphate at high phosphate concentration.[67]

These observations on phosphate binding have led to the postulate[67] of a 'flip-flop' mechanism, in which half of the sites are active at any point in time. Thus under normal conditions, one of the enzyme subunits will be phosphorylated. The other subunit binds a substrate molecule which is then phosphorylated, as the first subunit releases its phosphate group. Thus each half of the enzyme alternates between phosphorylated and non-phosphorylated states, with appropriate conformational change.

An Mn^{2+}–phosphate distance of 7.3 Å has been calculated from NMR data on the inactive Mn(II) enzyme, in accord with preliminary crystallographic data which indicates that substrates cannot gain direct access to the metal site.[67a] It is possible that the Mg^{2+} serves to promote the deprotonation, and hence the nucleophilicity, of the water group or the hydroxyl group of the substrate which is to attack the phosphoryl group being transferred.

Miscellaneous zinc enzymes

Rabbit liver fructose-1, 6-biphosphatase is a zinc metalloenzyme that binds four zinc ions per mole, one per subunit.[68] In the presence of the substrate analogue, $(\alpha + \beta)$ methyl D-fructofuranoside-1,6-biphosphate, at a level where two analogue molecules are bound per mole of enzyme, a total of eight zinc ions bind. In contrast, fructose-1, 6-biphosphatase from rat liver appears to contain twelve Zn(II)-binding sites per mole, three per subunit.[69] The sites of lowest affinity may however correspond to the Mg^{2+}-binding sites, Mg^{2+} being an activating cation. In the case of the rate liver enzyme it appears that while binding of Zn^{2+} to the first set of sites gives an active enzyme, binding to the second set of sites results in inhibition. Thus[69] Zn(II) can function both as an activator and as a negative allosteric regulator of fructose-1, 6-biphosphatase. [31]P nuclear relaxation studies on the Mn(II) enzyme give an Mn–phosphate distance of 3.3 ± 0.3 Å.

β-Lactamase (II) is an enzyme produced by *Bacillus cereus* which hydrolyses the β-lactam ring of penicillins or cephalosporins, and at present appears to be the only β-lactamase that requires a metal ion for activity. The enzyme binds two zinc(II) ions, and the cobalt(II) enzyme is also active. Spectroscopic studies[70] on the Co(II) and Cd(II) enzymes have implicated the only cysteine residue in the protein in metal binding. The use of 270 MHZ ^{1}H NMR spectroscopy[71] shows that resonances attributable to three of the histidyl residues in the apoenzyme shift on the addition of one mole of Zn(II), while resonances attributed to a fourth histidine residue shift on further addition of zinc to the monozinc enzyme. Thus it is suggested that the one zinc is bound by three histidine groups and the second zinc is bound by one histidine group. The remaining ligands have not yet been identified.

The α-amylases, which catalyse the hydrolysis of glucosides, contain Ca^{2+} which is necessary for activity and structure maintenance. The enzyme from *B. subtilis* also contains zinc, which has been thought to form a cross-link between two monomeric units. It has now been shown[72] that the enzyme exists in a monomer–dimer equilibrium in the absence of Zn^{2+}. Studies on the binding of Zn^{2+} show it to be bound only to the dimeric form of the enzyme, so it appears that the earlier view of the role of Zn^{2+} in this enzyme is incorrect.

The hemorrhagic activity of a toxin found in the venom of the rattlesnake *Crotalus atrox* has been shown to correlate with its proteolytic activity and with its zinc content. When zinc was removed from this hemorrhagic toxin, the proteolytic and hemorrhagic activities were equally inhibited and in direct proportion to the extent of zinc removal. It appears therefore that the hemorrhagic proteins are zinc-containing proteases.[73]

MAGNESIUM-ACTIVATED ENZYMES IN PHOSPHATE TRANSFER AND HYDROLYSIS

Most enzymes associated with the hydrolysis of phosphates or the transfer of phosphoryl groups require a metal ion for activity. This is usually Mg^{2+}, but it may be replaced by Mn^{2+} while in some cases Mn^{2+} appears to be the native cation.

The phosphatases catalyse the hydrolysis of phosphates. Some act on specific substrates but others are rather unspecific, although they may only be effective in certain pH ranges. As noted for alkaline phosphatase, phosphatases are usually thought to function by the formation of a phosphoenzyme intermediate with concommittant hydrolysis of the substrate. The phosphoryl group is subsequently transferred to a water molecule. If the phosphoryl group is transferred to other nucleophilic groups then the enzyme is demonstrating phosphoryl transfer activity. Such phosphoryl-transfer reactions are extremely important, and may involve the transfer of simple or complex phosphoryl groups, as in kinases and synthetases respectively. Phosphoglucomutase catalyses the transfer of phosphate from one part of a molecule to another (glucose-6-phosphate → glucose-1-phosphate). It is an interesting feature of these phosphoryl-transfer enzymes that the phosphoenzyme intermediate is not hydrolysed. Kinases probably do not involve this phosphoenzyme intermediate.

Much work has been carried out on model compounds to explore the binding of metal ions to phosphates and the mechanism of phosphoryl transfer. The transfer of a phosphoryl group involves the nucleophilic attack of species Y on phosphorus with loss of species X. In theory this reaction can be discussed[74] in

$$X—PO_3^{2-} + :Y \rightarrow X: + YPO_3^{2-}$$

terms of dissociative (S_N1) and associative (S_N2) mechanisms, but it appears probable that enzymes utilize associative pathways as this gives greater control of the stereochemistry of the substitution, and the rates of such processes are accelerated more effectively by metal ions. Indeed protonation and coordination of the transferable phosphoryl group inhibit an S_N1 mechanism, but will accelerate the S_N2 reaction. The cation exerts such an effect through charge neutralization and possibly polarization of the phosphoryl group, or, more feasibly, activation of water. The metal ion could also accelerate the S_N2 pathway by binding to both entering group and the phosphoryl group.

These aspects of the mechanism could be elucidated by distance measurements between the metal and the phosphoryl group. This has been accomplished[74,75] by the use of enzymes containing paramagnetic Mn^{2+} at the metal site, and the study of the effect of this paramagnetic centre on the longitudinal relaxation rate ($1/T_1$) of a nearby magnetic nucleus in the substrate, for example ^1H, ^{31}P, or ^{13}C. This technique requires a fast exchange of the substrate into the paramagnetic complex. Provided measurements are taken at several magnetic fields these distances may be measured to ≤ 0.1 Å. They may then be compared with known distances in metal–phosphate systems. Sometimes when more than one substrate is bound to an enzyme it is necessary to use a paramagnetic analogue of one of them. Spin labels have been of value here, as has the use of CrATP, a paramagnetic, substitution-inert analogue of MgATP.[76] The accumulation of a number of distance measurements on an enzyme has allowed the setting up of models for the active site.

The paramagnetic effect of Mn^{2+} on the proton relaxation rate of coordinated water has been used by Cohn[77] in pioneering work on the nature of metal–enzyme–substrate ternary complexes. Mn^{2+}, with its high magnetic moment, has a large effect upon the proton relaxation rate of water. The displacement of the aquo groups by other ligands should decrease the effectiveness of Mn^{2+} in increasing the relaxation rate of water. However the complexing of Mn^{2+} by proteins can increase the relaxation rate, as the relative rotational motion of the Mn^{2+} ion and water is hindered by the macromolecule. Thus the enhancement of the proton relaxation rate is a sensitive indicator of changes in configuration and the nature of the ligands occurring at the active site.

Kinases

We shall look at this group in more detail. The kinases catalyse the transfer of the terminal phosphoryl group of ATP to an acceptor molecule, as represented below. Kinase reactions involve a direct transfer of the phosphoryl group from ATP to the receptor substrate.

$$ATP^{4-} + HX \overset{M^{2+}}{\rightleftharpoons} ADP^{3-} + XPO_3^{2-} + H^+$$

Cohn has classified kinases into two groups on the basis of the binding scheme of the ternary complex of enzyme, metal and substrate, namely: Type **I**, M–S–E, and Type **II**, S–M–E. In **I** the substrate bridges metal and enzyme, while in **II** the metal acts as a bridge. This work involved the use of Mn(II) as a probe, NMR and EPR techniques, and spin labels.

The existence of the first class has been demonstrated by EPR spectra. The Mn–ADP binary complex has an EPR spectrum which does not change on formation of the ternary complex with enzyme. This indicates that there is no change in the nature of the ligands bound to the metal ion when the ternary complex is formed. Clearly therefore the metal is bound to substrate only and cannot be bridging. Further support comes from kinetic studies,[78] using relaxation techniques, on the ADP–creatine kinase system using a range of metal ions as cofactors. Rate constants for the association and dissociation of M–S and enzyme are practically independent of the nature of the metal, implying that metal–ligand substitution is not occurring, and that therefore the metal ion is not bridging.

The existence of ternary complexes of the type S–M–E having bridging metal ion was shown by NMR studies on the proton relaxation rate of water bound to Mn^{2+}. A large enhancement of relaxation rate on formation of the metal–enzyme complex is followed by a de-enhancement upon binding of the substrate. This is consistent with substrate substitution into the coordination shell of Mn(II) with replacement of water, but it could be the result of a conformational change.

This division of the kinase into two groups is confirmed by studies on their activation by different metal ions. Certain kinases have a low metal ion specificity. These belong to type **I**, M–S–E, where the requirement for the metal ion is mainly one of charge neutralization only. Those that belong to type **II** (E–M–S) have a greater selectivity, again reasonably, as here there are extra requirements in terms of the role of the metal ion as a bridge.

Examples of type **I** kinases are creatine kinase, adenylate kinase, hexokinase, and phosphoglycerate kinase. Several approaches have shown that the activating metal in creatine kinase binds to ATP only. The enzyme is a dimer of molecular weight 82,000, with an active site on each monomer. A cysteine residue is associated with each active site, which has been spin labelled for distance measurements to Mn(II). The overall conclusion from NMR measurements is that Mn(II) binds to the leaving ADP group. The long creatine–Mn(II) distance (9.8 Å) did not differ in the enzyme–creatine–Mn–ADP complex suggesting that the Mn(II) remains bound to the α- and β-phosphorous atoms.

Phosphoglycerate kinase appears to be an atypical type **I** kinase. The enzyme catalyses the conversion of 1,3-diphosphoglycerate into the 3-phosphoglycerate with production of ATP from ADP. The binding of Mn(II) nucleotides to the enzyme[79] was subject to a synergism that is not typical of type **I** kinases, while the EPR spectrum of the Mn(II) in the ternary complex differs from those observed for other kinases. One suggestion is that in this case the metal may bind to the substrate and enzyme, but with maintenance of enzyme–substrate interactions. A preliminary structural determination showed that Mg(II) is bound to both phosphoryl groups of ADP. Recently a high-resolution structure has been reported,[80] which shows the enzyme to be a single polypeptide chain organized into two widely separated domains of almost equal size. The binding site for MgATP or MgADP has been established, but the only possibility for the phosphoglycerate site was in the N-terminal domain, 10 Å away. This leads to difficulties in explaining how the two substrates can come close enough to transfer a phosphoryl group, and how water could be excluded from the site to prevent hydrolysis during the catalytic reaction. One possibility is that a 'hinge-bending' conformational change occurs on binding of phosphoglycerate, thus bringing together the two substrates and expelling the solvent. A similar mechanism had been suggested for hexokinase.[81]

Pyruvate kinase, a type **II** kinase, catalyses the transfer of phosphate to pyruvate and requires a divalent and a monovalent cation, the distance between which was found for the Mn^{2+}, Tl^+ enzyme to be 5 Å and 8 Å in the presence and absence of phosphoenolpyruvate. The Mn^{2+} bridges enzyme and substrate, while K^+ is thought to bridge the carboxyl group of phosphoenolpyruvate and the enzyme. A considerable number of distance measurements have been made on this enzyme, by several techniques, including the use of CrATP. The distance between Mn(II) and Cr(III) in the ternary complex with CrATP is about 5.2 ± 0.9 Å, suggesting van der Waals contact between the hydration spheres of the two metals.[82] A model based on fifteen distance measurements on four active complexes of pyruvate kinase suggests contact between the phosphorus of the γ-phosphoryl group of ATP and the carbonyl oxygen of pyruvate. Thus these NMR measurements support direct phosphoryl-group transfer from ATP to substrate,[83] in accord with crystallographic studies at 6 Å on cat muscle pyruvate

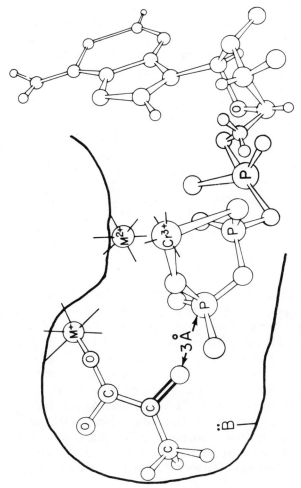

Figure 4.6 The active site of pyruvate kinase (Reproduced with permission from A. S. Mildvan, et al., *Biochemistry*, **18**, 4347 (1979)

kinase and its binary substrate complexes.[84] The present view of the active site in pyruvate kinase is depicted in Figure 4.6.

Phosphoglucomutase

This enzyme involves the formation of a phosphorylenzyme intermediate during the transfer of the phosphate group from position 6 to position 1 on the glucose molecule. The presence of phosphate bound to a serine residue at the active site has been studied by ^{31}P NMR spectroscopy.[85] The Mg(II) in phosphogluco-mutase is not bound to the phosphate group, while relaxation rates of the bound phosphate group, in the Ni(II) enzyme suggest the presence of a second-sphere complex of phosphate with the bound metal, i.e. enzyme–metal–(H_2O)–substrate. This leads to the correct positioning of the phosphoryl group for nucleophilic attack.[75]

DNA polymerase

Several DNA-dependent DNA polymerases and RNA polymerases and RNA-dependent DNA polymerases are zinc metalloenzymes. The DNA polymerases also require Mg^{2+} or Mn^{2+} for activity. The enzyme copies the DNA template, each step in the chain elongation involving the nucleophilic displacement of pyrophosphate on the appropriate deoxynucleoside triphosphate substrate by the 3′-OH group of the preceding sugar of the growing chain. Mildvan and his coworkers[86] have used Mn^{2+} as a paramagnetic probe and ^{31}P and proton relaxation rates to map the conformation of the substrates. In the binary complex all three phosphoryl groups are bound to Mn^{2+}, but on DNA polymerase only the γ-phosphoryl group remains coordinated. Distance measurements show that the resulting polyphosphate conformation is puckered. It is suggested that one role for the divalent cation is to assist the departure of the leaving pyrophosphate group by γ coordination, and possibly to assist nucleophilic attack on the α-phosphorus atom by strain and by hydrogen bonding through a coordinated water ligand.

SOME CALCIUM-DEPENDENT HYDROLYTIC ENZYMES

Reference has already been made to thermolysin and the α-amylases, while the Mg^{2+}–Ca^{2+}–ATPase will be discussed in a later chapter. Several phospho-lipases A_2 from different sources catalyse the specific hydrolysis of fatty acid ester bonds at the 2 position of 3-sn-phosphoglycerides. The enzyme has a requirement for Ca^{2+}. Ba^{2+} and Sr^{2+} bind and act as competitive inhibitors but Mg^{2+} does not bind at all. It appears that there is a second metal ion site, also specific for Ca^{2+}, which controls the way in which the enzyme interacts with the lipid at the lipid–water interfaces at higher pH values.[87] Activation by Ca^{2+} of proteases has been reported.[88]

Trypsin binds one Ca^{2+} per mole, which protects the enzyme from autolysis and from denaturation. The Ca^{2+}-binding site in trypsin has been studied[89] by several techniques, including the use of fluorescent Tb^{3+} and paramagnetic Gd^{3+}. In the latter case, proton relaxation rate measurements show that six water molecules are released from the aquo-Gd^{3+} ion on binding of trypsin, while recent X-ray studies have shown the Ca^{2+} site to involve two bidentate carboxylate ligands (Glu-70 and Glu-80) and the peptide carbonyl oxygens of Asn-72 and Val-75.[90]

REFERENCES

1. M. L. Bender and B. W. Turnquest, *J. Amer. Chem. Soc.*, **79**, 1889 (1957).
2. M. D. Alexander and D. H. Busch, *J. Amer. Chem. Soc.*, **88**, 1130 (1966).
3. D. A. Buckingham, D. M. Foster, and A. M. Sargeson, *J. Amer. Chem. Soc.*, **91**, 4102 (1969); D. A. Buckingham, D. M. Foster, L. G. Marzilli, and A. M. Sargeson, *Inorg. Chem.*, **9**, 11 (1970).
4. D. A. Buckingham, C. E. Davis, D. M. Foster, and A. M. Sargeson, *J. Amer. Chem. Soc.*, **92**, 5571 (1970); D. A. Buckingham, D. M. Foster, and A. M. Sargeson, *J. Amer. Chem. Soc.*, **92**, 6151 (1970).
5. M. A. Wells and T. C. Bruice, *J. Amer. Chem. Soc.*, **99**, 5341 (1977).
5a. J. T. Groves and R. M. Dias, *J. Amer. Chem. Soc.*, **101**, 1033 (1979).
6. J. McB. Harrowfield, V. Norris, and A. M. Sargeson, *J. Amer. Chem. Soc.*, **98**, 7282 (1976).
7. J. M. Pesando, *Biochemistry*, **14**, 681 (1975).
8. A. E. Dennard and R. J. P. Williams, *Transition Metal Chemistry* (Ed. R. Carlin) Vol. 2, Arnold, 1966.
9. J. E. Prue, *J. Chem. Soc.*, **1952**, 2331.
10. A. E. Dennard and R. J. P. Williams, *J. Chem. Soc. (A)*, **1966**, 812.
11. For reviews see M. F. Dunn, *Structure and Bonding*, **23**, 62 (1975); H. F. Bundy, *Comp. Biochem. Physiol.*, **57B**, 1 (1977); P. Wyeth and R. H. Prince, *Inorg. Persp. Biol. Med.*, **1**, 37 (1977); T. H. Maren, *Physiol. Rev.*, **47**, 595 (1967).
12. A. Liljas, K. K. Kannan, P. -C. Bergsten, I. Waara, K. Fridborg, B. Strandberg, U. Carlbom, L. Järup, S. Lövgren, and M. Petef, *Nature New Biol.*, **235**, 131 (1972); K. K. Kannan, B. Nostrand, K. Fridborg, S. Lövgren, A. Ohlsson, and M. Petef, *Proc. Natl. Acad. Sci. U.S.A.*, **72**, 51 (1975); L. E. Henderson, D. Hendriksson, and P. O. Nyman, *J. Biol. Chem.*, **251**, 5457 (1976).
13. K. K. Kannan, M. Petef, K. Fridborg, H. Cid-Dresdner, and S. Lövgren, *FEBS Letters*, **73**, 115 (1977).
14. K. K. Kannan, I. Waara, B. Nostrand, A. Borell, K. Fridborg, and M. Petef, *Proc. on Drug Action at the Molecular Level* (Ed. G. C. K. Roberts), McMillan, London, 1976.
15. P. J. Stein S. P. Merrill, and R. W. Henkens, *J. Amer. Chem. Soc.*, **99**, 3194 (1977).
16. R. Reipe and J. H. Wang, *J. Amer. Chem. Soc.*, **89**, 4229 (1967).
17. S. Lindskog, *J. Biol. Chem.*, **238**, 945 (1963); *Biochemistry*, **5**, 2641 (1966).

122

18. J. C. Kernohan, *Biochim. Biophys. Acta*, **81**, 346 (1964); **96**, 304 (1965).
19. P. Woolley, *Nature*, **258**, 677 (1975).
20. R. G. Khalifah, *J. Biol. Chem.*, **246**, 2561 (1971).
21. G. S. Brown, G. Navon, and R. G. Shulmann, *Proc. Natl. Acad. Sci. U.S.A.*, **74**, 1794 (1977).
22. R. L. Ward, *Biochemistry*, **8**, 1879 (1969); **9**, 2447 (1970); *Arch. Biochem. Biophys.*, **150**, 436 (1972).
23. R. L. Ward and P. L. Whitney, *Biochem. Biophys. Res. Comm.*, **51**, 343 (1973).
24. R. L. Peterson, T. Y. Li, J. T. McFarland, and K. L. Watters, *Biochemistry*, **16**, 726 (1977).
25. I. Bertini, C. Luchinat, and A. Scozzafava, *Inorg. Chim. Acta*, **22**, L23 (1977); *J. Amer. Chem. Soc.*, **99**, 581 (1977). See also *J. Chem. Soc. (Dalton)*, **1978**, 1269; *Inorg. Chim. Acta*, **23**, L15 (1977).
26. P. W. Taylor, R. W. King, and A. S. V. Burgen, *Biochemistry*, **9**, 2638, 3894 (1970).
27. J. F. Hower, R. W. Henkens, and D. B. Chesnut, *J. Amer. Chem. Soc.*, **93**, 6665 (1971).
28. B. P. N. Ko, A. Yazgan, P. L. Yeagle, S. C. Lottich, and R. W. Henkens, *Biochemistry*, **16**, 1720 (1977).
29. C. F. Chignell, D. K. Stavkweather, and R. H. Ehrlich, *Biochim. Biophys. Acta*, **271**, 6 (1972).
30. Y. Pocker and D. W. Bjorkquist, *Biochemistry*, **16**, 5698 (1977).
31. D. Demoulin, A. Pullman, and B. Sarkar, *J. Amer. Chem. Soc.*, **99**, 8498 (1977).
32. I. D. Campbell, S. Lindskog, and A. I. White, *Biochim. Biophys. Acta*, **484**, 443 (1977); R. K. Gupta and J. M. Pesando, *J. Biol. Chem.*, **250**, 2630 (1975).
33. M. E. Fabry, *J. Biol. Chem.*, **253**, 3568 (1978).
34. J. E. Coleman, *J. Biol. Chem.*, **242**, 5212 (1967); *Nature*, **214**, 193 (1967); S. Lindskog and J. E. Coleman, *Proc. Natl. Acad. Sci. U.S.A.*, **70**, 2505 (1973).
35. F. A. Quiocho and W. N. Lipscomb, *Adv. Protein Chem.*, **25**, 1 (1971); W. N. Lipscomb, J. A. Hartsuck, G. N. Reeke, F. A. Quiocho, R. H. Bethge, M. L. Ludwig, T. A. Steitz, H. Muirhead, and J. C. Coppola, *Brookhaven Symp. Biol.*, **21**, 24 (1968); *Phil. Trans. Roy. Soc. Lond.*, **B257**, 159 (1970).
36. R. K. Scheule, H. E. Van Wart, B. L. Vallee, and H. A. Scheraga, *Proc. Natl. Acad. Sci. U.S.A.*, **74**, 3272 (1977).
37. G. M. Alter, D. L. Leussing, H. Neurath, and B. L. Vallee, *Biochemistry*, **16**, 3663 (1977); C. A. Spilburg, J. L. Bethune, and B. L. Vallee, *Biochemistry*, **16**, 1142 (1977); *Proc. Natl. Acad. Sci. U.S.A.*, **71**, 3922 (1974).
38. R. Breslow and D. L. Wernick, *Proc. Natl. Acad. Sci. U.S.A.*, **74**, 1303 (1977).
39. B. L. Vallee, R. J. P. Williams, and J. E. Coleman, *Nature*, **190**, 633 (1961).
40. B. L. Vallee, T. L. Coombs, and F. L. Hoch, *J. Biol. Chem.*, **235**, 45 (1960).
41. D. S. Auld and B. Holmquist, *Biochemistry*, **13**, 4355 (1974).
42. R. G. Shulman, G. Navon, B. J. Wyluda, D. C. Douglass, and T. Yamane, *Proc. Natl. Acad. Sci. U.S.A.*, **56**, 39 (1966).
43. B. L. Vallee and J. F. Riordan, *Brookhaven Symp. Biol.*, **21**, 91 (1968).
44. B. L. Vallee, J. F. Riordan, D. S. Auld, and S. A. Latt, *Brookhaven Symp. Biol.*, **21**, 215 (1968).
45. H. M. Kagan, *Fed. Proc.*, **27**, 455 (1968).
46. F. W. Carson and E. T. Kaiser, *J. Amer. Chem. Soc.*, **88**, 1212 (1966).
47. S. Scheiner and W. N. Lipscomb, *J. Amer. Chem. Soc.*, **99**, 3466 (1977).
48. D. M. Hayes and P. A. Kollman, *J. Amer. Chem. Soc.*, **98**, 3335, 7811 (1976).
49. S. W. Melbye and F. H. Carpenter, *J. Biol. Chem.* **246**, 2459, 2463 (1971).
50. V. Kettman, *Ergeb. Exp. Med.*, **24**, 103 (1977).
51. F. Jurnak, A. Rich, L. Van Loon-Klassen, H. Bloemendal, A. Taylor, and F. H. Carpenter, *J. Mol. Biol.*, **112**, 149 (1977).

52. J. Feder, I. R. Garrett, and B. S. Wildi, *Biochemistry*, **10**, 4552 (1971).
53. P. M. Colman, J. N. Jansonius, and B. W. Matthews, *J. Mol. Biol.*, **70**, 701 (1972).
54. B. W. Matthews and L. H. Weaver, *Biochemistry*, **13**, 1719 (1974).
55. G. Voordouw and R. S. Roche, *Biochemistry*, **13**, 5017 (1974).
56. A. Fontana, C. Vita, E. Boccu, and F. M. Veronese, *Biochem. J.*, **165**, 539 (1977).
57. W. R. Kester and B. W. Matthews, *Biochemistry*, **16**, 2506 (1977); *J. Biol. Chem.*, **252**, 7704 (1977).
58. D. J. Plocke, C. Levinthal, and B. L. Vallee, *Biochemistry*, **1**, 373 (1962); D. J. Plocke, and B. L. Vallee, *Biochemistry*, **1**, 1039 (1962).
59. R. T. Simpson and B. L. Vallee, *Biochemistry*, **7**, 4343 (1968).
60. W. F. Bosron, R. A. Anderson, M. C. Falk, F. S. Kennedy, and B. L. Vallee, *Biochemistry*, **16**, 610 (1977).
61. J. F. Chiebowski and S. Mabrey, *J. Biol. Chem.*, **252**, 7042 (1977).
62. A. D. Sherry, S. Au-Young, and G. L. Cottam, *Arch. Biochem. Biophys.*, **189**, 277 (1978).
63. J. L. Bock and A. Kowalsky, *Biochim. Biophys. Acta*, **526**, 135 (1978).
64. R. A. Anderson and B. L. Vallee, *Biochemistry*, **16**, 4388 (1977).
65. H. Szajn and H. Csopak, *Biochim. Biophys. Acta*, **480**, 143 (1977).
66. J. F. Chiebowski, I. M. Armitage, and J. E. Coleman, *J. Biol. Chem.*, **252**, 7053 (1977).
67. C. Lazdunski, D. Chappelet, C. Petitclerc, F. Leterrier, P. Douzou, and M. Lazdunski, *Eur. J. Biochem.*, **17**, 239 (1970); **20**, 124 (1971).
67a. J. R. Knox and H. W. Wyckoff, *J. Mol. Biol.*, **74**, 533 (1973).
68. A. Benkovic, C. A. Caperelli, M. de Maine, and S. J. Benkovic, *Proc. Natl. Acad. Sci. U.S.A.*, **75**, 2185 (1978).
69. F. O. Pedrosa, S. Pontremol, and B. L. Horecker, *Proc. Natl. Acad. Sci. U.S.A.*, **74**, 2742 (1977).
70. R. B. Davies and E. P. Abraham, *Biochem. J.*, **143**, 129 (1974).
71. G. S. Baldwin, A. Galdes, H. A. O. Hill, B. E. Smith, S. G. Waley, and E. P. Abraham, *Biochem. J.*, **175**, 441 (1978).
72. R. Tellam, D. J. Winzor, and L. W. Nichol, *Biochem. J.*, **173**, 185 (1978).
73. J. B. Bjarnason and A. T. Tu, *Biochemistry*, **17**, 3395 (1978).
74. A. S. Mildvan and C. M. Grisham, *Structure and Bonding*, **20**, 1 (1974).
75. A. S. Mildvan *Acc. Chem. Res.*, **10**, 246 (1977).
76. J. I. Legg, *Coord. Chem. Revs.*, **25**, 103 (1978).
77. M. Coh, *Quart. Rev. Biophys.*, **3**, 61 (1970); *Acc. Chem. Res.*, **4**, 214 (1971).
78. G. G. Hammes and J. K. Hurst, *Biochemistry*, **8**, 1083 (1969).
79. B. E. Chapman, W. J. O'Sullivan, R. K. Scopes, and G. H. Reed, *Biochemistry*, **16**, 1005 (1977).
80. R. D. Banks, C. C. F. Blake, P. R. Evans, R. Haser, D. W. Rice, G. W. Hardy, M. Merrett, and A. W. Phillips, *Nature*, **279**, 773 (1979).
81. W. S. Bennett and T. A. Steitz, *Proc. Natl. Acad. Sci. U.S.A.*, **75**, 4848 (1978).
82. R. K. Gupta, *J. Biol. Chem.*, **252**, 5183 (1977).
83. A. S. Mildvan, D. L. Sloan, C. H. Fung, R. K. Gupta, and E. Melamud, *J. Biol. Chem.*, **251**, 2431 (1976); D. Dunaway-Mariano, J. L. Benovic, W. W. Cleland, R. K. Gupta, and A. S. Mildvan, *Biochemistry*, **18**, 4347 (1979).
84. D. K. Stammers and H. Muirhead, *J. Mol. Biol.*, **95**, 213 (1975).
85. W. J. Ray, A. S. Mildvan, and J. B. Grutzner, *Arch. Biochem. Biophys.*, **184**, 453 (1977).
86. D. L. Sloan, L. A. Loeb, A. S. Mildvan, and R. J. Feldmann, *J. Biol. Chem.*, **250**, 8913 (1975); L. A. Loeb and A. S. Mildvan, *Crit. Rev. Bioch.*, **6**, 219 (1979).
87. R. Franson and M. Waite, *Biochemistry*, **17**, 4029 (1978); A. J. Slotboom,

E. H. J. M. Jansen, H. Vlijm, F. Pattus, P. Soares de Araujo, and G. H. de Haas, *Biochemistry*, **17**, 4593 (1978).

88. J. L. Dimicoli and J. Bieth, *Biochemistry*, **16**, 5532 (1977); D. R. Phillips and M. Jakabová, *J. Biol. Chem.*, **252**, 5602 (1977).
89. M. Epstein, J. Reuben, and A. Levitzki, *Biochemistry*, **16**, 2449 (1977).
90. W. Bodl and P. Schwager, *FEBS Letters*, **56**, 139 (1975); W. Bodl and P. Schwager, *J. Mol. Biol.*, **98**, 693 (1975).

Added in proof

91. I. Bertini, E. Borghi and C. Luchinat, *J. Amer. Chem. Soc.*, **101**, 7069 (1979).
92. M. W. Makinen, L. C. Kuo, J. J. Dymowski and S. Jaffer, *J. Biol. Chem.*, **254**, 356 (1979). This shows that the carboxypeptidase-catalysed hydrolysis of an ester substrate proceedes via a mixed anhydride intermediate, deacylation of which is catalysed by a zinc bound hydroxide group.
93. A. Nakano, M. Tasumi, K. Fujiwara, K. Fuwa and T. Miyazawa, *J. Biochem. (Tokyo)*, **86**, 1001 (1979). Mn(II) coordinated water in MnCPA is not displaced by the binding of Gly–Tyr substrate.

CHAPTER FIVE

THE TRANSITION METALS IN BIOLOGICAL REDOX REACTIONS

The transition metals are involved in a range of biological redox processes. Iron and copper are of wide-ranging significance in this context while cobalt and molybdenum are also important. In these reactions the redox change in the transition metal ion is associated with the catalytic process in question, either in electron transfer or in oxidation or reduction of a substrate. In a few cases non-transition metals are associated with redox enzymes, in which case their function will be a structural or organizational one. Examples are the enzymes alcohol dehydrogenase (Zn^{2+}) and aldehyde dehydrogenase (K^+).

A number of well-known transition-metalloproteins act as storage and/or transport systems. The dioxygen carriers hemoglobin, hemerythrin, and hemocyanin, and the dioxygen storer myoglobin will be discussed in Chapter 7. Specific transport and storage proteins for the transition metals have been characterized in some cases. Those for iron, which are best understood, are described in Chapter 8, while those for copper will be referred to briefly in this chapter.

While this chapter will be concerned mainly with the biochemistry of iron, cobalt, copper, and molybdenum, some background information will be given on biological redox reactions, redox potentials in model systems, and mechanisms of electron transfer.

TYPES OF BIOLOGICAL OXIDATION

Biological oxidations may be brought about by one of the following processes (S = oxidized substrate, SH_2 = reduced substrate).

  (a) electron transfer with oxygen (or another species) as the terminal electron acceptor,

  (b) hydrogen atom abstraction, by (5.1) or (5.2)

$$2SH_2 + O_2 \rightarrow 2S + 2H_2O \tag{5.1}$$

$$SH_2 + O_2 \rightarrow S + H_2O_2 \tag{5.2}$$

  (c) hydride group transfer,

  (d) oxygen atom incorporation, and

  (e) hydroxyl group incorporation.

The free energy liberated in these oxidative processes may be utilized in the production of ATP. Hydrolysis of ATP will then, in turn, provide the energy for a number of important processes, such as muscle contraction or synthesis.

Equations (5.1) and (5.2) represent an overall reaction. In practice, the reaction may proceed *via* one or more intermediate hydrogen carriers, with oxygen as the terminal hydrogen acceptor. Such a scheme is usually represented by (5.3), where A^n and A^nH_2 represent oxidized and reduced states of a series of acceptors. This scheme is very important as in effect we have an electron transport chain.

$$SH_2 \diagdown \nearrow A^1 H_2 \diagdown \nearrow A^2H_2 \diagdown \cdots \diagup A^nH_2 \diagdown \nearrow H_2O_2$$
$$S \diagup \diagdown A^1 \diagup \diagdown A^2 \diagup \quad \diagdown A^n \diagup \diagdown O_2 \tag{5.3}$$

Electron-transfer chains

It is possible to write schemes similar to (5.3) involving electron transfer or oxygen transfer in addition to hydrogen transfer. Some typical carriers that might be involved in such a chain of electron acceptors have already been referred to in Chapter 1. Examples include the flavin nucleotide coenzymes, the nicotinamide coenzymes, the cytochromes, and the iron–sulphur proteins. The latter two examples are associated with a $Fe(II)–Fe(III)$ redox change. Oxygen is a very well-known physiological electron acceptor, but often this (or other naturally occurring electron acceptors) may be replaced by artificial electron acceptors. Thus if it is desired to examine an oxygen-utilizing species under anaerobic conditions, it is usually possible to replace oxygen as the terminal electron acceptor by the dye methylene blue. Many other dyes have also been used. Significantly, these may usually be autoxidized readily.

One important example of electron-transfer systems is that of linked reactions. Here one carrier intervenes between two separate dehydrogenase using two different substrates SH_2 and $S'H_2$ (Equation 5.4). Here then, instead of the

$$SH_2 \diagdown \nearrow C_{red} \diagdown \nearrow S'H_2$$
$$S \diagup \diagdown C_{ox} \diagup \diagdown S' \tag{5.4}$$

reduced carrier being reoxidized by a second carrier (and ultimately by a terminal electron acceptor), it is reoxidized by a molecule of a second substrate.

Of fundamental importance in this discussion are electron-transfer processes in mitochondria and photosynthesis. In the former case[1] electrons are transferred *via* a series of membrane-bound acceptors to dioxygen as the terminal electron acceptor. At specific points electron transport is coupled to the synthesis of ATP, a topic that has been at the centre of much discussion. It is currently believed that these two processes are linked by proton gradients, that cytochrome c oxidase, for example, also pumps H^+ across the mitochondrial membrane to build up a proton concentration gradient, which can then drive the synthesis of ATP. This theory of chemiosmosis is discussed in Chapter 9. The components of the electron-transfer chain in mitochondria are shown in Figure 5.1, on which probable points of synthesis of ATP are noted. The data in Figure 5.1 represent current views and may have to be corrected at a future time. Its conclusions are

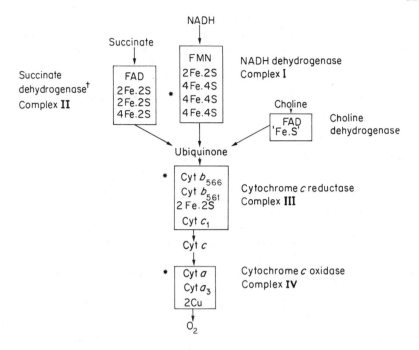

Figure 5.1 Some electron-transfer components of the respiratory chain.

*Sites of synthesis of ATP.
†See *Proc. Natl Acad. Sci., USA*, **76**, 3805 (1979)

drawn largely from visible and EPR data (for cytochromes and iron–sulphur proteins respectively). The components of this chain are arranged in sequence according to their redox potentials. Thus different substrates may interact at different parts of the redox chain.

The mechanism of electron transfer will be discussed later but it is appropriate to refer again to the reasons why these metalloproteins are so well suited to participate in the fast transfer of electrons, and also are able to provide a range of redox potentials.

The copper enzymes particularly involved in electron-transfer reactions are the blue proteins where the copper is in the copper(II) state. The Franck–Condon principle, as outlined in Chapter 2, requires that for fast electron-transfer reactions the geometries of the reactant and product complexes should easily be made equivalent. This then means that the energy involved in the promotion of the complex to an appropriately vibrationally excited state is small, otherwise there would have been a large additional contribution to the activation energy for electron transfer. As was discussed in detail in Chapter 2, copper(I) and copper(II) normally favour rather different stereochemistries. It may well be that the irregular structure imposed upon the copper in the copper blue proteins by the protein is one that lies between the favoured structures for the two oxidation states and therefore one that implies fast electron transfer.

The way in which the redox potentials of iron–sulphur proteins are controlled presents a particularly fascinating problem, and reflects a subtle role for the protein. These proteins are also well suited for fast electron transfer as they offer a delocalized, highly polarizable redox centre. The distortions present in the clusters are probably of significance.

Both Fe(II) and Fe(III) complexes prefer octahedral stereochemistries, but bond lengths will be slightly different. In heme systems the heme group provides an inflexible set of in-plane ligands, but does allow some flexibility in terms of the Fe–axial ligand bond lengths. Here again there is the possibility that fast electron transfer is favoured by the metal–ligand bond length lying in an intermediate position between the characteristic values of Fe(II) and Fe(III) bond lengths. Depending upon the nature of the axial ligands the cytochromes will show a range of redox potentials. Thus an unsaturated ligand will prefer to bond to iron(II) and stabilize that state. Of greater importance, however, is the question of the spin state of the iron(III) system. It appears that there is an equilibrium between spin states in most examples of ferric hemeproteins. The redox potential of the Fe(III)–Fe(II) couple will depend upon the spin state. The spin state will,

TABLE 5.1 Redox potentials for some biochemical and chemical couples.

Protein	E_0' (mV)	Ligand	E_0' (mV)
$Fe^{3+}-Fe^{2+}$		$Fe^{3+}-Fe^{2+}$	
Hemoglobin	170	Water	770
Myoglobin	46	(1,10-Phenanthroline)$_3$	1100
HR peroxidase	-170	Dipyridyl	960
Cytochrome c	260	(Cyanide)$_6$	220
Cytochrome c_1	220	Oxalate	20
Cytochrome c_2	320	(Oxime)$_3$	-251
Cytochrome c_3	-205		
Cytochrome c_{350}	250		
Cytochrome c_5	320		
Cytochrome b	50		
Cytochrome b_5	20	$Cu^{2+}-Cu$	
Cytochrome $P450$	-400	Water	167
Rubredoxin	-57	Imidazole	255
Adrenodoxin	-270	(Imidazole)$_2$	345
HiPIP (*chromatium*)	350	(NH$_3$)$_2$	340
Ferredoxin (*chromatium*)	-490	2-Methylthioethylamine	187
		(2-Methylthioethylamine)$_2$	243
$Cu^{2+}-Cu^+$		Pyridine	197
Azurin	328	(Pyridine)$_2$	270
Plastocyanin	370	(1,10-phenanthroline)	174
Rusticyanin	680	(2-Cl-1,10-phen)$_2$	400
Laccase	415	(2,9-Me$_2$-1,10-phen)$_2$	594
		(Ethylenebisthio-glycolic acid)$_2$	
NAD$^+$-NADH	-320	(Ethylenediamine)$_2$	-380
Riboflavin	-210	(Glycine)$_2$	-160
		(Alanine)$_2$	-130

in turn, depend upon the axial ligands. Any effect which prevents the close approach of the axial ligand will result in a high-spin complex, as also will the presence of weak ligands. It may be readily seen, therefore, why the redox potentials of the cytochrome vary with change of axial ligand.

The redox potentials (E'_0) of systems of biochemical importance are defined at pH7. Some values are listed in Table 5.1 together with data for other couples for comparison. As has been noted, it may be seen that certain groups of metalloproteins show a wide range of redox potentials.

Dehydrogenases

There are three main groups, all catalysing dehydrogenations, but utilizing different electron acceptors.

(1) Copper enzymes

These are of high redox potential and therefore always use the oxygen molecule as the electron acceptor. It appears that oxygen oxidizes copper(I) to copper(II). This latter fact may also be associated with the high affinity of copper proteins for oxygen. Copper enzymes will therefore be involved in the oxidation of substrates of high redox potential, such as the hydroquinones. They are often found as the terminal member of respiratory chains.

(2) Flavoproteins

These are of lower redox potential than the copper proteins and accordingly catalyse the oxidation of substrates of intermediate redox potentials and may interact with oxygen or with other electron acceptors.

(3) Nicotinamide coenzyme-dependent dehydrogenases

These are usually involved in the initial dehydrogenation of substrates such as amines and alcohols, the coenzyme (to which initial transfer from substrate occurs) having a redox potential of -0.32 V. They cannot be linked directly to oxygen therefore, but are often associated with flavoproteins. They are well suited to act in the initial dehydrogenation of many substrates.

Ferredoxins may remove hydrogen in addition to acting as electron-transfer agents.

The interrelationships between these three groups of dehydrogenases is summarized in Figure 5.2.

It may be noted that in general these dehydrogenases contain a number of metal atoms. Thus ascorbic acid oxidase contains eight atoms of copper per mole of protein while xanthine oxidase contains eight atoms of iron and two of molybdenum per mole. These are termed oxidases because O_2 is utilized as an electron acceptor, but they are dehydrogenases. In the first example, the product

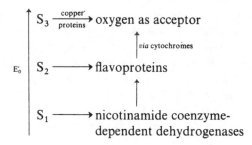

$$S_3 \xrightarrow[\text{proteins}]{\text{copper}} \text{oxygen as acceptor}$$

$$\uparrow \textit{via} \text{ cytochromes}$$

$$E_0' \quad S_2 \longrightarrow \text{flavoproteins}$$

$$\uparrow$$

$$S_1 \longrightarrow \text{nicotinamide coenzyme-}$$
$$\text{dependent dehydrogenases}$$

Figure 5.2 The relation between redox potential of substrate and electron transport.

of hydrogen atom abstraction will be the ascorbate free radical. The hydrogen atom must then pass to the oxygen molecule which will be bound at another site on ascorbic acid oxidase. However, we are dealing with a non-complementary reaction. Oxygen is reduced to water (Equation 5.1), a four-electron change. A slow reduction of oxygen would generate free-radical species of some lifetime, allowing the possibility of direct reaction with the ascorbate radical. The presence of a number of metal centres may be associated with the necessity of providing a rapid release of electrons to the oxygen molecule to prevent such a situation occurring. This would be followed by addition of H^+.

Enzymes catalysing oxygen atom incorporation

These include oxygenases (which catalyse the introduction of both atoms of molecular oxygen into a substrate) and hydroxylases, which catalyse the introduction of only one atom of a molecule of oxygen, and hence the formation of a monohydroxy derivative of the substrate. In both cases the introduction of oxygen may be demonstrated by the use of isotopically labelled oxygen.

The activity of oxygenases may be understood in terms of the intermediate formation of the dihydroxy derivative of the substrate. Thus a typical reaction catalysed by an oxygenase is the cleavage of an aromatic ring (Reaction 5.5).

$$\text{(5.5)}$$

There is good evidence for the involvement of iron(II) in many oxygenases. Thus tryptophan oxygenase (Reaction 5.6) has a heme prosthetic group. Activity is lost when this is removed and regained when it is restored. In general, the iron

$$\text{(5.6)}$$

and copper enzymes active in catalysing the direct action of oxygen have an open structure or ligand groups that may easily be replaced, so favouring oxygen binding. Copper appears to be present as copper(I) and there is evidence for the reaction of copper(I) with oxygen. The action of hydroxylases may be illustrated by Reaction (5.7), where it may be noted that oxygen-18 has appeared in the product water and that there is a requirement for a second oxidizable substrate, $S'H_2$.

$$SH_2 + {}^{18}O_2 + S'H_2 \xrightarrow[\text{hydroxylase}]{} SH^{18}OH + S' + H_2{}^{18}O \qquad (5.7)$$

The hydroxylase is therefore acting as an hydroxylase and as a dehydrogenase. In many examples, there is a requirement for heme prosthetic groups and in some cases for copper. Monophenol oxidase and dopamine hydroxylase are examples of copper enzymes, while peroxidase is a well-known example of a heme enzyme that catalyses the hydroxylation of certain aromatic and heterocyclic compounds by hydrogen peroxide.

The following scheme has been suggested[2] for the catalysis of the hydroxylation of dopamine (SH_2) by dopamine hydroxylase. Here ascorbate is required as the second substrate ($S'H_2$).

$$Cu^{2+}.\text{protein} + S'H_2 \rightarrow Cu^+.\text{protein} + S'H + H^+$$
$$Cu^+.\text{protein} + O_2 \rightarrow [Cu.\text{protein}.O_2]$$
$$S'H + SH_2 \rightarrow S'H_2 + SH\cdot$$
$$[Cu.\text{protein}.O_2] + SH\cdot \rightarrow [Cu.\text{protein}.O_2.SH\cdot]$$
$$[Cu.\text{protein}.O_2.SH\cdot] + S'H_2 \rightarrow Cu^{2+}.\text{protein} + SOH + H_2O + S'$$

In an earlier section it was noted that fast electron transfer to oxygen was necessary to prevent the build up of oxygen free-radical species and their attack upon an organic radical. However, there are examples of oxidases where this does occur and where there is no binding of oxygen by the enzyme. In this case the redox metal will be in a higher oxidation state and will directly reduce the bound substrate (5.8).

$$M^{n+} + S^- \rightarrow M^{n+}S^- \rightleftharpoons M^{(n-1)+}S \qquad (5.8)$$

$$(M = Fe, \text{ possibly } Cu)$$

The free-radical substrate then adds on oxygen, eventually giving the appropriate products. Interactions of this type between redox metal ions and organic groups are particularly important in the flavoproteins, mentioned under dehydrogenases.

Normally, however, in oxygenase and hydroxylase activity the oxygen or hydrogen peroxide and oxidized substrate are both bound at the protein. A number of important and interesting features associated with this behaviour will be discussed later in the chapter.

REDOX POTENTIALS OF MODEL SYSTEMS

The value of the redox potential for any transition metal couple in a metalloenzyme system will determine the role it may play in electron transport or, for example, whether a metalloprotein will carry oxygen or be oxidized by it. The protein ligand groups will largely control the redox potential by providing a site symmetry and metal-binding groups that will favour the two oxidation states to different extents. In this section it is hoped to examine the way in which the redox potential of metals in model complexes respond to variation in ligand and then to apply this to enzyme systems.[3]

A fair amount of information is available[4,5] for complexes of copper and iron, so allowing a study of the redox potentials for Cu(II)–Cu(I) and Fe(III)–Fe(II). Redox potentials are listed in Table 5.1 while formation constant data, contained in Table 3.3, tells us the favoured binding groups for these metals in the upper and lower oxidation state. We will not attempt an examination of other transition metal systems, partly due to a paucity of data and partly as the systems are not so convenient for study.

Copper

The formation constant data for copper may be readily understood in terms of the hard–soft formalism of Pearson, as outlined in Chapter 3. Copper(I) is much softer than copper(II), in accord with the fact that sulphur donor ligands are bound more strongly by copper(I) than copper(II), as are all unsaturated ligands such as o-phenanthroline and bipyridyl. Both nitrogen and sulphur donor ligands bind more strongly to copper(II) and copper(I) than do oxygen donors. For unidentate ligands, considerations of this type only are important. For polydentate ligands, however, a stereochemical effect will be superimposed upon this, as copper(I) and copper(II) have very different behaviour in terms of preferred coordination number and stereochemistry. Copper(I) is often linear, two-coordinate, but may increase its coordination number to four, in which case the geometry will be a tetrahedral one. Copper(II), in contrast, prefers to be six-coordinate, most complexes having a tetragonally distorted structure. For 1:2 stoichiometries, therefore, with ligands such as bipyridyl which, in addition to being soft, cannot give a tetragonally distorted octahedral structure, it is quite clear that copper(I) complexes will be stable.

The data in Table 5.1 allow us to observe the effect on redox potential for Cu(II)–Cu(I) of successively replacing aquogroups by other ligands. Replacement of one and then two water molecules by nitrogen donor ligands results in a more positive potential. Increase in the number of nitrogen donor atoms beyond this will not further stabilize the copper(I) state, however, as this will involve the formation of a complex with a coordination number higher than the favoured one. The binding of groups such as the aliphatic α-amino acids and ethylenediamines results in negative potentials for the couple. They prefer to be bound to copper(II) and will, in fact, cause disproportionation of copper(I) to copper metal and copper(II). Cyanide and mercaptide ligands correspondingly favour copper(I) to such an extent that they will reduce copper(II) to copper(I).

James and Williams have commented[4] on the importance of steric hindrance in determining the values of redox potentials. 2,9-Substituted phenanthrolines introduce a substantial degree of steric hindrance in four-coordinate square-planar copper(II) complexes ML_2. Four-coordinate copper(I) complexes are tetrahedral and will not show steric hindrance. Thus the high value of E'_0 in this case reflects the destabilization of the copper(II) state. Similarly the larger halide ions favour copper(I) rather than copper(II), reflecting the smaller size of the copper(II) ion.

Iron

The preferred stereochemistry for both Fe(II) and Fe(III) complexes is octahedral. We are not therefore concerned with major stereochemical problems as in the case of Cu(II)–Cu(I). However, it has been noted already that, in heme proteins, while there will be a preferred metal–axial ligand bond length for fast electron transfer, there is also another effect associated with the question of spin state of the ferric and ferrous ions.

Some redox potentials are listed in Table 5.1. As for copper there is a stabilization of the lower oxidation state by unsaturated ligands, with resulting high potentials. In the low-spin state, interaction with the ligands will become greater, and so the effect of unsaturated ligands should be more pronounced. Increase in basicity of the ligand (i.e. σ-electron donation) will obviously favour the ferric state, with a decrease in E'_0 Ultimately, however, this will cause a change from a high-spin to a low-spin state. But the energy involved in this electronic rearrangement is smaller for Fe(II) than for Fe(III) and the rearrangement will therefore occur first in the former case. This means that we have to compare the effect of σ donation on Fe(III) high spin with Fe(II) low spin. When both these oxidation states have become low spin (i.e. at a sufficiently high pK_a of ligand), then once again we are comparing similar spin states and ferric ion will be favoured by further increase of pK_a with a corresponding fall in redox potential. It is difficult to assess the relative effect of increased basicity on low-spin Fe(II) and high-spin Fe(III) but Williams[3] has shown that for iron porphyrin complexes, where there is a spin-state change, increase of ligand pK_a results in a *rise* in redox potential, when Fe(II) only is low spin, followed by a fall when both oxidation states are low spin. The stabilization of Fe(II) when low spin can be understood in terms of ligand field stabilization energy changes.

The effect of protein upon redox potential

A feature of iron and copper proteins is that they may show high redox potentials compared to those usually observed for model compounds, i.e. the lower oxidation state is stabilized with respect to the higher one. This may be understood quite readily for copper proteins. The copper(I) ion is much less specific in its steric requirements than the copper(II) species. The protein is therefore able to satisfy the steric requirements of the former species more readily, and probably with donor groups that bind more strongly to copper(I) than to copper(II). The steric effect is much more important for protein systems than for model complexes.

Of the protein groups available, it is probable then that the 'hard' groups, carboxylates, phenolates, and amino groups will be specific for copper(II) while sulphur donors, RSSR, R_2S, and particularly RS^-, will be bound to copper(I). However, under biological pH values, the proton competes effectively with the metal for the $-NH_2$, phenolate, and sulphide groups and their availability will be reduced. In practice, very little is known about the nature of metal-binding groups in copper proteins.

Any protein system that involved copper bound to, say, sulphide groups (i.e. copper(I) specific groups) would seem unlikely to be involved in fast electron transfer as the copper(II) would be generated in an environment characteristic of copper(I). The redox potential may be very high, in addition. It seems reasonable to suggest, therefore, that for copper proteins involved in electron transfer, at least some of the binding groups should be groups which are not particularly specific for either copper(I) or copper(II). It would also seem reasonable to predict that the copper in the oxygen carrier hemocyanin *would* be bound by sulphur groups to prevent irreversible oxidation.

The varying potentials of the iron heme enzymes will depend upon the protein binding groups in the fifth and sixth positions. The redox potentials vary from -0.4 V to $+0.1$ V. The low potential could result from carboxylate being bound to iron, i.e. favouring Fe(III) as the harder species. High potentials would result from unsaturated soft donor groups favouring Fe(II), or also from a steric misfitting which would destabilize the iron(III) state.

Thus in order to favour fast electron transfer in terms of the requirements of the Franck–Condon principle, the ferric–axial ligand bond length would be longer than that normally observed in ferric complexes. This would result in a greater tendency for high-spin iron(III) than that seen in model complexes. This implies a high redox potential for the heme enzyme. Again, this fits in with the fact that electron-transfer proteins have high redox potentials.

For non-heme iron proteins, there is also a range of redox potentials. These are more difficult to explain at present, as proteins with very different redox potentials appear to have identical iron–sulphur clusters. These differences must result from a more subtle effect of the protein.

A further complication results from the nature of the metal-binding site on the protein, as this will probably be hydrophobic in character. This will result in higher values of redox potentials compared to those observed in aqueous solution. This arises as the higher oxidation state of the metalloprotein will usually be charged while the lower one will not. The hydrophobic environment will thus stabilize the lower oxidation state. Furthermore, this allows the control of the redox potential by the protein depending upon the extent to which the metal centre is exposed to the aqueous solvent. Thus the 'protein' may serve to 'tune' the redox properties of the iron heme. This has been demonstrated[6] for heme proteins for which high-resolution crystallographic data is available. The redox potential decreases in proportion to the fraction of heme surface exposed to the aqueous solvent, and indeed is expressed by the empirical relationship $E_0' = -15\% (\% \text{ exposure}) + 345$ mV. This could also reflect the ease of access of the heme to dioxygen.

'Valence-state' determination

The oxidation state of copper, in particular, in a protein is not always determined easily. EPR measurements should show the presence of copper(II), but the absence of a signal may be attributed to factors other than the absence of copper(II). Sometimes magnetic susceptibility measurements may help, but the possibility of the presence of organic free radicals is a complicating factor. Visible spectra should allow a distinction to be made between copper(I) and copper(II) proteins. However, in practice, charge-transfer bands occur at quite high wavelengths in copper(I) complexes and difficulty may be experienced in assigning the bands in the visible spectrum of a copper protein to charge-transfer or $d \rightarrow d$ transitions. The important guide-line here is the question of the intensity of the bands. If the intensity is low, they are probably $d \rightarrow d$ bands.

The information contained in the previous sections allows a chemical approach to the study of oxidation state. The extent to which added groups are bound by the enzyme at the metal ion will allow a distinction to be made between valence states. Obviously the very strongly binding 'valence-specific' reagents are of little value as these will either cause disproportionation of copper(I) or reduction of copper(II) to give the favoured oxidation state anyway. The disproportionation of Cu^+ is represented below.

$$2\,Cu^+ \rightleftharpoons Cu^{2+} + Cu, \qquad E_0^v = 0.167\,V$$

Clearly, normally copper(I) is very unstable. The equilibrium may be displaced in either direction. Thus the soft ligands CN^-, I^-, R_2S all reduce copper(II) to copper(I) while ethylenediamine, for example, will convert a copper(I) complex to a copper(II) ethylenediamine complex. An examination of the available data for a range of ligands shows that the equilibrium is dependent both upon the nature and the preferred stereochemistry of the ligand, as indicated earlier.

Provided the reduction of copper(II) by the protein-binding groups is not possible, 2,2'-biquinoline is specific for copper(I), as also is the binding of oxygen or carbon monoxide. Chloride and bromide are bound more strongly by copper(II) than copper(I). This is probably due to a steric effect. Again, 2,2'-bipyridyl is specific for iron(II), although this is a kinetic phenomenon. Great care must be exercised in the use of this approach.

Spin-state determination

From the previous discussion it is clear that the matter of spin-state determination is also an important one. Iron(II) low-spin systems will be diamagnetic, while high-spin complexes will have high moments. Iron(III) systems may have either one or five unpaired electrons. In the latter case magnetic susceptibilities will clearly demonstrate in broad terms the composition of a complex in terms of its spin state. It is much more difficult to say whether a complex is completely low or high spin. Thus the magnetic moment of a low-spin complex may be raised above the spin-only value either by orbital contributions or by the presence of a small amount of the high-spin form and it will be difficult to rule out the latter explanation authoritatively. Another problem with measurement of magnetic moments is associated with the very high diamagnetic correction for heme groups.

ELECTRON TRANSPORT IN REDOX METALLOPROTEINS

Electron transfer proceeds at a very fast rate in biological processes with a low activation energy estimated to be of the order of 7 kcal mole^{-1}. The inorganic reactions discussed in Chapter 2 usually proceed at a very much lower rate with activation energies of 15–20 kcal mol^{-1}. The highly organized arrangement of electron carriers in the mitochondrial membrane is evidently associated with this phenomenon, providing as it undoubtedly must, highly conjugated metal–ligand systems, and allowing maximum orbital overlap of the constituents of the electron-transport chain.

For electron transport to occur between metal atoms in metalloproteins there must be electron transfer from one metal to a neighbouring atom of the protein binding group through overlap of orbitals. This electron must then be transported away before recombination with the positive 'hole' of the metal atom can occur. The importance of conjugated polarizable protein groups is plain in this connection and it is noteworthy that considerable attention has been given to the semiconducting properties of proteins in terms of band theory.

One general problem associated with electron transport is the nature of electron transport, whether it be by actual transfer of electrons or whether by group transfer, particularly that of hydrogen atoms. Electron transfer and group transfer are known for model systems, although sometimes it is not possible immediately to identify which is occurring. Thus the ferric–ferrous ion exchange, which one might expect to be by direct electron transfer, is catalysed by hydroxide ions. This suggests the following reaction pathway, in which electron transfer has really involved H atom transfer. However, does this involve direct atom transfer or electron transfer followed by proton transfer? One conclusion is that it is important to check the dependence of redox potential upon pH.

$$[Fe^{III}(H_2O)_5(OH)]^{2+} + [Fe^{II}(H_2O)_6]^{2+}$$
$$\rightarrow [(H_2O)_5Fe^{2+}\!-\!OH\cdots H\cdots OHFe^{2+}(H_2O)_5]$$
$$\rightarrow [Fe^{II}(H_2O)_6]^{2+} + [Fe^{III}(H_2O)_5OH]^{2+}$$

As a preliminary step before considering possible schemes for biological electron transport it is appropriate to summarize the various criteria for rapid electron transfer that we have discussed so far in this chapter and in Chapter 2. They are:

(a) that the requirements of the Franck–Condon principle be easily met,
(b) that electron spin be conserved, fastest electron transfer occurs when the net electron spin of the reactants is the same as that of the products,
(c) that the rates of outer-sphere reactions are faster when polarizable ligands are bound to the metal,

(d) reactions tend to be faster when complementary redox species are involved.

Condition (d) is satisfied in a cytochrome chain, when both reactants undergo one-electron changes. A problem arises in the interaction of dioxygen with the terminal member of the respiratory chain, as this is a non-complementary reaction. This is probably overcome by rapid electron release from a multienzyme system.

Models for electron transport

As models we may consider either reactions in solution or in the solid state. The problem with solution studies here is that condition (a) may not be readily met and measured activation energies will then also reflect the raising of molecules to vibrationally excited states. The protein system is more ordered than a solution system, being designed to overcome this problem. It may then be more appropriate to consider as models, electron transfer in the solid state as in semiconductors and photoconduction. This is explained either by the delocalized band model, in which electrons are conducted in an upper excited state, or by the localized hop model, which is related to electron transfer between redox centres in solution. The delocalized band model is not compatible with the properties of proteins and so will not be considered further.

Metal phthalocyanines have been used to model electron transport. Phthalocyanine (Figure 5.3) and the porphyrins are both 18 π-electron aromatic systems, and have similar structures except that phthalocyanine has aza links.

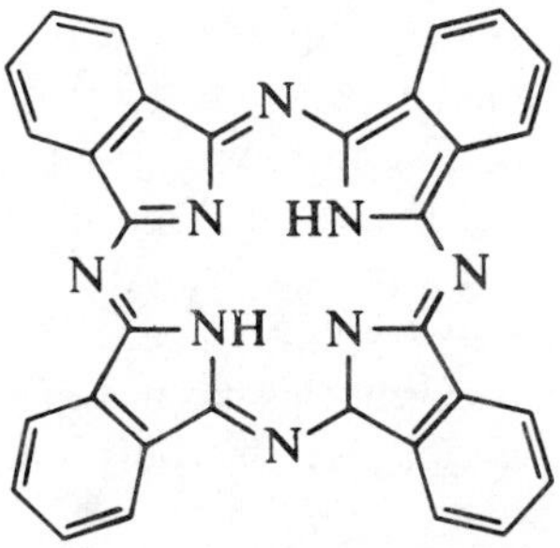

The structure of phthalocyanine.

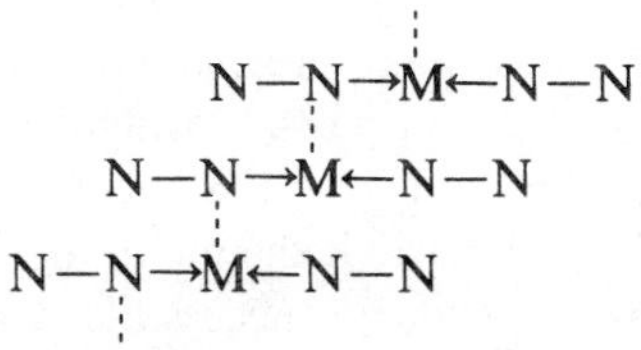

$$N-N\rightarrow\overset{\vdots}{M}\leftarrow N-N$$
$$N-N\rightarrow\overset{\vdots}{M}\leftarrow N-N$$
$$N-N\rightarrow\overset{\vdots}{M}\leftarrow N-N$$

Figure 5.3 Schematic representation of the structure of metal phthalocyanines.

Metal phthalocyanines show conductivity properties which are enhanced considerably by adding traces of an electron-acceptor impurity such as *o*-chloranil, with a reduction of activation energy. In the undoped samples thermal energy is necessary to produce carriers. This is eliminated by the addition of an electron acceptor. Here the conductivity is associated with the mobile π-electrons of the phthalocyanine ring. The crystal structure of metal phthalocyanines is schematically represented in Figure 5.3. The structure involves parallel sheets with interaction between neighbouring sheets. Even so, the activation energy for conduction is still high and this certainly suggests that the mechanism involved in these cases could not adequately account for longer-range electron transfer. The activation energy in this particular example could well be lowered by the presence of more polarizable ligands.

In solution redox reactions we have two general mechanisms; electron or group transfer *via* a bridge linking two metal centres (inner-sphere), or electron transfer *via* overlap of ligand orbitals (outer-sphere) in which case reaction is accelerated by the presence of polarizable ligands. It is interesting to consider the actual act of the electron transfer in the bridged system. Depending upon the timing of electron transfer to and from the bridging atom X (before group transfer), it is possible to have no electron concentration in this group or to generate a species, $M^{n+}.X^-.M^{n+}$. For either of these to be extended to polyatomic bridges it is essential to have a mobile conjugated electron system such as the porphyrin group. A large polarizable, bridging ligand, such as a porphyrin, also has the advantage of providing a long pathway for electron transport (some 15 Å in this case). Delocalization of charge onto the ligand also means that smaller changes in bond lengths are necessitated on redox changes.

The above points are all relevant to electron-transfer proteins, but complications arise in that channels are necessary to allow the electron to enter and leave the redox centre. These channels may help to control the direction and rate of electron transport. The presence of a porphyrin group will facilitate electron transport through the initial part of the channel, but it appears that the 'hop' mechanism must also be involved, in which electron transfer proceeds *via* amino acid residues. Phenolate, thiol, and tryptophan groups are suggested to be appropriate residues for this function, but this has now been queried.[3] Residues with S—S groups may still be suitable for this purpose. Quinone and flavin groups could also function in this way.

The ligand which receives the electron from the metal must have an appropriate low-energy empty orbital to accept the electron, and be able to transfer the electron away before it returns to the metal. In addition, if the ligand is a good σ donor the positive charge on the metal will be reduced, so allowing the *d*-electron density to expand out from the nucleus. This too will favour electron transfer to the ligand, particularly if the metal ion is in a low-spin state, as then the electron density is particularly 'accessible' to the ligand. In this context Williams has drawn attention to EPR studies[7] on a range of transition metal complexes with the ligand maleonitriledithiolate which show that here the unpaired electron density of the metal is largely associated with the sulphur ligand. The significance lies in the fact that the properties of this ligand are very much like those of the porphyrin group and demonstrate that ligands with the properties discussed above are able in effect to abstract an electron from the metal as a first step in electron transfer.

Kinetic studies on electron-transfer proteins

This work has followed two general courses. One approach has involved the study of the reactions of isolated electron-transport proteins (such as the cytochromes, blue copper proteins, and iron–sulphur proteins) with simple inorganic reagents whose general redox behaviour was well characterized. These reagents have included Cr(II), Fe(II), $[Ru(NH_3)_6]^{2+}$, tris (1,10-phen) Co(III), and $[Fe(CN)_6]^{6-}$. An elaboration of this work has involved the study of reactions between isolated electron-transfer proteins, e.g. cytochromes and blue copper proteins. The second approach has been concerned with the direct study of mitochondrial and photosynthetic electron transport in intact or partially disrupted preparations. This will be considered later.

The relevance of reactions between proteins and inorganic redox agents to reactions between electron-transfer proteins is open to question. Nevertheless they appear to have contributed to an understanding of a complex subject. Some reactions with inorganic complexes include reduction by the outer-sphere reagent $[Fe(EDTA)]^{2-}$ of ferricytochrome c,[8] trifluoroacetylated cytochrome c,[9] azurin, plastocyanin, stellacyanin, and laccase;[10] reduction of cytochrome c by the inner-sphere reagent Cr(II);[11] oxidation by tris (1,10-phenanthroline) cobalt(III) and/or other cobalt(III) complexes of ferrocytochrome c,[8,12] cytochrome c_{551},[12] stellacyanin, plastocyanin, and azurin;[13] and oxidation by various Co(III) complexes of 2Fe ferredoxins.[14] Examples of reactions between isolated electron-transfer proteins are cytochrome c–azurin,[15,16] cytochrome c–plastocyanin, cytochrome c_{551}–azurin,[16] cytochrome c–cytochrome c,[17] cytochrome c_{551}–cytochrome c_{551},[18] and cytochrome c–Cocytochrome c.[19]

The reactions of electron-transfer proteins with outer-sphere reagents or with other electron-transfer proteins are usually examined quantitatively in terms of Marcus theory. This theory correlates the rate constant of such a reaction (k_{12}) with the electron self-exchange rate constants of the two reactants, k_{11} and k_{22}, and the equilibrium constant for the electron-transfer reaction K, via the equation (expressed in its simplest form).

$$k_{12} = (k_{11}k_{22}K)^{\frac{1}{2}}$$

The two electron self-exchange rate constants are characteristic of the activation process that the individual electron-transfer proteins/inorganic reagents must undergo to transfer an electron. These parameters will be constant if the electron-transfer process is independent of the nature of the reactants. This was found to be the case[16] for the redox reactions of cytochromes c_{553}, c_{551}, and c with a range of blue copper proteins. The self-exchange rate constants for cytochromes c_{551} and c_{553} (from prokaryotic organisms) are similar but that of horse heart cytochrome c is much lower. The lower reactivity of cytochrome c may reflect the lower accessibility of the heme crevice, and its high positive charge relative to the other cytochromes. In contrast, the two prokaryotic cytochromes have a second opening to the heme crevice, with a different electrostatic and amino acid side-chain environment for protein–protein electron transport.

Reaction of cytochrome c with $[Fe(EDTA)]^{2-}$, $[Ru(NH_3)_6]^{2+}$, $[Co(phen)]^{3+}$, and $[Fe(CN)_6]^{3-}$ has been assessed for deviation from strict Marcus theory in terms of hydrophobic character and d-orbital availability of the small complex. This is suggested to be consistent with an outer-sphere mechanism at the heme edge of cytochrome c that is exposed at the protein surface. Thus the most reactive substrate[8] was $[Fe(CN)_6]^{3-}$ which is assumed to penetrate the protein surface through one of its monodentate cyanide ligands, possibly allowing direct Fe–CN–heme π overlap. With $[Co(phen)_3]^{3+}$ it is assumed that contact occurs between the heme and phenanthroline edges, a view that allowed an interpretation of the reactivity of a range of substituted 1, 10-phenanthroline complexes in terms of variation in heme–phenanthroline edge distances.[12]

Reduction of ferricytochrome c by Cr(II) proceeds[11] *via* a rate-determining cleavage of the iron–sulphur bond (axial Met group), and attack by Cr(II) with reduction *via* the inner-sphere mechanism. Anion-catalysed cytochrome c–Cr(II) reaction is faster than the iron–sulphur bond rupture. In the case of thiocyanate the mechanism changes to outer-sphere, with electron transfer taking place *via* the porphyrin ring, but chloride ion catalyses an inner-sphere process with a Fe—Cl—Cr transition state.

Recent kinetic studies on the oxidation and reduction of *Pseudomonas aeruginosa* azurin[13] by inorganic complexes have shown that protein–complex association occurs prior to electron transfer, and that pH effects give information about the nature and location of the binding sites. The presence of these protein–complex interactions may have implications in terms of the application of Marcus theory.

An alternative mechanism for electron transfer between proteins that is open to quantitative assessment is that of non-adiabatic electron tunnelling.[20] Rate constants for $^{Fe}cyt_c$–$^{Fe}cyt_c$ and $^{Co}cyt_c$–$^{Fe}cyt_c$ electron transfers calculated by this theory are in good agreement with experiment,[19] while experimental verification of electron tunnelling between bound cytochrome c and cytochrome c peroxidase has been presented.[21] The measured transfer distance is about 7 Å between heme edges, with transfer of about 15–20 Å between iron atoms.

THE HEME IRON PROTEINS

These are associated with dioxygen transport and storage (hemoglobin and myoglobin), electron transfer in cytochromes, catalysis of oxidations by dioxygen and hydrogen peroxide (oxidases and peroxidases), and with catalysis of the decomposition of hydrogen peroxide (catalases). The wide range of biological functions displayed by these heme proteins serves to emphasize the subtle way in which the properties of a series of apparently closely related compounds may be controlled by the protein. Before looking at these reactions we shall consider some general features of heme proteins, beginning with an outline discussion of physical and chemical aspects of the porphyrin ring system. A comprehensive review series on the porphyrins is available.[22]

The porphyrin system

All compounds are derived from porphin (Figure 5.4) by substitution of H atoms by side chains. All eight pyrrole carbon atoms are completely substituted in natural metalloporphyrins. Figure 5.4 also illustrates the nomenclature for numbering the molecule, while the substituents present in some important porphyrins are listed in Table 5.2. The nature of the side chains is important in providing additional stabilizing interaction between heme group and protein in hemoproteins. In some hemoproteins the substituents of the porphyrin group are also linked to the protein structure. In cytochrome c, for example, heme and protein groups are linked *via* thioether groups. There will also, of course, be direct iron–protein linkages.

The porphyrins are highly conjugated molecules. Their electronic spectra show a number of bands, which will be characteristic of a particular porphyrin as the spectrum is dependent upon the nature of the substituents at the pyrrole carbon atoms. A very intense band, around 25,000 cm^{-1} is the Soret band (or γ band) while there are two other bands of lower intensity, the α and β bands. These are close together in frequency. The presence of good σ- or π-donor axial ligands

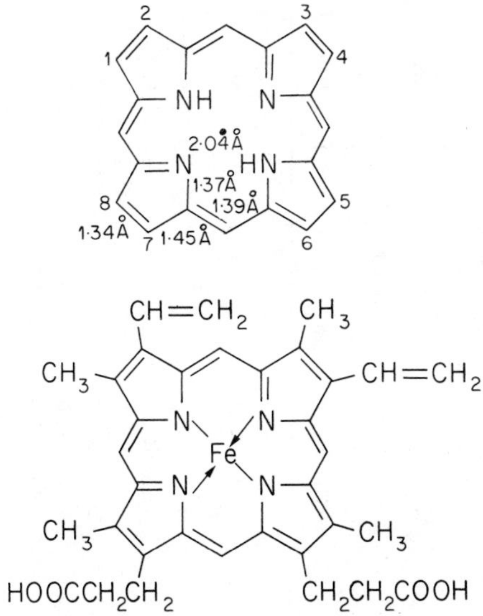

Figure 5.4 Porphin and iron (II) protoporphin.

TABLE 5.2 Some naturally occurring porphyrins.

Porphyrin	\multicolumn{6}{c}{Substituents}					
	1	2	3	4	5	6
Protoporphyrin	CH_3	$CH=CH_2$	CH_3	$CH=CH_2$	CH_3	CH_2CH_2COOH
Hematoporphyrin	CH_3	$CHOHCH_3$	CH_3	$CHOHCH_3$	CH_3	CH_2CH_2COOH
Etioporphyrin	CH_3	CH_2CH_3	CH_3	CH_2CH_3	CH_3	CH_2CH_3
Deuteroporphyrin	CH_3	H	CH_3	H	CH_3	CH_2CH_2COOH
Mesoporphyrin	CH_3	CH_2CH_3	CH_3	CH_2CH_3	CH_3	CH_2CH_2COOH
Chlorocruoroporphyrin	CH_3	CHO	CH_3	$CH=CH_2$	CH_3	CH_2CH_2COOH
Coproporphyrin	CH_3	CH_2CH_2COOH	CH_3	CH_2CH_2OH	CH_3	CH_2CH_2COOH
Rhodoporphyrin	CH_3	CH_2CH_3	CH_3	CH_2CH_3	CH_3	$COOH$

Porphyrin	\multicolumn{2}{c}{Substituents}		\multicolumn{4}{c}{$E_0 V^{20}$}			
	7	8	A	A'	B	C
Protoporphyrin	CH_2CH_2COOH	CH_3	0.015	0.137	−0.033	−0.183
Hematoporphyrin	CH_2CH_2COOH	CH_3	0.004	—	−0.099	−0.200
Etioporphyrin	CH_3	CH_2CH_3	−0.029	—	—	—
Deuteroporphyrin	CH_2CH_2COOH	CH_3	—	—	—	—
Mesoporphyrin	CH_2CH_2COOH	CH_3	−0.063	—	—	−0.229
Chlorocruoroporphyrin	CH_2CH_2COOH	CH_3	—	0.246	—	−0.113
Coproporphyrin	CH_2CH_2COOH	CH_3	−0.036	—	−0.010	−0.247
Rhodoporphyrin	CH_2CH_2COOH	CH_3	—	—	—	—

A, A': 5th and 6th ligands pyridine at pH9.6 and pH7.
B: 5th and 6th ligands α-picoline at pH9.6.
C: 5th and 6th ligands cyanide at pH9.6.

in metalloporphyrins will intensify the absorption of the α band. These bands are due to $\pi \rightarrow \pi^*$ transitions.

The two protons attached to nitrogen in a porphyrin are readily replaced by a metal ion. The structure of iron(II) protoporphyrin is shown in Figure 5.4. Here the metal is four coordinate with a square-planar environment. Cytochrome c involves the protoporphyrin structure with the addition of protein cysteine side chains across the vinyl double bonds in positions 2 and 4. Hemoglobin, myoglobin, plant peroxidases, and the catalases also involve the protoporphyrin group.

The porphyrin molecule is essentially planar, but two pyrrole rings are tilted up and two are tilted down so that the nitrogen atoms are slightly out of plane. The molecule is not completely rigid and the presence of bulky substituents may cause puckering. As indicated in Figure 5.6 the distance from the pyrrole nitrogen to the centre of the ring is 2.04 Å. The metal–nitrogen bond lengths in metalloporphyrins vary slightly (e.g. 2.10 Å for ferric porphyrins, 1.95 Å for nickel porphyrins), but not as much as a consideration of the ionic radii of the metal ions would suggest. This is a reflection of the inability of the porphyrin ring system to adjust itself to fit the metal ion in the metalloporphyrin.

Most metalloporphyrins involve four-coordinate metal ions. Others involve five-coordinate metal ions with square-pyramidal structures, or six-coordinate metal ions with distorted octahedral geometry. These are formed by adding one or two axial ligands to the metal ion. Some five-coordinate metalloporphyrins are known with a tetragonal-pyramidal stereochemistry. One example is provided by the high-spin iron(III) porphyrins, where the iron(III) lies above the plane of the porphyrin group. A further example involves vanadium in the vanadyl etioporphyrin. In contrast, the low-spin iron(III) porphyrin, bis(imidazole)iron(III) tetraphenylporphyrin, has the iron lying in the porphyrin plane. Much discussion has centred on the importance of 'out-of-plane' iron in porphyrins, particularly in dioxygen carriers. Resonance Raman spectroscopy has been used to assess this displacement in several porphyrins.[23]

Studies in solution

Naturally occurring porphyrins tend to be insoluble in acid solution, and to associate extensively in basic solution. These problems have meant that either non-aqueous solvents have been used or special, soluble porphyrins have been synthesized. These soluble porphyrins have included sulphonato or carboxylato species, but even these are not always monomeric. While some forms of aggregation result from dipole–dipole interaction (and depend upon the presence of electron-withdrawing groups in the porphyrin) specific monomer–dimer equilibria have been characterized involving bridging oxo and hydroxo groups. The dissociation of these dimeric species follows a complex mechanism.[24]

Iron(II) porphyrins are readily autoxidized to the iron(III) species. The formation of an intermediate five-coordinate dimeric structure with a bridging oxo group is well established, but more recently it has been shown that a dimeric structure with a bridging dioxygen group (or peroxo group) is formed in an earlier stage of the reaction.[25] A mechanism is suggested below, P-porphyrin.

$$PFe \xrightarrow{\ O_2\ } PFeO_2 \xrightarrow{\ PFe\ } PFe-OO-FeP$$
$$PFeOFeP \xleftarrow{\ PFe\ } PFeO$$

The iron(II) heme proteins are also readily autoxidized, but in this case the mechanism will be different, depending on the possibility of dimerization occurring (e.g. if the heme protein is five coordinate). Iron(III) heme proteins require an extra anion for neutralization. If it is chloride or hydroxide the heme is known as a chlorohemin or a hematin respectively.

The reduction of ferriporphyrins has been considered using powerful, single-electron reducing agents such as Cr(II) and V(II), and multielectron reducing agents. In the latter case it appears that electron transfer occurs after the formation of ferriporphyrin dimeric species (for tetra(4-N-methylpyridyl)porphyrinatoiron(III)).[24] The reduction of tetra (p-sulphonatophenyl)porphinatocobalt(III) by Cr(II) has been interpreted in terms of electron transfer to the cobalt centre *via* the porphyrin π cloud, and not *via* the fifth and sixth ligand positions.[26]

Substituent effects on porphyrin rings

These are of great importance. There are a great number of closely related iron hemes having slightly different redox properties. The presence of different substituents on the porphyrin ring could contribute to these finer points of difference. It does appear, though, that when major changes in property are required, other factors in the heme proteins such as the nature of the fifth and sixth ligand or the function of distant groups on the protein are important. Thus myoglobin, hemoglobin, peroxidase, catalase, and certain cytochromes, e.g. b_5, all have the protoporphyrin heme structure, but their physiological properties are quite different, as are their redox potentials.

It has already been noted that the bands in the spectra of the porphyrins are substituent sensitive. This will be carried through to the spectra of the iron hemes and the substituents may well affect the $d \rightarrow d$ and porphyrin $\rightarrow$ metal charge-transfer bands. The overall situation is complex. Nevertheless, spectroscopic changes resulting from the presence of substituents forms the basis of the cytochrome classification scheme.

The presence of electron-withdrawing substituents on the porphyrin decreases the basicity of the porphyrin donor atoms. This is reflected in the decreased rate of incorporation of metal ions into the porphyrin. It is probable that the ability of the porphyrin to undergo deformation is also important in binding metal ions, as this enhances the coordinating ability of the nitrogen lone pairs by directing them away from the porphyrin cavity. It may also increase the acidity of the porphyrin

NH groups by destroying the tautomerism. The ability to undergo this deformation may be dependent on the substituents.

Electron-withdrawing substituents on the porphyrin also affect the binding of the fifth and sixth ligands to the metal. The binding of these ligands to porphyrins and heme proteins has been much studied.[27] Thus the binding of some substituted pyridines to zinc tetraphenylporphyrin has been monitored[28] by ^{15}N NMR. The porphyrin ^{15}N chemical shift correlates well with the electron-donating ability of the pyridine substituent. A comparison of the Mössbauer spectra[29] of bis(piperidine) and carbonylpiperidine iron(II) porphyrins shows that the axial carbonyl group affected the iron–porphyrin bonding. The carbonyl group is a π acceptor, and it appears that the porphyrins are affected by this. Evidently, as these studies shown, the σ- and π-bonding characteristics of the porphyrin can be modified to accommodate the electronic requirements of the axial ligands. This, clearly, is of significance in understanding the relationship between the biological function of a heme protein and the nature of the axial ligand to iron.

A further extension of the study of axial binding groups involves the use of substituted tetraphenylporphyrins with 'tails' that can coordinate to the axial position. This is illustrated by recent work on porphyrins with 'tails' having sulphur donor groups. These model binding interactions in cytochrome c, which contains the Met-80 sulphur group as an axial ligand.[30] An example of a tail-base porphyrin is shown below.

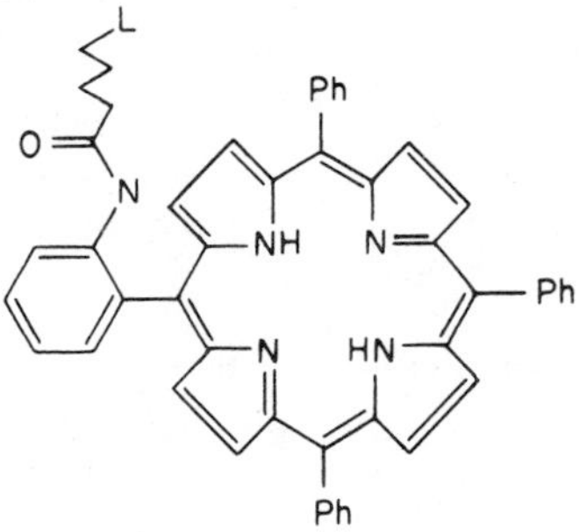

The lability of axial nitrogen bases in six-coordinate synthetic iron(II) porphyrins has been examined[31] by NMR techniques. This shows that ligand exchange occurs by a dissociative mechanism with a five-coordinate transition state.

Redox properties of metalloporphyrins

Some redox potentials are given in Table 5.2. Due to dimerization and solubility problems these are often difficult to measure. In practice, it is often easier to do this for low-spin complexes. Nevertheless, it is important that a study of these redox potentials be made as a foundation for the understanding of the redox potentials of the cytochromes and other electron-transfer catalysts. The presence of the porphyrin ligand will stabilize the higher oxidation state compared to the aquo couple. The presence of electron-withdrawing groups on the porphyrin will decrease the basicity of the nitrogen donors (i.e. σ donors). This will result in a destabilization of the iron(III) state.

The axial ligands will have a very significant effect upon the redox potential of the iron porphyrin. The experimental study of iron protoporphyrin and other metalloporphyrins has confirmed the picture developed for simple model complexes. Strong σ donors stabilize the iron(III) state, but this may also result in a change from high spin to low spin, which will vary for iron(III) and iron(II). When high-spin Fe(III) is converted to low-spin Fe(II), the Fe(II) complex will be stabilized relative to Fe(III). On further increase of ligand σ-donor strength, however, so that both redox states are low spin, the iron(III) complex will once more be stabilized. Spectral studies show that in iron(III) hemes there is a delicate balance between high- and low-spin types. Changes in the nature of the axial ligands in particular and also in the porphyrin substituents may exert a considerable effect.

Many heme systems involve the coordination of imidazole and water in the fifth and sixth positions. In view of this, it might be particularly useful to examine the behaviour of metalloprophyrins with one strong-field and one weak-field ligand in the axial positions. A further important area involves the study of the effect of axial ligands on the redox properties of porphyrins with a range of metals. In the case of the complex Ru(OEP)(CO)(py), where OEP is the octaethylporphyrin anion and py is pyridine, electrochemical oxidation takes place in two stages. One electron is removed from the porphyrin ring and the second is removed from the metal ion.[32] In the corresponding osmium complex the stages are reversed.

Metal ion incorporation into porphyrins

Enzyme systems have been identified which catalyse the incorporation of iron into the porphyrin ring. These only utilize ferrous iron, and accordingly the insertion reaction is inhibited completely by the presence of oxygen.[33]

Prosthetic groups of heme proteins

Some important porphyrins were listed in Table 5.2 along with the redox potential of the corresponding heme. The most important heme is undoubtedly protoheme. It may occur free in tissues and is the prosthetic group of hemoglobin, myoglobin, erthyrocruorin, peroxidases, catalases, and the b cytochromes. Chlorocruoroporphyrin is the prosthetic group of chlorocruorin. However, recombination experiments have shown that the structure of the prosthetic group is relatively unimportant in determining the properties of certain heme proteins. Thus apohemoglobin may be recombined with a range of iron porphyrins with retention of oxygen-carrying properties, although, not unexpectedly, the details of the behaviour, such as the oxygen uptake curve, will be affected.

The cytochromes

These are hematin compounds which are involved in electron-transfer chains in the mitochondria, in which electron transfer is associated with the presence of the iron(III)–iron(II) redox couple. Terminal members of the cytochrome chain must also have the property of reacting directly with oxygen, so for these there will be extra requirements. Cytochromes are also involved in aspects of the nitrogen cycle and in enzymic reactions associated with photosynthesis.

Some fifty cytochromes have been characterized, but the most studied example is cytochrome c. The classification of the cytochromes is based upon the nature of the porphyrin ring system and in turn upon spectral data for the pyridine derivative. Type a have the α band in the porphyrin spectra at longer wavelengths than 570 nm; type b have the α band in the range 555–560 nm and have vinyl side chains on the porphyrin ring; while type c have the α band in the range 548–552 nm and have porphyrin rings in which the vinyl groups have been saturated, as in cytochrome c where these are associated with the thioether connection with the protein. In general the type 'a' cytochromes form a less cohesive group. The b and c types in general have two strong field donor groups in the fifth and sixth coordination position and are usually low-spin complexes. These cannot normally combine with small molecules. Cytochrome a_3, on the other hand, the terminal member of the cytochrome chain, has water as one of the axial ligands (with imidazole as the other) and so binds oxygen.

The study of the cytochrome chain sequence involves its breakdown to single cytochromes or small complexes such as cytochrome c oxidase which contain a number of groups. The redox potential of single cytochromes may then be measured by standard experimental methods. It is important, however, to show that these will still react with the other components of the cytochrome chain as before and much effort has gone into attempts to reconstitute the cytochrome sequence.

Cytochrome c

This is one of the most studied of all biological molecules.[34] This cytochrome has been isolated from many sources, and contains a heme group that is covalently bound to protein *via* thioether groups. X-ray analysis shows that the axial ligands are His-18 and Met-80 in both oxidation states.

Cytochrome c is a member of the mitochondrial respiratory chain found in all eukaryotes. It accepts an electron from cytochrome c_1 and then transfers it to the cytochrome oxidase complex (Figure 5.1). A central problem in this molecule is the way in which the electron moves from the surface to the heme group and then out again. The possibility that this involves conformational changes in the molecule, as the iron cycles between Fe(II) and Fe(III), has been explored through detailed X-ray structural data at 2 Å resolution on oxidized and reduced tuna cytochrome c.[35] This shows beyond reasonable doubt the absence of any conformational change in going from one redox form to the other. This result is of the utmost importance, and also precludes any but the smallest changes in metal–ligand bond lengths as the reaction proceeds. The absence of conformational change in cytochrome c is also demonstrated by NMR studies on diamagnetic Co(III) cytochrome c and diamagnetic Fe(II) cytochrome c.[36]

Horse heart cytochrome c has a molecular weight of around 12,400 and involves 104 amino acid residues in the protein chain. Both methionine residues have been identified[37] by ^{13}C NMR, and the role of Met-80 as a ligand confirmed. However it appears that Met-60 is sensitive to the change in oxidation state. Spin-label studies have indicated that this residue is probably close to the membrane-binding site of cytochrome c.

The peptide chain of cytochrome c is wrapped around the heme so that it provides a heme-containing crevice, in which the heme group has one edge exposed to the solvent. The heme crevice is however blocked by a phenylalanine residue, Phe-82. The heme is surrounded by hydrophobic side chains, while the polar side chains lie on the outside of the molecule. The hydrogen bonding has been analysed in detail,[35] and shows the important involvement of six of the ten glycine residues in this and the several interactions with a propionic acid residue.

One feature of the structure is the presence of two channels which lead from the surface of the molecule to the heme group. The main channel is lined with the side chains of hydrophobic amino acid residues, as is a smaller channel, which is not solvent accessible. These are suggested to function as pathways for exit and entry of an electron, respectively. Thus one pathway is suggested to involve entry of an electron via Tyr-74, Trp-59, Tyr-67, heme, and Fe. However, in cytochrome c_{550}, the residue Tyr-74 is replaced by a leucine residue, so this pathway now seems unlikely.

Another feature of these channels is that their outlets are associated with lysine residues. It has long been suggested that these represent a binding site for another electron-transfer protein such as the neighbouring cytochrome oxidase complex. This proposition has been explored[38] through the chemical modification of these lysine residues. Trifluoroacetylation of lysine residues 55 and 99 had no effect on reaction with cytochrome oxidase, but modification of lysine residue 13 did. This residue lies at the top of the heme crevice. This result appears to support a binding role for this residue.

A further approach to the study of the role of the protein in cytochrome c has involved the preparation of heme-containing fragments by peptic hydrolysis[39] or cyanogen bromide cleavage[40,41] of cytochrome c. The former approach gives an undecapeptide containing residues 11–21 of native cytochrome c, while the latter cleavage produces a heme-containing fragment made up of residues 1–65. The porphyrin–protein thiol links are at residues 14, 17 (cysteine). The cyanogen bromide fragments 1–65 and 66–104 have been complexed and the peptide link resynthesized[41] to give a product indistinguishable from native cytochrome c. Furthermore this has been repeated with a synthetic fragment 66–104. This opens up a useful method for studying the role of particular amino acid residues through the synthesis and incorporation of a modified fragment into the protein.

Several mechanisms have been proposed for electron transfer in cytochrome c. A protein-hopping mechanism seems unlikely in view of the data described above, but electron tunnelling through the hydrophobic channel is still possible.[3]

The work on model compounds has emphasized the possibility that an electron can enter the protein through the exposed edge of the heme. This provides an attractive alternative mechanism for cytochrome c. It is noteworthy however, that the activation energy for the reaction is lowered in view of the constant Fe—L bond lengths generated by the protein.

The interaction of cytochrome c with cytochrome oxidase has attracted much attention (cytochrome oxidase will be discussed separately). This work has profited from the use of metal-substituted cytochrome c, as indeed has the study of cytochrome c itself. Thus Sn(IV) and Zn(II) cytochromes c have been used[42] to explore the interaction of cytochrome c with cytochrome oxidase and with mitochondria, into which it may be reincorporated. The fluorescence yields of these two cytochromes c fall on binding. This has led to the conclusion that the distance between the porphyrin rings of cytochrome c and cytochrome a is 35 Å. Furthermore, measurements of the polarized emission of metal-free cytochrome c when bound to orientated layers of cytochrome c oxidase indicate that the porphyrin is bound obliquely to the plane of the oxidase layers with an angle of about 70° from heme plane to membrane plane. This observation may well be of considerable significance in electron-transfer mechanisms. Copper cytochrome c exhibits luminescence characteristics of copper porphyrins; this property may be a useful indicator to monitor interaction with cytochrome oxidase.[43] Kinetic studies on the reaction of cytochrome c with cytochrome oxidase are consistent with the assumption that binding is the dominant parameter in reactivity.[44]

The presence of the axial ligand Met-80 in cytochrome c is probably an important factor in controlling its electron-transfer properties. This is supported by recent studies on model compounds.[45] Low-spin iron(II) and iron(III) complexes of tetraphenylporphyrin with thioether axial ligands have almost identical Fe—S bond lengths (Fe^{II}—S = 2.34 Å, Fe^{III}—S = 2.33, 2.35 Å). Thus any increase in Fe—S bond length on oxidation of iron is off-set by the increased charge attraction of Fe(III) for its ligands. The presence of methionine as an axial ligand in cytochrome c may thus be a particularly appropriate one for constant Fe—S bond length and rapid electron transfer. These workers have also prepared tail-base porphyrins, with a porphyrin-linked imidazole axial ligand and a thioether ligand. Electrochemical studies on this and the bis(imidazole) complex (cf. cytochrome c_3) shows that substitution of one imidazole by the thioether results in a shift in the Fe(II)/Fe(III) redox potential of about $+ 167$ mV. However the difference between cytochromes c_3 and c is about 460 mV. The authors attribute this partly to the replacement of histidine by methionine (about 160 mV), with the remainder coming from environmental effects. This latter destabilization of the Fe(III) oxidation state reflects the inability of cytochrome c to delocalize the positive charge generated by oxidation of the iron. It is of considerable significance therefore that the heme propionic acid side chains of cytochrome c are buried in the protein, but that this is not the case with c_3. Thus it appears that control of ionization of propionic acid groups is a crucial factor in determining redox potentials of cytochromes.

Some bacterial cytochromes

This section will include reference to work on isolated cytochromes from bacteria, and is intended to illustrate the range of cytochrome types found in this area.

Cytochromes c_3 are hemoproteins that contain a number of heme groups, and have a very negative redox potential at pH 7 (Table 5.1). A number have been isolated from *Desulphovibrio* species, and have molecular weights in the range 13,000–16,000, with four heme groups. They act as an electron carrier that couples the hydrogenase and, for example, the thiosulphate reductase present in *D. desulphuricans* and *D. gigas*. However in this case the two cytochromes c_3 are not interchangeable.[46] It appears reasonable that the four heme units in this cytochrome act as a heme cluster, with heme–heme interaction, but this has not yet been resolved conclusively. NMR studies show that for cytochrome c_3 from *D. vulgaris*, resonances of three different oxidation states appear to develop in turn as oxidation proceeds. Oxidation states I, II, and III are assumed to represent oxidation of one, two, and four hemes respectively. When the oxidized form is more than one third of cytochrome c_3 then all three states coexist.[47] This shows that electron exchange between the reduced state and the three oxidized states is too slow to cause exchange averaging of the NMR spectra even though the hemes are close together (see below). The reduction titration of cytochrome c_3 with dithionite shows that the hemes are reduced in two-electron steps.[48] Electrochemical techniques show the electrode reactions of cytochrome c_3 to be four one-electron processes and that the hemes are non-equivalent and non-interacting.[49]

A model for the structure of cytochrome c_3 has been proposed[48] on the basis of sequence results and NMR data. The primary structure of the protein contains two Cys-x-x-Cys–His and two Cys-x-x-x-x-Cys–His regions, which provide the heme-binding sites, while four invariant His residues have been assigned as axial ligands from EPR and NMR data, i.e. one for each iron. The model also invokes two pairs of heme groups, roughly planar to each other and about 10 Å apart.

Cytochrome c_{550} has been isolated[50] from *Thiocapsa roseopersicina*, and was shown to be heat stable, as it was capable of reducing sulphide even after heating to 80–100°C for some minutes.[51] Cytochrome c_{552} isolated from the thermophilic bacterium *Thermus thermophilus* is also thermo-resistant.[52]

A number of resonance Raman and NMR studies on bacterial cytochromes have now been published. Recent papers have been reviewed.[53]

The 2 Å resolution structure of cytochrome c_{551} from *Pseudomonas aeruginosa* involves the same folding pattern and hydrophobic heme environment as cytochromes c, c_2, and c_{550}. Cytochrome c_{551} has 82 amino acid residues compared to 104 for cytochrome c, sequences 38–57 being missing. These observations fit in well with experiments[40] that show that this region is not vital to the formation of an ordered structure in cytochrome c.

The b *cytochromes*

This group of cytochromes contain protoheme, which is not covalently bound to the protein *via* heme substituents. Cytochromes b_6 and b_{559} (and *f*) are found in chloroplasts and undergo light-induced redox reactions, in accord with a role in photosynthetic electron transport.[54] Cytochromes b_6 and b_{559} have molecular weights of 40,000 and 110,000 respectively. In the case of b_{559} it is possible that non-protein constituents of the molecular weight are artefacts of the isolation procedure.

Cytochrome b_5 is a membrane-bound protein that catalyses electron transport in, for example, the hydroxylation of steroids. It contains 93 amino acid residues and one heme group, which is linked to the protein by two histidine residues that act as axial ligands. An unusual feature is that a hydrophilic edge is exposed to the solvent.[55] Despite the two strong axial ligands, there is a pH-dependent equilibrium between high- and low-spin forms. This has led to the suggestion that the iron ligands are able to alter their bond distances rapidly, and so facilitate electron transfer. The X-ray studies show a hydrophobic groove leading to the heme, which could provide a path for electrons. The structure of cytochrome b_{562} from *E. coli* has been determined at 25 Å resolution.[56] The axial ligands are histidine and methionine.

Cytochrome P450

This cytochrome occurs as a complex with an iron–sulphur protein and a reductase. The complex catalyses a wide range of hydroxylations. Cytochrome *P*450 is so named as its reduced complex with carbon monoxide shows maximum absorption at 450 nm, the 'anomalous' Soret band. The enzyme cycle involves the formation of a substrate–Fe(III) cytochrome *P*450 complex, which is reduced to the Fe(II) form before binding dioxygen. The dioxygen reacts with substrate to give the hydroxylated product, water, and Fe(III) *P*450. Binding of substrate promotes a change from a low-spin to a high-spin heme, with a change in coordination number from 6 to 5. Thus cytochrome *P*450 shows electron transfer and dioxygen binding properties. The formation of dioxo Fe(II) species has been claimed from stopped-flow studies,[57] while the binding of dioxygen has been probed by carbon monoxide binding.[58] This shows that the Fe—CO bonding is similar to that found for hemoglobin and myoglobin.

The nature of the axial ligands is of importance in view of the remarkable way in which they allow the protoheme complex to carry out a range of functions. There is growing evidence that the axial ligand common to five- and six-coordinate Fe(III) *P*450 is a deprotonated cysteine residue. The identity of the sixth ligand has not been established, but histidine, lysine, and cysteine have all been suggested. A range of Fe(II) and Fe(III) complexes have been prepared, in an endeavour to mimic the physical and spectroscopic properties of *P*450 at the different stages of its catalytic cycle.[59] These have shown that the 450 nm band is shown in porphyrin complexes with mercaptide anion in the axial position.[60]

Heme enzymes—peroxidase and catalase

Peroxidases are heme proteins that catalyse oxidations by hydrogen peroxide. Catalases catalyse the disproportionation of hydrogen peroxide, and so may be regarded as a special example of peroxidase activity in which the substrate oxidized by hydrogen peroxide is another molecule of hydrogen peroxide. It is not surprising therefore, that the behaviour of peroxidase and catalase show a number of similarities, including the presence of iron in oxidation state IV during the enzyme cycle.

$$H_2O_2 + SH_2 \xrightarrow{\text{peroxidase}} 2H_2O + S$$
$$H_2O_2 + H_2O_2 \xrightarrow{\text{catalase}} 2H_2O + O_2$$

These two enzymes are very widespread, and protect against the build up of dangerous concentrations of hydrogen peroxide in living systems as a consequence of only partial reduction of dioxygen. Another product of partial reduction of dioxygen is superoxide ion, O_2^-. As discussed later in this chapter, it is possible that this is removed *via* the activity of the enzyme superoxide dismutase.

Peroxidases[61]

All peroxidases isolated so far from plant sources contain the prosthetic group ferriprotoporphyrin IX (Table 5.2). This is also present in catalases. Horse radish peroxidase, HRP, is the most well-studied peroxidase, but those from other plant sources are now receiving attention. A number of peroxidases from animal tissue have also been studied, including thyroid peroxidase, which is fairly similar to HRP, and lactoperoxidase, myeloperoxidase, and glutathione peroxidase. Lactoperoxidase and myeloperoxidase probably contain a formyl-type porphyrin, while glutathione peroxidase is unique in that it contains one atom of selenium per subunit. Cytochrome *c* peroxidase from baker's yeast also contains ferriprotoporphyrin IX. Many of these enzymes are mixtures of isoenzymes of differing physical but similar catalytic properties. Thus isoenzymes A and C of HRP both contain Ca^{2+}, which appears[62] to be involved in the stabilization of the protein conformation, as is evidenced by the lower thermal stability of the calcium-free isoenzyme. These isoenzymes differ substantially in their ability to exchange Ca^{2+}.

Horse radish peroxidase has long been thought to involve a histidine residue and an aquo group as the fifth and sixth ligands to iron(III), while a second histidine residue may be hydrogen bonded to the aquo group. In solutions of low pH, HRP is high spin, but a low-spin species is formed at high pH with a pK_a for the equilibrium between the two forms of about 11. Several approaches suggest[63] that the complex may be five-coordinate under acidic and neutral conditions, and that the alkaline form results from coordination of another protein ligand to give a six-coordinate species. This view is not unanimously accepted at present.[64]

Oxidation of Fe(III)HRP with hydrogen peroxide gives a green product, compound **I**, which has two oxidizing 'equivalents' above iron(III). On slow decomposition compound **I** is converted into a red product, compound **II**, which has one oxidizing equivalent above iron(III). These observations were demonstrated originally by titration methods.

A number of techniques have shown[65] that both compounds **I** and **II** involve iron(IV). The extra electron removed from compound **I** is assumed, therefore, to come from the protein or the porphyrin, with the generation of a radical species. The addition of one electron to compound **I** thus gives a non-radical, iron(IV) compound **II**. Recent NMR work[66] has led to the suggestion that compounds **I** and **II** are high- and low-spin iron(IV) compounds respectively, and that the additional oxidizing equivalent in compound **I** is not associated with the heme group. Earlier studies on compound **I** from cytochrome *c* peroxidase had suggested that the additional oxidizing equivalent was associated with an aromatic amino acid residue of the protein. Compound **I** of cytochrome *c* peroxidase is red rather than green.

HRP may be reduced by dithionite to give iron(II) peroxidase, which reacts with dioxygen to give an inactive species (compound **III**). The reduced enzyme has also been obtained from Co(II)HRP. The ligand hyperfine structure of the EPR spectrum of this enzyme confirms the axial ligand to be a nitrogen donor.[67]

The reduction of compounds **I** and **II** in the enzyme cycle of HRP is brought about by the abstraction of electrons from oxidizable substrates. This means that many peroxidase reactions involve the formation of free radicals, which then react further, for example with dioxygen to give superoxide ion. Dioxygen is known to intervene in other HRP-catalysed oxidations of substrates by hydrogen peroxidase.[68]

Catalases

These have molecular weights of about 240,000, made up of four identical subunits, each containing one heme group, with high-spin iron(III). Chemical modification techniques have implicated histidine and tyrosine residues in the activity of the enzyme, while the fifth and sixth ligands are an amino acid residue and an oxygen donor, presumed to be water. CD and MCD spectra suggest that histidine is probably not a ligand.[69]

Catalase reacts with hydrogen peroxide giving compound **I**, which is then able to oxidize hydrogen peroxide. Loss of one oxidizing equivalent from compound **I** gives compound **II**. Catalase is also able to act as a peroxidase in the catalysis of the oxidation of certain substrates by hydrogen peroxide. These supply electrons for the reduction of catalase compounds **I** and **II** as in peroxidase. Oxidation of compound **II** by hydrogen peroxide gives compound **III** which appears to have three oxidizing equivalents above Fe(III). As in the case of peroxidase, compound **I** is accepted to be Fe(IV) with a radical grouping in the protein or porphyrin.

It was once thought that a catalase–H_2O_2 complex occurred as an intermediate in the formation of compound **I**. However it is now known that compound **I** can be generated from catalase by a variety of oxidizing agents. Nevertheless it is possible that binding of hydrogen peroxide by Fe(III) occurs at an early stage in the mechanism.

Models for catalase and peroxidase

A considerable amount of work has been carried out[70] on the oxidation of simple inorganic compounds by compounds **I** and **II**, but a greater interest has been shown in the use of simple ions and complexes in the catalysis of decomposition of hydrogen peroxide. Iron and copper complexes have been well studied.[71] The simple Fe(II) and Fe(III) aquo cations are effective at pH < 2 but their tendency to hydrolysis limits studies in the pH region relevant to catalase. The Fe(III) hemin compound is effective, but this is limited by its tendency to form oxo-bridged dimers. Clearly this cannot happen in catalase itself.

Some success has been achieved recently[72] with an Fe(III) macrocycle, shown below. This compound catalyses the decomposition of hydrogen peroxide with a

rate constant $\sim 10^5$ dm^3 mol^{-1} s^{-1} at 25 °C compared with values for catalase in the range $10^6 - 10^7$ dm^3 mol^{-1} s^{-1}. The mechanism of this reaction appears to be a free-radical one rather than one involving iron in higher oxidation states.

IRON–SULPHUR PROTEINS

Iron–sulphur proteins are involved in a wide range of biological processes, including electron transfer in photosynthesis, nitrite reduction, steroid hydroxylation, and oxidative phosphorylation in mitochondria. The first example, ferredoxin, was discovered in dinitrogen-fixing systems, and they are now known to be widespread in biology. Some examples are listed in Table 5.3. Examination of this table will indicate the diverse range of organisms in which they have been found.

The simplest class of iron-sulphur proteins is the rubredoxins, which contain one iron atom per mole of protein. The iron is bound by four cysteine residues. All other iron–sulphur proteins contain 'acid-labile' sulphur, giving H$_2$S on treatment with mineral acid. They are of low molecular weight and contain two, four, or eight iron atoms per mole. More complex enzymes such as xanthine oxidase and nitrogenase also contain iron–sulphur clusters. The iron–sulphur clusters in these proteins may be extruded and incorporated into other proteins,

TABLE 5.3 Iron–sulphur proteins.*

Protein	Source	Mol. wt.	E_0' (mV)
Rubredoxin (1Fe)	*C. pasteurianum*	6000	− 570
2Fe.2S	Spinach	10,600	− 420
	Microcystis	10,300	
	Azotobacter	21,000	− 350
(Adrenodoxin)	Pig adrenals	13,000	− 270
	Mitochondria	30,000	280
	E. coli	12,600	− 360
	Halobacterium	14,000	− 345
(Putidaredoxin)	*P. putida*	12,500	− 240
4Fe.4S	*Bacillus*	8000	− 380
	Desulphovibrio	6000	− 330
(HiPIP)	*Chromatium*	9650	350
8Fe.8S	*Clostridium*	6000	− 395
	Chromatium	10,000	− 490
	Rhodospirillum rubrum	13,000	
Succinate dehydrogenase 8Fe.8S; FAD		200,000	
NADH dehydrogenase 28Fe.28S; FMN			
Xanthine oxidase 8Fe.8S; 2FAD; 2Mo		30,000	

*A 3Fe protein has been reported: *J. Biol. Chem.*, **255**, 1797 (1980).

while valuable contributions to the understanding of these structures have come from the synthesis and study of some excellent inorganic analogues. Furthermore certain of these synthetic analogues have been shown to replace ferredoxin in the mediation of electron transfer to hydrogenase; a feature which serves to emphasize their value as models.

Iron–sulphur proteins show a remarkably wide range of redox potentials (+ 350 to − 600 mV), a property which is central to their essential role in biological electron transport. This feature has attracted much attention, particularly as these ranges of redox potentials are associated with proteins having very similar Fe—S clusters. A review series has been devoted to iron–sulphur proteins,[73] and many other reviews are available.[74,75]

Rubredoxin

This protein was originally isolated from *Clostridium pasteurianum*, and has now been found in a number of organisms. All examples have one iron atom and four cysteine residues per mole of protein, with molecular weights around 6000 and some 54 amino acid residues. The oxidized protein is red and the reduced protein is colourless, the iron cycling between Fe(III) and Fe(II).

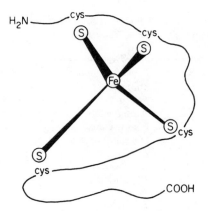

Figure 5.5 Schematic representation of the structure of
rubredoxin.

X-Ray diffraction techniques have shown that the iron is bound by the four cysteine sulphur atoms in an approximately tetrahedral stereochemistry (Figure 5.5). Recent EXAFS measurements[76] have confirmed this structure and have shown the equivalence of the four Fe—S bond lengths (average value 2.267 ± 0.003 Å). The structure of rubredoxin from *Desulphovibrio vulgaris* has also been determined, and found to be similar to the *Clostridium* species,[77] with mean Fe—S bond length 2.29 Å. The only observable difference between the two enzymes was the nature of the hydrophilic groups on the surface of the molecule.

It had been suggested, prior to the X-ray study, on chemical evidence[78] that cysteine groups were involved in coordination. The apoenzyme may be prepared. The number of free—SH groups has been found to be 2.7 by titration with *p*-mercuribenzoate. This is lower than the expected value of 4 probably due to the oxidation of the —SH group during the isolation of the apoenzyme. However, the metalloprotein cannot be reconstituted by the addition of iron to a solution of the aporubredoxin at pH 7. This will only take place at higher pH in the presence of 2-mercaptoethanol. Again ^{59}Fe exchange with the iron in rubredoxin will only occur in the presence of 2-mercaptoethanol. All this is suggestive of a role for cysteine in metal binding.

A number of physical measurements have been carried out on both oxidized and reduced froms of rubredoxin. The two forms differ in ORD, optical, and EPR spectra. The last-mentioned technique showed an absorption at $g = 4.3$ for the oxidized form, but no absorption for the reduced form. A more recent report[79] quotes magnetic moments (by the NMR technique) of 5.85 and 5.05 BM for oxidized and reduced forms respectively, implying high-spin Fe(III) and Fe(II). This has been confirmed by Mössbauer spectra, while the NMR results suggest a disordered tetrahedral array of ligands. Calculations on model compounds also suggest both redox forms to be high spin.[80]

The near-infrared circular dichroism spectrum of the Fe(II) rubredoxin has been reported.[81] This shows only one band in the range $9100-5550 \text{ cm}^{-1}$ (at

$6250\ cm^{-1}$). Circular dichroism studies reveal a very large anisotropy factor for this band and hence suggest it to be $d \to d$ in origin. This may be assigned as a $^5E \to ^5F$ transition for high-spin Fe(II) in a tetrahedral field, so confirming the maintenance of this stereochemistry on reduction. This result also allows an estimation of $10\,Dq = 6250\ cm^{-1}$ for the —SH group.

One recent result[82] has also confirmed the tetrahedral structure. The laser Raman spectrum of oxidized C. *pasteurianum* rubredoxin in aqueous solution shows bands at 368 and $314\ cm^{-1}$. The spectrum of crystalline rubredoxin shows bands at 365 and $311\ cm^{-1}$. These may be assigned to ν_{Fe-S} modes of the FeS_4 tetrahedron. Depolarization studies confirm that these bands are the ν_3 and ν_1 fundamentals of a tetrahedral complex. The results are also important in that they show that the environment around the metal ion is the same in the crystalline state and in solution.

Many complex enzymes contain one or more iron–sulphur centre. It would be helpful to be able to characterize these different systems *in vivo*. To this end the magnetic circular dichroism spectrum of oxidised C. *pasteurianium* rubredoxin has been studied.[83] This technique is more discriminating than absorption spectroscopy. The spectrum shows the presence of two one-electron charge-transfer transitions (S $\to$ Fe) in the region 15,000 to $28,000\ cm^{-1}$. This approach has been extended to other iron–sulphur proteins and will be discussed later.

2Fe.2S ferredoxins

These proteins are usually isolated from plant sources, notably spinach and also alfalfa and parsley. They are found in chloroplasts and their important role in photosynthesis is well established.[73] A 2Fe.2S protein from non-plant sources is putidaredoxin, of molecular weight 12,500, isolated[84] from *Pseudomonas putida*, that acts as a specific one-electron transfer agent in the enzyme complex that catalyses the hydroxylation of camphor. The sulphur atoms can be replaced by isotopically enriched sulphur or by selenium.[85] An example of a 2Fe.2S ferredoxin from an animal source is adrenodoxin, which is isolated from the adrenal cortex.[86] This is part of a multienzyme system that catalyses the hydroxylation of steroids; the following electron-transfer scheme has been suggested.[87] NADPH $\to$ flavoprotein $\to$ adrenodoxin $\to$ cytochrome $P450 \to O_2$. One half of the iron in adrenodoxin is reduced, giving rise to an EPR signal at $g = 1.94$ which is characteristic of reduced ferredoxins (Table 5.4).

The structures around the redox sites in these 2Fe.2S proteins have not yet been determined by X-ray techniques, although preliminary data is available on two proteins (from *Spirulina platensis*[88] and *Aphanothece sacrum*[89]). Nevertheless a fairly clear picture has been built up by the use of various instrumental techniques. The interpretation of their magnetic properties has caused problems in the past, because the oxidized form is diamagnetic and EPR inactive, while the reduced form is paramagnetic and usually has the $g = 1.94$ EPR signal (Table 5.4). The only diamagnetic form of iron is low-spin octahedral Fe(II), of d^6 configuration. If this is thought to be present in the oxidized

TABLE 5.4 'g' values for oxidized and reduced iron–sulphur proteins.

Protein	Reduced	Oxidized
Clostridial ferredoxin (8Fe.8S)	$gx = 1.89$ $gy = 1.96$ $gz = 2.00$	
Azotobacter (2Fe.2S)	$gx = 1.93$ $gy = 1.94$ $gz = 2.00$	
Pseudomonas putidaredoxin (2Fe.2S)	$g = 1.94$	
Spinach ferredoxin (2Fe.2S)	$gx = 1.89$ $gy = 1.96$ $gz = 2.00$	
Adrenodoxin (2Fe.2S)	$\begin{cases} g_1 = 1.94 \\ g_{11} = 2.01 \end{cases}$	
Rubredoxin		$g = 4.43$

ferredoxin, then it implies that the reduced form contains low-spin Fe(I), d^7, an unlikely situation. On the other hand, rubredoxin is paramagnetic in the oxidized form, and its redox properties may be simply interpreted in terms of the reduction of high-spin Fe(III) to high-spin Fe(II). Clearly we should not attempt to understand the magnetic data for 2Fe.2S proteins on the basis of redox changes in isolated iron atoms. It may be done for rubredoxin as in this case there is only one atom of iron per mole of protein.

The diamagnetic nature of the oxidized two-iron ferredoxins may be understood by postulating antiferromagnetic interaction between the two iron atoms, that is some type of dimeric structure, allowing the effective pairing of electrons giving a resultant spin of $S = 0$. On reduction, one electron is accepted, formally giving one Fe(III) group ($S = 5/2$) and one Fe(II) group ($S = 2$), which interact giving a value for the dimer of $S = \frac{1}{2}$. In other words, one electron has been added to a diamagnetic system. Much attention has been paid to the source of the $g = 1.94$ EPR signal, but this may be accounted for by antiferromagnetic coupling.

Much attention has been paid to the electronic state of the iron atoms in 2Fe.2S proteins and the interaction between them, as these are the simplest iron–sulphur systems in which this occurs. Magnetic susceptibility[90], ENDOR,[91] and Mössbauer[92] measurements have all been carried out on various 2Fe.2S proteins. Thus the Mössbauer spectra of[57] Fe-enriched putidaredoxin give parameters for the iron atoms in the reduced, paramagnetic protein which are typical of high-spin Fe(III) and high-spin Fe(II). Mössbauer spectra measured in magnetic fields show conclusively that antiferromagnetic coupling takes place to give a total spin of $S = \frac{1}{2}$, and that the Fe(III) atom has a ligand arrangement of relatively low symmetry and strong Fe—S covalency. The parameters for the ferrous iron indicate a distorted tetrahedral symmetry, while the oxidized form of the protein contains high-spin Fe(III) coupled to give $S = 0$. Furthermore Mössbauer and ENDOR spectroscopy have shown that the reduced form contains localized Fe(II) and Fe(III) oxidation states,[73] rather than a 'valence delocalized' system.

It appears then, at this point in time, that the structure of the 2Fe.2S proteins can best be regarded[93] as a bitetrahedral one, with the iron atoms bridged by two sulphide groups and bound additionally by two protein cysteine groups per iron atom, as represented schematically below:

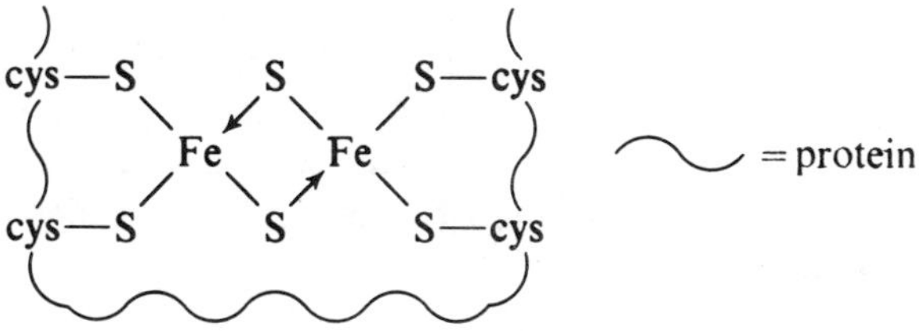

This structure is strongly supported by the study of model compounds, some excellent examples being available. Potassium dithioferrate $KFeS_2$ is a linear polymer of iron atoms bridged by two sulphur atoms. The Fe—Fe distance[94] is 2.7 Å, a value that implies some degree of Fe—Fe interaction. Magnetic susceptibility studies confirm[93] $KFeS_2$ to be an antiferromagnet, with the exchange coupling constant J approximately -200 cm^{-1}, a value that compares well with $J = -183$ cm^{-1} for spinach ferredoxin[90] so supporting the analogy between model and protein. This is confirmed by EPR studies on a single crystal of $KFeS_2$ which shows that there is delocalization of spin on to the sulphur atoms, as previously observed in EPR studies on several 2Fe.2S proteins.

Several other sulphur-bridged binuclear complexes of iron have been considered as models for assessing the electronic structure of the 2Fe.2S chromophore,[95,96] including the following species whose structural, spectroscopic and redox properties show it to be similar to that of the iron–sulphur protein.[97] This

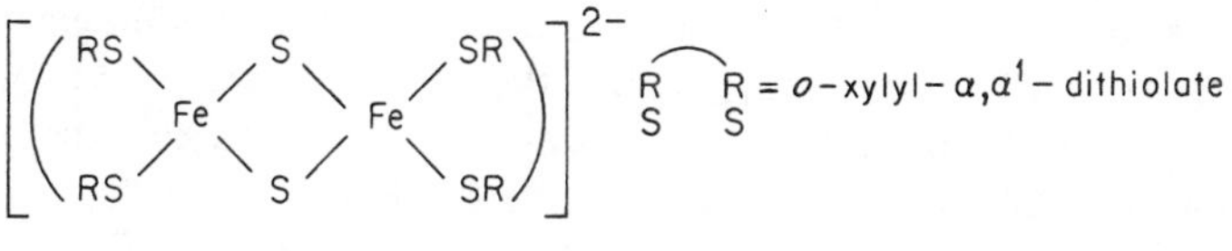

compound involves distorted tetrahedra around the iron, with bond lengths Fe—Fe $= 2.70$ Å, Fe—S $= 2.21$ Å, and Fe—SR $= 2.30$ Å. Calculations[98] on the compound $[Fe_2S_2(SH)_4]^{2-}$ suggest that the antiferromagnetism of the 2Fe ferredoxins is due to superchange rather than direct Fe—Fe interaction. It has been suggested that redox changes involving Fe(II)/Fe(III) dimeric systems may be accompanied by bridge angular deformation, and that the redox potential may reflect the ability of the system to undergo deformation.

The electron-transfer reactions of reduced spinach ferredoxins with horse heart cytochrome c, metmyoglobin, [Fe(EDTA)]$^-$, and related compounds have been studied.[99] One conclusion from the application of Marcus theory to these observations is that the redox centre of the protein is inaccessible.

4Fe.4S and 8Fe.8S proteins

These proteins appear to contain a common iron–sulphur prosthetic group, in which iron and labile sulphide are arranged in a distorted cubic structure, with each iron atom linked to the protein *via* cysteine residue. The four iron atoms and four sulphur atoms in the Fe_4S_4 cube form interpenetrating and concentric tetrahedra. This structure has been shown[100] in both oxidized and reduced forms of high-potential iron protein from *Chromatium*, and is represented schematically below. It may be seen that the structure can be considered as a pair of interacting Fe_2S_2 groups.

(protein not shown)

Peptococcus aerogenes ferredoxin (or clostridial ferredoxin, Fd) of molecular weight 6000, is an 8Fe.8S iron–sulphur protein. This contains[101] two Fe_4S_4 cubes separated from each other by 12 Å. In this case the structure of the oxidized form only was determined. The striking similarity in Fe—S chromophore in these two proteins is of particular interest in view of their wide difference in redox potential, the values for high-potential iron protein (HiPIP) and the clostridial ferredoxin (Fd) being $+0.3$ V and -0.4 V respectively, as it is difficult to conceive of this difference being due only to differences in the protein environment of the two clusters. One most interesting feature that emerges from the comparison of the structural data for the clusters is that bond lengths and angles in the HiPIP$_{red}$ and Fd$_{ox}$ Fe_4S_4 clusters are the same within the root mean square deviation of each mean value, while the HiPIP$_{ox}$ cluster is smaller by $0.1–0.2$ Å in certain dimensions. In view of this it has been suggested that the large difference in redox potentials between HiPIP and Fd may be explained on the assumption that the Fe_4S_4 cluster has three oxidation states available, rather than two, HiPIP cycling between the upper two and Fd between the bottom two. HiPIP$_{ox}$ and Fd$_{red}$ clusters would be paramagnetic, while the intermediate state (common to HiPIP$_{red}$ and Fd$_{ox}$) would be diamagnetic, through antiferromagnetic interactions and an even number of electrons. HiPIP$_{red}$ and Fd$_{ox}$ do have identical magnetic states.[102] These proposals are represented schematically below:

Formal oxidation states	State	HiPIP	Fd
$3Fe^{2+}/Fe^{3+}$ $[Fe_4S_4(S—Cys)_4]^{3-}$	C^{3-} (paramagnetic)	(Super-reduced)	Reduced
$2Fe^{2+}/2Fe^{3+}$ $[Fe_4S_4(S—Cys)_4]^{2-}$	C^{2-} (diamagnetic)	Reduced	Oxidized
$Fe^{2+}/3Fe^{3+}$ $[Fe_4S_4(S—Cys)_4]^{-}$	$C^{=}$ (paramagnetic)	Oxidized	(Super-oxidized)

If there are two possible redox equilibria available and the $HiPIP_{red}$ and Fd_{ox} are experimentally indistinguishable in terms of their iron–sulphur groups, then it is difficult to see why the $HiPIP_{red}$ cluster cannot be reduced further, or why the Fd_{ox} cluster cannot be oxidized further. These two possibilities are shown above by 'super-reduced HiPIP' and 'super-oxidized Fd'. This must reflect a subtle effect of the protein which prevents further oxidation or reduction of each cluster, so only allowing the transfer of a single electron by each cube. This important role of the protein has been confirmed by the observation[103] that HiPIP could be reduced to the 'super-reduced' form in the presence of 80% dimethylsulphoxide, which caused the unfolding of the protein. Later it was shown that ferredoxin could be oxidized to the 'super-oxidized' form by ferricyanide, and that the EPR signal of this species was identical to that of oxidized HiPIP.[104] These important results do not however explain how the protein is able to control the redox potential in this way. Carter[105] has presented a very detailed comparison of the active sites of HiPIP and Fd, and has drawn attention to the fact that they may be brought into equivalent orientations by assuming that the Fe_4S_4 clusters belong to the point group C_s. This results in polypeptide segments and particular tyrosyl residues becoming analogous. He suggests that interactions with the polypeptide backbone and tyrosine hydroxyl groups places the iron–sulphur clusters in diastereomeric environments, and that this accounts for the difference in their redox behaviour.

Synthetic and extruded analogues of the Fe_4S_4 cluster

The Fe_4S_4 cubes described above are reproduced very closely by synthetic analogues of the type $[Fe_4S_4(SR)_4]^{2-}$. These may be prepared by reaction of ferric chloride, sodium methoxide, sodium hydrosulphide, and a mercaptan.[106] These analogues have been developed largely by Holm and his group.[96] They have also shown that the active site cores of the iron–sulphur proteins may be extruded or displaced intact. This is accomplished by the use of a solvent of 80% dimethylsulphoxide and 20% water, to unfold the protein, and benzenethiol to act as the extruding reagent[107] by displacing the native cysteine thiol group. Thus the iron–sulphur centres may be extruded to give identifiable, synthetic analogues, in both 4Fe.4S and 2Fe.2S cases. Thus the extrusion of the 8Fe Fd_{ox} from *C. pasteurianum* in 80% hexamethylphosphoramide gives[108] two Fe_4S_4

cores, as anticipated. The same technique applied to the hydrogenase (Mol. wt. 60,500) from *C. pasteurianum* showed that it contained three Fe_4S_4 groups. The application of this technique to nitrogenase is discussed in Chapter 6.

A range of Fe_4S_4 analogues has been synthesized, including $[Fe_4S_4Cl_4]^{2-}$ by reaction[109] of benzoyl chloride with $[Fe_4S_4(SR)_4]^{2-}$. The dianion $[Fe_4S_4(SR)_4]^{2-}$ undergoes a one-electron reduction to give $[Fe_4S_4(SR)_4]^{3-}$. These anions correspond to 4Fe Fd_{ox} (HiPIP$_{red}$) and 4Fe Fd_{red} (HiPIP$_{superox}$) respectively. At present there is no analogue for 4Fe $Fd_{superox}$ (HiPIP$_{ox}$)—this would be the monocharged anion. Solid-state magnetic susceptibility studies on the dianion confirm the presence of an antiferromagnetically coupled Fe_4 unit. A range of techniques suggest[110] that all four iron atoms are equivalent in the cluster, and that it is more realistic to regard the Fe_4S_4 core as a fully delocalized system rather than in terms of Fe(II) and Fe(III). There is a difference in this respect between the Fe_2S_2 and Fe_4S_4 proteins.

Specific studies on the activity of analogue clusters in enzyme systems has been hindered by the insolubility of most clusters in water and by the hydrolysis of soluble clusters.[111] However such clusters can be stable in aqueous solution in the presence of an excess of the appropriate thiol. The clusters shown below are examples of soluble alkyl thiolate and peptide clusters which are soluble and which may be stabilized in aqueous solution under anaerobic conditions. They can replace[112] ferredoxin, in a hydrogen-evolving system utilising *Spirulina maxima* ferredoxin as electron mediator from dithionite to *C. pasteurianum* hydrogenase, confirming their value as models.

$$[Fe_4S_4(SCH_2CH_2OH)_4]^{2-} \, ; \, [Fe_4S_4(Ac.Gly_2.Cys.Gly_2.Cys.Gly_2NH_2)_2]^{2-}$$

One problem with these analogues is that their redox potentials determined in non-aqueous solvents are considerably more negative than those of the proteins determined in aqueous solution. However the redox potentials of the water-soluble clusters $[Fe_4S_4(S(CH_2)_nCO_2)_4]^{6-}$ are similar to those for the ferredoxins: cluster, $n = 2$, -580 mV; ferredoxin -400 mV. The potential of the analogue with $n = 3$ is little different from the $n = 2$ analogue, showing that the negative charges on the thiols are not important. The only feature common to the protein and the analogue in aqueous solution, and which is not present for the analogues in non-aqueous solvents, is hydrogen bonding. It is suggested therefore that this is important in controlling the redox potentials of protein-bound clusters,[113] a view already mentioned earlier.

Iron–sulphur proteins in electron-transfer chains

Membrane-bound iron–sulphur proteins play key roles in electron transport in mitochondria, particularly in complexes **I**, **II**, and **III** (as shown in Figure 5.1), and also in chloroplasts and chromatophores in photosynthetic plants and bacteria respectively.[73] The cytochromes in such electron-transfer chains may be characterized by UV–visible spectroscopy, but while iron–sulphur proteins are usually brown or red brown in colour, the intensities of their absorptions in this already complicated spectral region are too low to be of diagnostic value in characterizing particular clusters. It appears that MCD is able to distinguish between isolated 2Fe.2S and 4Fe.4S clusters and this technique[83,114] may become useful, but of greater value at present is low-temperature EPR spectroscopy. Using this technique, iron–sulphur clusters may be distinguished by their g values, temperature dependence of the spectrum, midpoint redox potential, and their response to various substrates and reducing agents.

Important results have been obtained[115] from EPR studies on submitochondrial particles prepared from the yeast *Candida utilis* grown in a medium enriched with ^{57}Fe. Hyperfine interactions of ^{57}Fe atoms in the EPR spectrum gave information on 4Fe.4S and 2Fe.2S clusters—so signals were identified from 4Fe.4S clusters 2, 3, and 4 of NADH dehydrogenase (complex I), and centre 3 of the succinate dehydrogenase (complex II). Between 50 and 80 K only the spectra of the 2Fe.2S proteins could be observed, centre 1 in complex I, centres 1 and 2 in complex II and the Rieske centre of complex III.

There has been some uncertainty as to the exact number of iron–sulphur centres in these complexes. Thus complex I was thought once to contain seven centres and then five. This 'fifth' centre is now suggested to be a 4Fe.4S. FAD complex, which has a role as an electron carrier in fatty acid oxidation.[116]

Other matters of importance include the location of the phosphorylation sites in the electron-transfer chain. The 4Fe.4S centre 2 in complex I (from *Paracoccus denitrificans*) is thought to be such a site.[17] Furthermore it is now thought that certain of these iron–sulphur proteins could also act as proton pumps and so could have a role in energy transduction in the context of chemiosmotic theory (see Chapter 9). The Rieske centre has been implicated as a proton carrier under certain conditions.

Membrane-bound iron–sulphur proteins are also involved in the electron-transfer chains of photosynthetic dioxygen-evolving organisms, and indeed possibly to be involved with primary photochemical events. At least three such proteins have been characterized *via* EPR, two of them being low-potential centres in photosystem I, while a third is similar to the Rieske centre of mitochondrial electron transport. This iron–sulphur protein is thought to function between plastoquinone and cytochrome f. These have been reviewed recently.[75]

SOME NON-HEME IRON ENZYMES

Several oxygenases and hydroxylases contain non-heme iron, although they have not usually been well studied. Protocatechuate-3,4-dioxygenase from *Pseudomonas aeruginosa* catalyses the intradiol cleavage of protocatechuic acid with the insertion of two atoms of dioxygen to give β-carboxy-*cis, cis*-muconic acid.

The enzyme contains eight atoms of Fe(III) per mole (molecular weight 700,000), and is made up of eight identical units, each of which is made up of two pairs of non-identical subunits.[118] The red colour of the enzyme results from an absorption at 450nm which is attributed to Fe(III) interaction with an amino acid residue. Addition of substrate in the absence of dioxygen shows an increase in intensity at 480 nm, while a new band appears at 520 nm on admission of dioxygen to the system. These observations are explained in terms of enzyme substrate and enzyme–substrate-O_2 complex fromation. After reaction the original spectrum is reformed. The resonance Raman spectrum of the enzyme showed[119] four lines at 1177, 1265, 1505, and 1605 cm^{-1}, while the spectrum of the iron(III) p-cresol complex showed four lines at 1180, 1222, 1488, and 1618 cm^{-1}. A similar spectrum has been recorded for iron(III) transferrin, and have been assigned to vibration of phenolate coordinated to Fe(III). Thus it appears that Fe(III) is coordinated by a tyrosyl residue in protocatechuate-3, 4-dioxygenase, a notable achievement for the technique in view of the size of the enzyme. The other ligands in the coordination shell are not known, but conflicting reports are available.[120]

Non-heme iron is also involved in lipooxygenase[121] (high-spin iron(III)) and ω-hydroxylase from $P.$ $oleororans$. The latter enzyme contains one iron atom and one cysteine residue per polypeptide chain (molecular weight 40,800).[122] An interesting example is ribonucleotide reductase (from $E.$ $coli$) which contains two non-identical subunits. One of these contains two non-equivalent high-spin Fe(III) centres in an antiferromagnetically coupled complex and a radical tyrosine residue in the polypeptide chain.[123]

COPPER PROTEINS

These are involved in a range of biological processes, from electron transport to oxidation of a range of substrates. Hemocyanins, the dioxygen-carrying proteins will be discussed in Chapter 7, while two copper proteins, superoxide dismutase and cytochrome oxidase will be considered in separate sections. Examples of copper proteins are listed in Table 5.5. They often have high redox potentials and sometimes a high (but even) number of copper atoms. Some copper proteins still have no known physiological function. Several reviews are available.[124-128]

It has been customary to classify copper species present in proteins into three types. While this should not be regarded as a rigid classification, it has served a useful purpose in the discussion of these proteins. It should be noted that the original definition applied only to copper is the blue proteins. The types may be defined as follows:

Type 1 Cu. These blue species have an intense absorption band near 600 nm (and bands at about 450 and 750 nm), an EPR spectrum with an unusually small hyperfine coupling constant, and high redox potentials. The absorption at 600 nm has an extinction coefficient about 100 times larger than those observed for Cu(II) complexes. The type 1 Cu thus stands out as a unique site, and much

TABLE 5.5 Some copper proteins.

Protein	Mol. wt	Mole Cu/mole protein	E_0' (mV)	Colour/function
Plastocyanin	10,500	1		Blue/electron transport
Stellacyanin	20,000	1	415	Blue/electron transport
Azurin	16,000	1	328	Blue/electron transport
Umecyanin	14,000	1		Bule/electron transport
Plantacyanin	8000	1		Blue/electron transport
Rusticyanin	16,500	1	680	Blue/electron transport
Laccase	60,000–120,000	4	450	Blue/oxidase
Ascorbate oxidase	150,000	8		Blue/oxidase
Ceruloplasmin	130,000	(8)	390	Blue/oxidase
Dopamine-β-monooxygenase	290,000	2		Non-blue/oxidase
Galactose oxidase	68,000	1		Non-blue/oxidase
Dopamine oxidase				Non-blue/oxidase
Benzylamine oxidase		2		Non-blue/oxidase
Polyphenol oxidase	119,000	4		Non-blue/oxidase
Cytochrome c oxidase	100,000	2,2Fe		Electron transport
Superoxide dismutase	32,000	2,2Zn		O_2^- disproportionation
Human serum albumin				Cu(II) transport

effort has been devoted to reproducing these features in model complexes. Blue copper proteins may contain type 1 Cu only, or they may contain copper of all three types, as in blue oxidases. Blue proteins with a single type 1 Cu (or two independent type 1 centres) are usually associated with electron transfer.

Type 2 Cu. This Cu centre, which is present in all the blue multicopper oxidases, has spectroscopic properties similar to complexes of Cu(II). This type is EPR detectable.

Type 3 Cu. This is also present in all multicopper oxidases, and appears to consist of two Cu^{2+} ions which are diamagnetic as a result of strong antiferromagnetic interactions. It can thus act as a two-electron donor/acceptor centre, and is essential to the reduction of dioxygen. The type 3 site has a strong absorption at around 330 nm, with $\varepsilon = 3000$–$5000\,dm^3\,mol^{-1}\,cm^{-1}$.

In addition to these types 1, 2, and 3 Cu found in blue copper proteins, there are non-blue copper proteins. The copper sites in these non-blue copper proteins should not strictly be compared with those of type 2 Cu.

Blue electron-transfer proteins with a single type 1 Cu

Certain blue proteins isolated from several bacterial sources (*Pseudomonas aeruginosa*, *Ps. fluorescens*, *Ps. denitrificans*, and *Bordetella pertussis*, for example) appear to have very similar properties, and are now termed azurins. These contain one copper ion (type 1) per mole of protein (molecular weight ~ 16,000). Another much studied protein is plastocyanin. This is found in plant chloroplasts and other photosynthetic organisms and is an essential component of the photosynthetic electron-transfer chain. Plastocyanins from different sources show only minor differences. Other blue proteins are listed in Table 5.5 Rusticyanin is of interest in view of its high potential, but this discussion will be restricted to plastocyanin and azurin. All of these proteins cycle between Cu(I) and Cu(II).

An important advance in the study of blue copper proteins is the determination of the structures of oxidized poplar plastocyanin[129] (at 2.7 Å resolution) and *Ps. aeruginosa* azurin at 3.0 Å resolution.[130] The copper-binding site of plastocyanin is made up of the imidazole groups of His-37 and His-87 and the sulphur atoms of Cys-84 and Met-92. These ligand groups represent a compromise between the requirements of Cu(I) and Cu(II), suggesting an enhanced ease of electron transport. The geometry around the copper is distorted tetrahedral, the bond angles differing by up to 50° from tetrahedral values. This again suggests enhancement of electron transfer, as it is intermediate between the square-planar and tetrahedral geometries favoured by Cu(II) and Cu(I) respectively.

The structure of plastocyanin offers several hydrophobic channels for quantum mechanical tunnelling of electrons, but an outer-sphere mechanism seems particularly attractive as the coordinated imidazole group of His-87 is exposed to the surrounding medium. The exterior of the protein shows hydrophilic and hydrophobic areas which may provide specific receptor sites for interaction between plastocyanin and its redox partners.

The structure of azurin is probably similar to plastocyanin in terms of X-ray evidence[130] and this conclusion is also supported by EXAFS data[131] on the oxidized protein, which shows the presence of a very short Cu—S distance of 2.10 ± 0.02 Å, while two nitrogen ligands (imidazole) are 1.97 Å away. It was suggested that the fourth ligand was a sulphur donor, Cu—S ~ 2.25 ± 0.05 Å (methionine). It is noteworthy that stellacyanin does not contain[132] a methionine residue, so it appears that the ligands of the type 1 Cu site are not invariant. This may be relevant to the control of the redox potential.

Models of the type 1 Cu site

Many workers had previously suggested that the coordination geometry around Cu in type 1 sites would be distorted tetrahedral. Thus on the basis of a comparison with the properties and spectra of model compounds, the EPR,[133] CD and MCD spectra[134] and the high redox potentials[4,135] of blue proteins were interpreted.

Many predictions had also been made concerning the nature of the ligand groups on the basis of chemical methods and instrumental techniques. The ^{1}H NMR spectra of azurin and plastocyanin were interpreted in terms of cysteine and methionine ligands,[136] while the same technique allowed the postulate of two histidine ligands.[137] The presence of cysteine and methionine ligands has also been shown by the analysis of the charge-transfer bands of Ni(II), Mn(II), and Co(II) azurin.[138] Bonding *via* methionine was supported by resonance Raman studies,[139] while X-ray photoelectron spectroscopy had been used[140] to support bonding *via* cysteine, although there appears to be some uncertainty over the exact interpretation of this data. Despite some incorrect predictions that are not discussed here, there has clearly been some excellent work on modelling the type 1 site.

Kinetic studies on blue electron-transfer proteins

Some references to the reaction of these proteins with small redox complexes[10,13] and other electron-transfer proteins[15,16] have already been given. Some of this material has been reviewed.[128] Kinetic parameters for the reduction of azurin and plastocyanin with Fe(EDTA)$^{2-}$ have been interpreted in terms of long-distance transfer of an electron to a buried metal centre. In contrast the oxidation pathway involves conformational changes which allow contact of [Co(phen)$_3$]$^{3+}$ with the copper ligands. It is noteworthy that the structure of oxidized plastocyanin confirms the accessibility of His-87. Type 1 Cu in stellacyanin seems to be equally accessible to both types of redox reagent, but this does differ in at least one ligand in the coordination sphere of Cu.

Blue oxidases with types 1, 2, and 3 Cu

Among the most studied examples of blue oxidases are laccase,[128] ceruloplasmin, and ascorbate oxidase. Laccase, which is usually prepared from *Rhus vernicifera* or *Polyporus versicolor*, catalyses the oxidation by dioxygen of diphenols and aromatic diamines, in which the dioxygen undergoes a four-electron reduction to water. As it is a water-soluble enzyme and easily purified, it has thus proved to be a particularly convenient system for studying the problems of multielectron redox reactions. Laccase contains one each of type 1 and type 2 Cu and two moles of antiferromagnetically linked type 3 Cu. This pair of EPR non-detectable Cu ions is commonly supposed to function as a cooperative two-electron unit. This results in a two-electron reduction of dioxygen, which bypasses the formation of superoxide. This view of the two type 3 Cu ions is not universally accepted, and it has been suggested[141] that they usually act as independent single-electron donors. It is useful, however, to note the analogy with hemocyanins where dioxygen is bound by a pair of copper ions, and to suggest that this is also the role of the two type 3 Cu ions in laccase.

Ceruloplasmin, the major copper-containing protein of mammalian blood plasma, also catalyses the four-electron reduction of dioxygen to water with concomitant oxidation of substrate. The physiological role of ceruloplasmin is thought to be the oxidation of iron(II), while it may also function as a copper-transport protein. Ceruloplasmin probably has seven or eight copper atoms per mole (eight seems reasonable) and so far two type 1, one type 2, and four type 3 sites have been found.

Intermediates have been detected in the reaction of dioxygen with reduced laccase[142] and with reduced ceruloplasmin.[143] Rapid-quench EPR experiments[144] with reduced laccase and $^{17}O_2$ have shown that the $H_2^{17}O$ thus formed is bound equatorially at the type 2 Cu ion, suggesting that this is close to the dioxygen-reducing site in the enzyme, and confirming the view that water is released from type 2 Cu. Other experiments have shown that on oxidation with dioxygen the type 1 Cu and the type 3 Cu became reoxidized but that the type 2 Cu remained reduced.[145] Thus three electrons have been transferred to dioxygen. This confirms that one-electron transfers do take place after the initial two-electron transfer, and implies that the O^- radical is formed. The existence of an oxygen radical has been shown by EPR spectroscopy.

Pulsed EPR spectroscopy[146] of laccase and ceruloplasmin has led to the conclusion that imidazole is a ligand at both type 1 and type 2 sites: indeed a later report suggests that the type 2 Cu in ceruloplasmin contains three nitrogen donor ligands.[147]

Irradiation at 77 K of oxidized ceruloplasmin and laccase[148] at the 330 nm absorption (i.e. at the type 3 chromophore) leads to the reduction of type 1 Cu, as demonstrated by the loss of the intense absorption at 610 nm. Type 2 Cu was unaffected. This demonstrates the possibility of direct energy transfer between type 1 and type 3 sites in the blue copper oxidases, in accord with the role of these sites for substrate and dioxygen binding respectively. It will be interesting to receive the results of experiments such as these on laccases which have had the type 2 Cu removed.[146]

Several model systems have been reported for type 3 Cu. There are usually macrocycles that bind two copper(II) ions. Thus the ligand shown below (L) forms[149] a complex with azide and Cu(II) of stoichiometry $Cu_2(N_3)_4L$, with a

Cu. . .Cu separation of 5.145 Å, and yet is completely diamagnetic as shown by susceptibility measurements between 4.2 and 390 K. Many binuclear copper(II) complexes exhibit antiferromagnetic behaviour, but very few are diamagnetic at room temperature and these have smaller Cu. . .Cu separations. The electronic spectrum of the complex shows bands at 380 nm ($\varepsilon = 3000 \text{ dm}^3 \text{ mol}^{-1} \text{ cm}^{-1}$) and 650 nm (350).

Non-blue copper proteins

Galactose oxidase[128] catalyses the oxidation of primary alcohols to the corresponding aldehyde with reduction of dioxygen to hydrogen peroxide. It has a molecular weight of 68,000 and contains one copper atom per molecule. As such it is the simplest non-blue copper protein. EPR studies suggest that the Cu(II) ion is bound by at least two nitrogen ligands, probably imidazole groups, while water has also been suggested as a ligand. There is some evidence for a role for a further histidine residue in the active site, while an indole group of tryptophan is also implicated. The fluorescence of some of the tryptophan residues is quenched by the Cu^{2+}, which is calculated to be less than 12 Å away.[150] The activity of the enzyme is cut dramatically when two out of the eighteen tryptophan residues are oxidized by N-bromosuccinimide,[151] and this modification also results in a decrease in the optical rotation at 314 nm and a change in the EPR spectrum, due, it is suggested, to a change in stereochemistry around Cu^{2+} from pseudo square-planar towards tetrahedral.

A number of amine oxidases are copper proteins that utilize the two-electron reduction of O_2 to H_2O_2. Benzylamine oxidase contains two Cu^{2+} ions per mole, and evidence has been produced for the participation of Cu^{2+} in the catalytic cycle. The enzyme is suggested to oxidize the substrate and is then reoxidized itself by reaction with O_2, which is converted to H_2O_2.[152]

Cytochrome c oxidase

Cytochrome c oxidase[153] is the terminal member of the electron-transport chain in mitochondria and so interacts with dioxygen and cytochrome c. It contains two type a cytochromes (a and a_3) and two copper ions. Thus this complex contains four metal centres, the same number found in the blue copper oxidase laccase. In contrast to laccase, cytochrome c oxidase is difficult to study as it is a high molecular weight, membrane-bound protein that requires solubilization. The spectra of the copper components are masked by those of the heme groups, which overlap themselves. Not surprisingly in an enzyme of this complexity there are still major points of disagreement about the mechanism of its action.

Arguments based upon the magnetic properties of the components of cytochrome c oxidase in both oxidized and reduced forms lead to the conclusion that one heme centre (cyt a_3) and one copper centre are antiferromagnetically coupled. Thus only one copper centre has been detected by EPR spectroscopy. The other is undetectable by both EPR and electronic spectroscopy. Cytochrome a_3 is believed to be high spin in the fully oxidized cytochrome c oxidase, as addition[154] of cyanide causes a change in the electronic spectrum that is consistent with a change from a high-spin to a low-spin heme. MCD measurements have also been useful[155] in studying the magnetic properties of cytochrome c oxidase. The temperature dependence of the paramagnetic susceptibility of oxidized cytochrome c oxidase has been studied in the range 200–7 K[156] and 4.2–1.5 K.[157] The results are consistent with a model in which two isolated $S = \frac{1}{2}$ centres and a spin-coupled $S = 2$ centre are present. These correspond to low-spin Fe(III)(d^5), Cu(II)(d^9), and high-spin Fe(III). . .Cu(II) respectively. The conclusions from the study of the magnetic properties of the heme groups in the oxidized, half-reduced, and fully reduced forms of the enzyme are shown below and represent the scheme most generally accepted at present. Each heme undergoes two changes in spin state during reduction.

Form	cytochrome a_3	cytochrome a
Fully oxidized	High-spin Fe(III)	Low-spin Fe(III)
Half reduced	Low-spin Fe(II)	High-spin Fe(III)
Fully reduced	High-spin Fe(II)	Low-spin Fe(II)

A model has been proposed[156] for the antiferromagnetically coupled cytochrome a_3–Cu centre, in which the two metals are linked by an imidazolate bridge, similar to that known to be present in superoxide dismutase between Cu and Zn.

A (Fe^{2+}–bipyrimidine–Cu^{2+}) site has been synthesized, showing that such Fe–Cu binuclear centres are feasible.[158] A nitrogen ligand has been implicated in the coordination sphere of the EPR-detectable copper.[159]

Other areas of particular interest at present are the study of intermediates formed during the reaction of the enzyme with dioxygen, and the reaction of the enzyme with inhibitors such as NO, CN^-, N_3^-, and CO. Anaerobic potentiometric titrations of cytochrome oxidase in the presence of CO suggest that two electrons must be added before the reduced cytochrome a_3–CO complex is formed.[160] The second centre that is reduced is believed to be the undetectable $Cu(Cu_u)$ suggesting that it plays a key role in the organization of the active site of cytochrome oxidase.

A number of intermediates have been detected in the reaction between reduced cytochrome c oxidase and dioxygen; these include[161] a species with an absorption maximum at 428 nm ('oxygenated' cytochrome c oxidase), which was then converted to the fully oxidized form in a stepwise reaction involving at least three intermediates. Another approach has been to oxidize the carbon monoxide complex of cytochrome c oxidase with ferricyanide. This gives the mixed oxidation state cytochrome c oxidase, as the cytochrome a_3^{2+}. CO species is inert to oxidation. The CO may be removed by flash photolysis, so allowing the study of the oxidation of this 'partially reduced' form of cytochrome c oxidase with dioxygen. This also gives an oxygenated oxidase ($\lambda_{max} = 611$ nm) as the first product; i.e. $Cu_u^{2+}.a_3^{2+}.O_2$. Studies such as these at low temperature give information[162] on the sequence of oxidation–reduction of the various centres in the oxidase. It appears that electrons enter via cytochrome a and are transferred rapidly to the detectable Cu. The reaction of the fully reduced oxidase with O_2 is very rapid, the rate constant of the slowest step being $\sim 700\,s^{-1}$. Binding of dioxygen is thought to involve cytochrome a_3. The antiferromagnetically linked pair of cytochrome a_3^{3+} and Cu_u^+ then transfers two electrons to dioxygen to give peroxide, or alternatively it is thought that cytochrome a_3^{2+} transfers two electrons to O_2, prior to oxidation of Cu^+ to Cu^{2+}. Subsequent steps involve a two-electron or two one-electron reduction of peroxide to water.

Kinetic studies[163] on the reaction of reduced cytochrome c with cytochrome c oxidase in the fully oxidized form, as obtained from the preparation, and with cytochrome c oxidase oxidized from the reduced form by O_2, show that different results are obtained in the two systems. The product of oxidation with O_2 of the fully reduced cytochrome c oxidase apparently differs from the normal preparation, which is probably not the catalytically active species. These important differences are assumed to result from a different rate of intra-molecular electron transfer in the two types of oxidases.

Superoxide dismutase

Several copper proteins such as erythrocuprein, cerebrocuprein, and hepatocuprein are now known to be identical. This widespread copper protein is suggested to have the important protective function of catalysing the disproportionation of superoxide ion, O_2^-, and so has been named superoxide dismutase.[164] Superoxide ion is a toxic radical and is converted to dioxygen and peroxide in the enzyme-catalysed reaction. The hydrogen peroxide thus formed will be decom-

$$2O_2^- \longrightarrow O_2 + O_2^-$$

posed by the enzyme catalase, and so these two enzymes protect cells from the toxic byproducts of oxygen respiration. Aerobic organisms appear to contain high levels of superoxide dismutase in accord with this view. The enzyme has been purified from a diverse range of microorganisms, plants and animal organs. Enzymes obtained from eukaryotic sources contain copper and zinc, while those isolated from other sources may contain iron and zinc or manganese and zinc. A review[165] and a book[165a] are available on superoxide and superoxide dismutase.

The best studied dismutase is that isolated from bovine erythrocytes. It has a molecular weight of 31,400 and is made up of two identical subunits, each of which contains 151 amino acid residues, one copper, and one zinc. The geometry around the copper[166] is distorted square planar, with four imidazole ligands from His-44, His-46, His-61, and His-118, while it is possible that water could bind in one unprotected axial position. The zinc ion has a tetrahedral environment with three nitrogen donors (His-61, His-69, and His-78) and one oxygen donor (Asp-81). His-61 is a common ligand to both metals, which are thus bridged by an imidazolate group, as shown below. The 270 MHz NMR spectrum of this enzyme is in accord with the crystal structure data.[167]

Both Cu and Zn may be removed readily by dialysis to give the inactive apoenzyme. Zn(II) may be replaced by Co(II), giving an enzyme that shown antiferromagnetic interaction between the two centres.[168] Stepwise addition of Cu(II) to Cu-free superoxide dismutase shows that the enzyme with only one subunit containing Cu is more active than the enzyme having Cu bound to both subunits.[169] This demonstrates that there is some sort of cooperative interaction between the two subunits.

The Zn(II) ion almost certainly has a structural role in the formation and stabilization of the active site.[170] The superoxide ion probably interacts with the Cu centre, undergoing the reactions shown below.

$$ECu(II) + O_2^- \longrightarrow ECu(I) + O_2$$
$$ECu(I) + O_2^- + 2H^+ \longrightarrow ECu(II) + H_2O_2$$

Models for superoxide dismutase

A number of copper complexes catalyse the disproportionation of superoxide ion. Indeed the aquocopper(II) complex is more effective than the enzyme in catalysing this reaction, although it is not possible to make this comparison under biological conditions. Lippard and his coworkers have synthesized and studied some interesting models for the imidazolate-bridged bimetal centre.[171] These include imidazolate-bridged dicopper(II) species. This grouping has also been incorporated into the macrocycle 1,4,7,13,16,19-hexaza-10,22-dioxatetra-cosane shown below. Each Cu of the Cu_2 (imid)$^{3+}$ ion is bound to one set of three nitrogen atoms at the end of the cavity and to a neutral imidazole group, giving five-coordinate species. These models allow the study of bridge cleavage reactions as in this compound the two metal centres will not be able to move away from each other when the bridge is cleaved.

MOLYBDENUM ENZYMES

Some seven molybdenum enzymes are known at present (Table 5.6). Dinitrogen fixation and nitrate reduction in microorganisms require molybdenum and the appropriate enzymes will be discussed in Chapter 6. Nitrate reductases are also found in plants. Another microbial molybdenum enzyme is xanthine dehydrogenase, while other dehydrogenases from *E. coli* may require Mo. Three molybdenum enzymes are known in mammalian systems; xanthine oxidase, aldehyde oxidase, and sulphite oxidase. Xanthine dehydrogenase and xanthine oxidase are very similar but utilize $NADH^+$ (or ferredoxin) and dioxygen as electron acceptors respectively. Discussion in this section will be limited to the mammalian molybdenum enzymes. Each will be described briefly prior to a

TABLE 5.6 Soe molybdenum enzymes.

Enzyme	Source	Mol. wt.	Stoichiometry
Xanthine oxidase	Bovine milk	275,000	2Mo; 8Fe, 8S, 2FAD
Xanthine dehydrogenase	*Micrococcus*	250,000	2Mo; 8Fe, 8S, 2FAD
Aldehyde oxidase	Hog liver	270,000	2Mo; 8Fe, 8S, 2FAD
Sulphite oxidase	Bovine liver	110,000	2Mo; 2heme
Nitrogenase	*A. vinelandii*	226,000	2Mo; 24Fe, 22S
Nitrate reductase	*N. crassa*	228,000	1–2Mo; cyt$_b$, FAD
Formate dehydrogenase			

general discussion of certain common features. Some molybdenum chemistry will also be presented. Some general reviews are available.[172]

The information given in Table 5.6 shows that flavin adenine dinucleotide is a cofactor for some of the molybdenum enzymes. This complicates the use of EPR techniques as free-radical semiquinones may be generated. Flavins may undergo one- or two-electron reductions giving semiquinones and quinones. Flavoproteins are often linked in pairs. Such a unit may be associated with redox reactions involving from one to four electrons.

Flavoquinone (F1)

Flavosemiquinone

Flavohydroquinone (F1H$_2$)

Aspects of molybdenum chemistry[173]

Molybdenum is an element of the second transition series, having a $4d^5 5s^1$ configuration. In its compounds it shows a range of oxidation states and stereochemistries. The biologically important oxidation states are $Mo(VI)(d^0)$, $Mo(V)(d^1)$, and $Mo(IV)(d^2)$; but Mo(III), cannot be completely discounted. The preferred coordination numbers for Mo(V) and Mo(VI) are five and six, and four and six respectively.

Mo(V) species are paramagnetic and will have an EPR spectrum. For ^{95}Mo and ^{97}Mo isotopes, $I = \frac{5}{2}$ and hence six satellite peaks will be observed. For the other isotopes $I = 0$, and one broad band will occur. This means that for naturally occurring Mo(V), as the natural abundance of the $I = \frac{5}{2}$ species is only about 25%, it will be difficult to observe the hyperfine splitting. The synthesis of $^{95}Mo(V)$-enriched compounds is sometimes necessary.

Mo(V) complexes are usually anionic, having oxygen, halogen, thiocyanate, or cyanide as ligands. Oxo species are particularly important and dimerization frequently occurs through oxo bridges. Under certain conditions this leads to diamagnetic species, through spin–spin interaction. Mo(V) compounds are usually prepared by reduction of molybdates or MoO_3 in acid solutions by chemical or electrochemical methods. Some oxo complexes with other ligands include $MoOCl_3(dipy)$, $Mo_2O_3Cl_4(dipy)_2$, and $\{MoO(oxalate)H_2O)_2O_2\}^{2-}$. The latter species is diamagnetic due to the presence of a Mo—Mo bond. An ethyl xanthate complex $[(C_2H_5OCS_2)_2MoO]_2O$ has a structure involving two distorted octahedrons linked by a Mo—O—Mo bond with *cis* double-bonded oxygen atoms on each molybdenum. Most of these Mo(V) complexes are not particularly stable with respect to oxidation. Some others, including those with EDTA and with 8-hydroxyquinoline-5-sulphonic acid, are quite stable.

Mo(VI), d^0, complexes are also usually anionic with oxygen as the preferred ligand. A variety of organic hydroxo complexes are known but these are not well defined. The tetrahedral oxo anion MoO_4^{2-} is well known, but in acid solution polymerization occurs giving a range of polymolybdate ions of some complexity. Amines give complexes with molybdates which are usually ill characterized, but with diethylenetriamine $MoO_3.dien$ is formed. Mo(VI) also forms various binuclear oxo complexes.

It is to be expected, therefore, that, in biological systems, molybdenum will be bound through carboxy groups or hydroxyl groups of tyrosine and serine residues. It is clear, however, from studies with—SH binding agents, that cysteine is a likely binding site of the metal ion.

It is important to ask why molybdenum is required in these biological processes, and to attempt to identify any unique property of Mo that could account for this. The dominant property of Mo is its ready formation of oxo compounds. This appears to be significant, as most of the reactions catalysed by molybdenum enzymes involve the transfer of an oxygen atom. This should be borne in mind when considering mechanisms.

Xanthine oxidase[174]

The enzyme catalyses the oxidation of xanthine to uric acid. The oxygen incorporated into the product does not come from dioxygen but from water, in

$$\xrightarrow[\substack{H_2O \\ enzyme}]{O_2}$$

accord with the fact that the enzyme can utilize other oxidizing agents. The enzyme is usually isolated from cows milk and contains two Mo atoms, two FAD groups, and two each of two different Fe_2S_2 clusters. It appears likely that there are two subunits in the enzyme, each involving one Mo, one FAD, and one each of the two Fe_2S_2 system. It has been postulated that the enzyme involves a dimolybdenum site, but this now seems unlikely, and it seems that the two Mo atoms are far apart. There is no evidence for a $[Mo(V)]_2$ dimer from EPR spectroscopy. The molybdenum is well established as the site for the binding and reduction of xanthine.

The enzyme may lose molybdenum or sulphur spontaneously during the extraction procedure. The formation of the desulpho form of the enzyme is also involved in the deactivation of the enzyme by cyanide, as CN^- will abstract sulphur readily to form thiocyanate. This has led to the suggestion that molybdenum is bound by a persulphide ligand ($-S-S^-$), but other sources of sulphur should not be excluded at present. (See note added in proof.)

The resting enzyme contains Mo(VI) and, as will be discussed later, it is reduced to Mo(IV) *via* Mo(V) during the enzymatic reaction. Reoxidation to Mo(VI) occurs through two one-electron transfers to the oxidant. The sequence of electron transfer was once thought to be the linear sequence below, which was established by kinetic studies utilizing EPR and electronic spectra. It is more

$$xanthine \rightarrow Mo \rightarrow FAD \rightarrow Fe_2S_2 - O_2$$

probable that the reduced Mo transfers electrons to both FAD and iron–sulphur clusters to an extent dependent upon their potentials, and both interact with dioxygen. The formation of intermediate Mo(V) species has been confirmed by EPR spectroscopy. The EPR spectrum is suggestive[175] of Mo—S binding in a distorted octahedral environment, but there are some uncertainties[172] in the interpretation of this data, partly because of the absence of monomeric model compounds of Mo(V), of known structure, for comparison purposes.

Another important feature of the Mo(V) EPR signal from xanthine oxidase is ligand hyperfine splitting arising from a proton.[176] In D_2O the doublet bands all coalesce into singlets showing that this proton is exchangeable. Furthermore the

(1)   (2)

pH dependence of this phenomenon indicates that the proton has an apparent pK_a of ~ 8. Studies with 8-deuteroxanthine (**I**) show that the proton originates in the 8 position of xanthine. A more recent investigation suggests that the Mo(V) form of the enzyme reacts with two enzyme-bound protons, one strongly coupled and one weakly coupled.[175]

The implication of Mo(IV) in the reduced form of xanthine oxidase is based largely upon the use[177] of the compound alloxanthine(**2**). This forms a stable complex with the reduced enzyme, so that excess reducing agent and alloxanthine can be removed. Subsequent titration with $[Fe(CN)_6]^{3-}$, after due allowance for the oxidation of the flavin and iron–sulphur centres, shows that two electrons may be removed from the Mo, showing it was present in the Mo(IV) state. It appears that in the xanthine oxidase-catalysed oxidation of xanthine that two electrons and at least one proton are transferred to the enzyme. Current views reject the possibility that these are transferred as a hydride species, and support the view that the proton is present on a protein atom bound to the molybdenum.

The desulpho xanthine oxidase has continued to attract attention. Thus reduction of this form of the enzyme to the Mo(V) state gives an EPR signal that differs from that obtained from the functional enzyme. An understanding of this phenomenon may well contribute information on the structure of the molybdenum centre. The desulpho enzyme has a Mo(V) EPR signal that shows the presence of a weakly coupled proton and a strongly coupled proton as for the normal enzyme. However the exchange rate constant of the strongly coupled proton differed dramatically in the two cases, being $0.40\,s^{-1}$ and $85\,s^{-1}$ for the desulpho and functional enzyme respectively (at pH 8.2 and 12°C).[178] It is suggested that many aspects of the differences between the two forms of the enzyme are consistent with E—Mo=S and E—Mo=O, protonation giving the EPR active forms E—Mo—SH and E—Mo—OH with pK_a values of 8 and 10. This view is compatible with the important role of oxo groups in molybdenum chemistry. Thus in the functional enzyme the proton transferred from the 8 position of the substate is accepted by the Mo=S group.

The xanthine molecule is probably bound to the Mo *via* the N-9 atom. Other ligands may include deprotonated groups, possibly cysteine. The proton from the C-8 position is transferred to an enzyme group, whose identity is not yet resolved. It may be Mo=S, or coordinated nitrogen as suggested by model studies.[179] A number of mechanisms have been proposed.[172,180] Features of these include coupled proton–electron transfer with nucleophilic attack of

hydroxide ion on the C-8 atom. The possibility of oxo transfer from an oxomolybdenum(VI) enzyme to the substrate has not been given much attention.

Aldehyde oxidase and sulphite oxidase

Aldehyde oxidase can be isolated from several sources. It catalyses the oxidation of aldehydes by dioxygen, but will also catalyse the oxidation of pyrimidines. The latter reaction may well be the physiological function of the enzyme, a suggestion of some interest as xanthine oxidase and aldehyde oxidase will then be purine and pyrimidine oxidases respectively in accord with their similar biochemistries. The electron-transfer chain in aldehyde oxidase is similar to that in xanthine oxidase, but may involve an additional electron carrier in coenzyme Q.[181]

Sulphite oxidase catalyses the oxidation of sulphite to sulphate, although again the source of oxygen is H_2O and not O_2. Oxo group transfer seems an attractive possibility here. Sulphite oxidase differs in its electron-transfer components, containing two b-type cytochromes. It is made up of two subunits each of which contains one Mo and one heme. EPR data is suggestive of the electron-transfer sequence $SO_3^{2-} \rightarrow Mo \rightarrow heme \rightarrow O_2$.[182] It has shown the proton ligand hyperfine splitting.

It is clear therefore, that these two enzymes have features in common with xanthine oxidase. They all seem to utilize the Mo(VI), Mo(V), and Mo(IV) oxidation states. The long-standing suggestion[183] that niobium, tantalum, and tungsten be used to probe the molybdenum sites has been justified by the preparation of a tungsten-substituted sulphite oxidase[184] by feeding rats a tungsten-containing diet. The tungsten enzyme is completely inactive, but is reduced to W(V) on treatment with dithionite. It is not however reduced further by excess dithionite. Thus the inactivity of the tungsten-substituted sulphite oxidase may reflect its inability to be reduced to W(IV).

The molybdenum cofactor

Evidence has been put forward to show that some of the molybdenum enzymes may have a common molybdenum-containing cofactor.[185] Thus nitrate reductase from the *nit 1* mutant *Neurospora crassa*, while possessing other reductase activity, cannot reduce NO_3^-. However full activity can be restored by treatment of *nit 1* extracts with the neutralized, acid-hydrolysis products of molybdenum enzymes. The implication is that an enzyme fragment containing molybdenum is donated, so giving the active nitrate reductase. This factor has now been isolated, although it is not certain at present that all molybdenum enzymes contain exactly the same cofactor. Nevertheless this is highly indicative of a common site around Mo and a common mechanism for the reactions of molybdenum enzymes.

THE COBALT B$_{12}$ COENZYMES

The biological activity of cobalt is largely confined to its role in the B_{12} series of coenzymes. The vitamin B_{12} molecule has been studied by X-ray diffraction,[186] and its structure has been shown in Chapter 3. It involves a conjugated corrin ring (usually buckled), which has a porphyrin-type structure apart from having one methine group less. The corrin system provides four nitrogen donor atoms, in a plane for cobalt as found in the porphyrins. The fifth and sixth coordination positions are occupied by the benzimidazole of the nucleotide group and a cyanide ion. The presence of cyanide is an artefact of the isolation procedure and is of no biological significance. The cyanide complex is not active as a coenzyme. This complex without cyanide is termed cobalamin, so vitamin B_{12} is cyanocobalamin. A whole range of other substituents may take the place of cyanide.

On hydrolysis the benzimidazole group is displaced by water, giving the cobinamide series of complexes, while a variety of ligands may occupy the sixth position. The fifth position may be occupied by OH^- or CN^- instead of water. The cobinamides and cobalamins are represented schematically in Figure 3.5. Of particular interest, both chemically and biochemically, is the case when the sixth ligand is a carbanion, that is where there is a Co—C bond. Until the demonstration of the direct cobalt–carbon σ bond linking an adenosyl group to cobalt in adenosylcobalamin, organocobalt complexes were regarded as rare, unstable species. Now a series of very stable organocobalt complexes has been prepared using model compounds based upon the B_{12} structure.

Some aspects of the biological function of the B_{12} coenzymes have already been given in Chapter 3. Some comments on model compounds were also made. The biochemical functions have been reviewed on a number of occasions.[187] In summary, it may be noted that these include methyl group transfer, skeletal rearrangements of molecules, and various reduction reactions. Alkylcobalamins are very susceptible to photolysis.

Models for cobalt B_{12} coenzymes

Much interesting inorganic chemistry has arisen in this context. The value as models of cobalt complexes of dimethylglyoxime, the 'cobaloximes' has already been noted in Chapter 3. The crystal structure[188] of a substituted alkylcobaloxime shows that the Co—N (in-plane) and Co—C bond lengths are very similar to those found for the coenzyme.

It appears that the underlying reason for the good correlation between the properties of the coenzyme and models is the presence of a strong planar ligand field of the correct strength. Thus a number of other ligands that satisfy this condition will also show model behaviour of varying appropriateness. One important difference between coenzyme and model compound, which probably contributes to the presence of a number of differences in behaviour, is the fact that there is a great deal of steric hindrance in the coenzyme which is not duplicated by the cobaloximes. This is important in Co—C homolysis.[189]

The redox behaviour of B_{12} coenzymes

The naturally occurring cobalamins (hydroxocoabalamin, methylcobalamin, and adenosylcobalamin) all contain coablt(III), but lower oxidation state species are important in their biological activity. They may be reduced in one-electron steps. The cobalt(II) cobalamin is low spin. EPR studies have shown the hyperfine structure due to coupling with the cobalt nucleus, and the ligand hyperfine structure due to interaction with the benzimidazole nitrogen atom. Further reduction gives the cobalt(I) cobalamin$_1$ which is unstable with respect to oxidation. These Co(III), Co(II), and Co(I) forms of aquocobalamin have been termed B_{12_a}, B_{12_r}, and B_{12_s}. B_{12_s} is a powerful nucleophile and is rapidly oxidized by both dioxygen and water at a pH below 8.

The nucleophilic properties of B_{12_s} result in its ready reaction with a range of alkylating agents to give alkylcobalamins. The cleavage of the cobalt–carbon bond of alkylcobalamin may involve as products Co(III) and a carbanion, Co(II)

$$OH-Cbl^I(B_{12_s}) + CH_3I \rightarrow CH_3Cbl$$

and an alkyl radical, or Co(I) and a carbonium ion. A discussion of methyl-transfer reactions catalysed by vitamin B_{12} coenzymes will be given in Chapter 10.

Adenosylcobalamin-dependent rearrangements

These rearrangements involve the migration of a hydrogen from one carbon to an adjacent one, while a group X migrates in the opposite direction. Group X is either a bulky alkyl or acyl residue or it is an electronegative group (OH or NH_2). Relatively little is known about the cobalamin-dependent enzymes involved in these reactions, and attention has been focussed on the role of the cobalamin. Some adenosylcobalamin (AdoCbl)-dependent enzymes are listed in Table 5.7. Further examples are known.[190]

TABLE 5.7 Adenosylcobalamin-dependent rearrangements.

Enzyme	Reaction catalysed	Group X $\left(\begin{smallmatrix} H & & H \\ \| & & \| \\ C-C & \rightleftharpoons & C-C \\ \| & & \| \\ X & & X \end{smallmatrix} \right)$
Gulatamate mutase	L-Glutamate → L-*threo*-β-methylaspartate	—CH(NH$_2$)COOH
Diol dehydrase	Ethylene glycol → acetaldehyde 1,2-Propanediol → propion-aldehyde	—OH
Glycerol dehydrase	Glycerol → β-hydroxypro-pionaldehyde	—OH
α-Lysine mutase	D-Lysine ⇌ 2,5-diamino hexanoic acid	—NH$_2$
Methylmalonyl-CoA mutase	2-Methylmalonyl-CoA ⇌ succinyl-CoA	—COSCoA

Several mechanisms have been put forward for these rearrangements. Abeles *et al.*[187] have suggested from studies on diol dehydrase that the Co—C bond of AdoCbl undergoes homolysis to give an adenosyl radical, which then removes a hydrogen atom from C-1 of a substrate molecule to give deoxyadenosine. The substrate radical (S·) then undergoes a rearrangement to a new radical (P·)

related to the product, which then reabstracts a hydrogen atom from the methyl group of the deoxyadenosine, regenerating the adenosyl radical.

$$S^{\cdot} \longrightarrow P^{\cdot} \xrightarrow{\text{deoxyadenosine}} PH + \text{adenosyl radical}$$

An alternative view (due to Schrauzer[187]) postulates heterolytic cleavage of the Co—C bond in AdoCbl to give cobalt(I) cobalamin and $4',5'$-anhydroadenosine as intermediates. The cobalt(I) cobalamin then reacts with the substrate to give an intermediate organocobalamin which decomposes *via* an intramolecular 1,2-hydride shift to cobalt(I) cobalamin and the product. The presence of Co(I) intermediates in the enzyme-catalysed reaction (diol dehydrase) is supported[191] by the fact that N_2O causes inhibition. This is thought to be a specific oxidant for cobalt(I) cobalamin.

Much less is known about the migration of the group X. Many model systems are available which involve such migration,[190] but it is difficult to extrapolate these conclusions to the enzyme-catalysed reaction.

A number of rearrangement reactions are also catalysed by methyl cobalamin.[190]

Studies on cobinamides and cobalamins with a range of axial ligands

Spectral studies on a range of complexes in which the nature of the axial ligand is changed show that there is dependence of the spectra on the ligand. The presence of carbanion ligands (particularly $C_2H_5^-$) with water as the other axial ligand, causes considerable changes in spectra. This latter effect may be understood in terms of an equilibrium between five- and six-coordinate species. Weak axial ligands in the cobinamide series (with H_2O as the fifth ligand) only give the six-coordinate species. Strong donors (e.g. $C_2H_5^-$) only give the five-coordinate species (with water displaced), while $C_2H_3^-$ and CH_3^- give a mixture of the two forms. Comparable changes have been seen for the cobalamin complexes, where in this case benzimidazole must be the leaving ligand for the cobalt complex to become five coordinate. Similarly, it has been shown that the coenzyme itself is an equilibrium mixture of five- and six-coordinate species with a fast rate of interconversion between these species.

The underlying chemical reason for the existence of the five-coordinate species is the well-known *trans* effect. However, while the cause is a well-known one, the effect is most unusual for cobalt(III). Five-coordinated Co(III) complexes are very rare. Another anomalous feature is the fast rate of interconversion of five- and six-coordinate species. Reactions of cobalt(III) complexes are usually much slower. In general, however, the rate of ligand exchange at cobalt in vitamin B_{12} is extremely fast.

It is noteworthy that both these features are well known for cobalt(II). Such complexes are labile and stable five-coordinate species are known. Here again we have a situation ideally designed for electron transfer, for cobalt(III) to be converted to cobalt(II). It is significant that the ligands best able to cause this situation, the carbanions, are themselves strong reducing agents.

REFERENCES

1. T. Ohnishi, *Eur. J. Biochem.*, **64**, 91 (1976).
2. M. Goldstein, E. Lauber, W. E. Blumberg, and J. Peisach, *Fed. Proc.*, **24**, 604 (1965).
3. G. R. Moore and R. J. P. Williams, *Coord. Chem. Rev.*, **18**, 125 (1976); *FEBS Letters*, **79**, 229 (1977) R. J. P. Williams, *R. I. C. Revs.*, **1**, 13 (1968).
4. B. R. James and R. J. P. Williams, *J. Chem. Soc.*, **1961**, 2007.
5. C. J. Hawkins and D. D. Perrin, *J. Chem. Soc.*, **1962**, 1351.
6. E. Stellwagen, *Nature*, **275**, 73 (1978).
7. A. H. Maki, N. Edelstein, A. Davison, and R. H. Holm, *J. Amer. Chem. Soc.*, **86**, 4580 (1964).
8. H. L. Hodges, R. A. Holwerda, and H. B. Gray, *J. Amer. Chem. Soc.*, **96**, 3132 (1974); S. Wherland and H. B. Gray, *Proc. Natl. Acad. Sci. U.S.A.*, **73**, 2951 (1976).
9. R. A. Holwerda, R. A. Read, R. A. Scott, S. Wherland, H. B. Gray, and F. Millett, *J. Amer. Chem. Soc.*, **100**, 5028 (1978).
10. R. A. Holwerda and H. B. Gray, *J. Amer. Chem. Soc.*, **97**, 6036 (1975); R. C. Rosenberg, S. Wherland, R. A. Holwerda, and H. B. Gray, *J. Amer. Chem. Soc.*, **98**, 6364 (1976); **97**, 5620 (1975).
11. T. J. Przystas and N. Sutin, *Inorg. Chem.*, **14**, 2103 (1975).
12. J. V. McArdle, K. Yocom, and H. B. Gray, *J. Amer. Chem. Soc.*, **99**, 4141 (1977).
13. J. V. McArdle, C. L. Coyle, H. B. Gray, G. S. Yoneda, and R. A. Holwerda, *J. Amer. Chem. Soc.*, **99**, 2483 (1977); M. G. Segal and A. G. Sykes, *J. Amer. Chem. Soc.*, **100**, 4585 (1978); A. G. Lappin, M. G. Segal, D. C. Weatherburn, and A. G. Sykes, *J. Amer. Chem. Soc.*, **101**, 2297 (1979); A. G. Lappin, M. G. Segal, D. C. Weatherburn, R. A. Henderson, and A. G. Sykes, *J. Amer. Chem. Soc.*, **101**, 2302 (1979).
14. F. A. Armstrong, R. A. Henderson, M. G. Segal, and A. G. Sykes, *J. C. S. Chem. Comm.*, **1978**, 1102; *J. Amer. Chem. Soc.*, **101**, 6912 (1979).
15. M. T. Wilson, C. Greenwood, M. Brunori, and E. Antonini, *Biochem. J.*, **145**, 449 (1975).
16. S. Wherland and I. Pecht, *Biochemistry*, **17**, 2585 (1978).
17. E. Oldfield and A. Allerhand, *Proc. Natl. Acad. Sci. U.S.A.*, **70**, 3531 (1973).
18. R. M. Keller, K. Wüthrich, and I. Pecht, *FEBS Letters*, **70**, 180 (1976).
19. J. C. W. Chien, H. L. Gibson, and L. C. Dickinson, *Biochemistry*, **17**, 2759 (1978).
20. J. J. Hopfield, *Proc. Natl. Acad. Sci. U.S.A.*, **71**, 3640 (1974); M. J. Potasek and J. J. Hopfield, *Proc. Natl. Acad. Sci. U.S.A.*, **74**, 221 (1977).
21. M. J. Potasek, *Science*, **201**, 151 (1978).
22. D. Dolphin (Ed.), *The Porphyrins*, Vols. 1–7, Academic Press, New York, 1978.
23. L. D. Spaulding, C. C. Chong, N. T. Yu, and R. H. Felton, *J. Amer. Chem. Soc.*, **98**, 2517 (1975).
24. F. L. Harris and D. L. Toppen, *Inorg. Chem.*, **17**, 74 (1978).
25. D. H. Chin, J. Del Gaudio, G. N. La Mar, and A. L. Balch, *J. Amer. Chem. Soc.*, **99**, 5486 (1977).

182

26. E. B. Fleischer, M. Krishnamurthy, and S. K. Cheung, *J. Amer. Chem. Soc.*, **97**, 3873 (1975).
27. S. J. Cole, G. C. Curthoys, and E. A. Magnusson, *J. Amer. Chem. Soc.*, **92**, 2991 (1970).
28. D. Gust and D. N. Neal, *J. C. S. Chem. Comm.*, **1978**, 681.
29. B. R. James, J. R. Sams, T. B. Tsui, and K. J. Reimer, *J. C. S. Chem. Comm.*, **1978**, 746.
30. D. A. Buckingham and T. B. Rauchfus, *J. C. S. Chem. Comm.*, **1978**, 705.
31. J. D. Saterlee, G. N. La Mar, and T. J. Bold, *J. Amer. Chem. Soc.*, **99**, 1088 (1977).
32. G. M. Brown, F. R. Hopf, T. J. Meyer, and D. G. Whitten, *J. Amer. Chem. Soc.*, **97**, 5385 (1975).
33. J. E. Falk and R. J. Porra, *Biochem J.*, **70**, 66 (1963).
34. F. R. Salemme, *Ann. Rev. Biochem.*, **46**, 299 (1977); R. E. Dickerson and R. Timkovich, in *The Enzymes* (Ed. P. D. Boyer), (3rd edn.) **XI**, 397 (1975).
35. R. Swanson, B. L. Trus, N. Mandel, G. Mandel, O. B. Kallai, and R. E. Dickerson, *J. Biol. Chem.*, **252**, 759 (1977); T. Takano, B. L. Trus, N. Mandel, G. Mandel, O. B. Kallai, R. Swanson, and R. E. Dickerson, *J. Biol. Chem.*, **252**, 776 (1977); N. Mandel, G. Mandel, B. L. Trus, J. Rosenberg, G. Carlson, and R. E. Dickerson, *J. Biol. Chem.*, **252**, 4619 (1977).
36. R. J. P. Williams, G. R. Moore, and P. E. Wright, in *Biological Aspects of Inorganic Chemistry* (Eds. A. W. Addison, W. R. Cullen, D. Dolphin, and B. R. James), Wiley, New York, 1977, p. 369.
37. A. Schejter, A. Lanir, I. Vig, and J. S. Cohen, *J. Biol. Chem.*, **253**, 3768 (1978).
38. N. Stauden Mayer and N. G. Siong, *Biochemistry*, **16**, 600 (1977).
39. M. T. Wilson, R. J. Ranson, P. Masiakowski, E. Czarnecka, and M. Brunori, *Eur. J. Biochem.*, **77**, 193 (1977); A. M. T. Jehanli, D. A. Stotter, and M. T. Wilson, *Eur. J. Biochem.*, **71**, 613 (1976).
40. R. R. Hantgen and H. Taniuchi, *J. Biol. Chem.*, **252**, 1367 (1977).
41. L. E. Barstow, R. S. Young, E. Yokale, J. J. Sharp, J. C. O'Brien, P. W. Berman, and H. A. Harbury, *Proc. Natl. Acad. Sci. U.S.A.*, **74**, 4248 (1977).
42. J. M. Vanderkooi, R. Landesberg, G. W. Hayden, and C. S. Owen, *Eur, J. Biochem.*, **81**, 339 (1977).
43. J. M. Venderkooi, B. Chance, and A. Waring, *FEBS Letters*, **88**, 273 (1978).
44. B. Errede and M. D. Kamen, *Biochemistry*, **17**, 1015 (1978).
45. T. Mashiko, J. -C. Marchon, D. T. Musser, C. A. Reed, M. E. Kastner, and W. R. Scheidt, *J. Amer. Chem. Soc.*, **101**, 3652 (1979). See *Inorg. Chem.*, **18**, 206 (1979).
46. M. Bruschi, C. E. Hatchikian, L. A. Golovleva, and J. Le Gall, *J. Bact.*, **129**, 30 (1977).
47. C. C. McDonald, W. D. Phillips, and J. Le Gall, *Biochemistry*, **13**, 1952 (1974).
48. C. M. Dobson, N. J. Hoyle, C. F. Geraldes, P. E. Wright, R. J. P. Williams, M. Bruschi, and J. Le Gall, *Nature*, **249**, 425 (1974).
49. K. Niki, T. Yagi, H. Inokuchi, and K. Kimura, *J. Amer. Chem. Soc.*, **101**, 3335 (1979).
50. U. Fisher and H. G. Treuper, *FEMS Microbiol, Letters*, **1**, 87 (1977).
51. K. Hon-ami and T. Oshima, *J. Biochem.* (*Tokyo*), **82**, 769 (1977).
52. C. Greenwood and D. Barber, *Inorganic Biochemistry* (Specialist Periodical Report of the Chemical Society, Ed. H. A. O. Hill), **1**, 233 (1979).
53. R. J. Almassy and R. E. Dickerson, *Proc. Natl. Acad. Sci. U.S.A.*, **75**, 2674 (1978).
54. D. B. Knaff, *Coord. Chem. Revs.*, **26**, 47 (1978).
55. F. S. Matthews, M. Levine, and P. Argos, *J. Mol. Biol.*, **64**, 449 (1972).
56. F. S. Mathews, P. Bethge and E. Czerwinski, *J. Biol. Chem.*, **254**, 1699, (1979).
57. E. Begard, P. Debey, and P. Douzou, *FEBS Letters*, **75**, 52 (1977).

58. D. H. O'Keefe, R. E. Ebel, J. A. Peterson, J. C. Maxwell, and W. S. Caughey, *Biochemistry*, **17**, 5845 (1978).
59. C. K. Chang and D. Dolphin, *J. Amer. Chem. Soc.*, **97**, 5948 (1975); J. O. Stern and J. Peisach, *J. Biol. Chem.*, **249**, 7495 (1974); S. Koch, S. G. Tang, R. H. Holm, R. B. Frankel, and J. A. Ibers, *J. Amer. Chem. Soc.*, **97**, 916 (1975); J. P. Collman, T. N. Sorrell, K. O. Hodgson, A. K. Kailshrestha, and C. E. Strause, *J. Amer. Chem. Soc.*, **99**, 5180 (1977).
60 J. P. Collman and T. N. Sorrell, *J. Amer. Chem. Soc.*, **97**, 4133 (1975).
61. H. B. Dunford and J. S. Stillman, *Coord. Chem. Revs.*, **19**, 187 (1976).
62. R. H. Haschke and J. M. Friedhoff, *Biochem. Biophys. Res. Comm.*, **80**, 1039 (1978).
63. S. Vuk-Pavlovic and Y. Siderer, *Biochem. Biophys. Res. Comm.*, **79**, 885 (1977); I. Morishama, S. Ogawa, T. Inubushi, T. Yonezawa, and T. Iizuka, *Biochemistry*, **16**, 5109 (1977).
64. D. Job, J. Ricard, and H. B. Dunford, *Arch. Biochem. Biophys.*, **179**, 95 (1977).
65. T. H. Moss, A. Ehrenberg, and A. J. Bearden, *Biochemistry*, **8**, 4159 (1969); M. J. Stillman, B. R. Hollebone, and J. S. Stillman, *Biochem. Biophys. Res. Comm.*, **72**, 554 (1976).
66. I. Morishima and S. Ogawa, *Biochemistry*, **17**, 4384 (1978).
67. M. -Y. R. Wang, B. M. Hoffman, and P. F. Hollenberg, *J. Biol. Chem.*, **252**, 6268 (1977).
68. G. Galliani, B. Rindone, and A. Marchesini, *J. C. S. Chem. Comm.*, **1976**, 782.
69. M. Kajiyoshi and F. K. Anan, *J. Biochem. (Tokyo)*, **81**, 1319 (1977).
70. E. Steiner and H. B. Dunford, *Eur. J. Biochem.*, **82**, 543 (1978); P. Jones and D. Mantle, *J. Chem. Soc. (Dalton)*, **1977**, 1849.
71. J. H. Wang, *J. Amer. Chem. Soc.*, **77**, 822, 4715 (1955); R. C. Jarnagin and J. H. Wang, *J. Amer. Chem. Soc.*, **80**, 786, 6471 (1958); H. Sigel, *Angew. Chem. Int. Ed.*, **8**, 167 (1969).
72. A. C. Melnyk, N. K. Kildahl, A. R. Rendina, and D. H. Busch, *J. Amer. Chem. Soc.*, **101**, 3232 (1979).
73. *Iron–sulphur proteins* (ed. W. Lovenberg). Vols. I, II, III, Academic Press, New York, 1973, 1977, 1977.
74. D. O. Hall, *Adv. Chem. Ser.*, Vol. 162 1976; D. O. HaLL, R. Cammack and K. K. Rao, *Chemie in unserer Zeit*, **1977**, 165.
75. R. Malkin and A. J. Bearden, *Coord. Chem. Revs.*, **28**, 1 (1979).
76. B. Bunker and E. A. Stern, *Biophys. J.*, **19**, 253 (1977).
77. E. T. Adman, J. C. Sieker, L. H. Jensen, M. Bruschi, and J. Le Gall, *J. Mol. Biol.*, **112**, 113 (1977).
78. W. Lovenberg and W. M. Williams, *Biochemistry*, **8**, 141 (1968).
79. W. D. Phillips, M. Poe, J. F. Weiker, C. C. McDonald, and W. Lovenberg, *Nature*, **227**, 574 (1970).
80. R. A. Bair and W. A. Goddard, *J. Amer. Chem. Soc.*, **100**, 5669 (1978).
81. W. A. Eaton and W. Lovenberg, *J. Amer. Chem. Soc.*, **92**, 7195 (1970).
82. T. V. Long, T. M. Loehr, J. R. Allkins, and W. N. Lovenberg, *J. Amer. Chem. Soc.*, **93**, 1909 (1971).
83. J. C. Rivoal, B. Briat, R. Cammack, D. O. Hall, K. K. Rao, I. N. Douglas, and A. J. Thomson, *Biochim. Biophys. Acta*, **493**, 121 (1977).
84. D. W. Cushman, R. L. Tsai, and I. C. Gunsalus, *Biochem. Biophys. Res. Comm.*, **26**, 577 (1967).
85. E. Münck, P. G. Debrunner, J. C. M. Tsibris, and I. C. Gunsalus, *Biochemistry*, **11**, 855 (1972).
86. K. Suzuki and T. Kimura, *Biochem. Biophys. Res. Comm.*, **19**, 340 (1963).
87. T. Kimura, *Structure and Bonding*, Vol. 5, Springer-Verlag, Berlin, 1969, p. 1.

184

88. K. Ogawa, T. Tsukihara, H. Tahara, Y. Katsube, Y. Matsura, N. Tanaka, M. Kukudo, K. Wada, and H. Matsubara, *J. Biochem.* (*Tokyo*), **81**, 529 (1977).

89. A. Kunita, M. Koshibe, Y. Nishikawa, K. Fukuyama, T. Tsukihara, Y. Katsube, Y. Matsura, N. Tanaka, M. Kakudo, T. Hase, and H. Matsubara, *J. Biochem.* (*Tokyo*), **84**, 989 (1978).

90. G. Palmer, W. R. Dunham, J. A. Fee, R. H. Sands, T. Iizuka, and T. Yonetani, *Biochim. Biophys. Acta*, **245**, 201 (1971).

91. J. Fritz, R. Anderson, J. Fee, G. Palmer, R. H. Sands, J. C. M. Tsibris, I. C. Gunsalus, W. H. Orme-Johnson, and H. Beinert, *Biochim. Biophys. Acta*, **253**, 110 (1971).

92. K. K. Rao, R. Cammack, D. O. Hall, and C. E. Johnson, *Biochem. J.*, **122**, 257 (1971); W. R. Dunham, A. J. Bearden, I. T. Salmeen, G. Palmer, R. H. Sands, W. H. Orme-Johnson, and H. Beinert, *Biochim. Biophys. Acta*, **253**, 134 (1971).

93. W. R. Dunham, G. Palmer, R. H. Sands, and A. J. Bearden, *Biochim, Biophys. Acta*, **253**, 373 (1971); W. R. Eaton, G. Palmer, J. A. Fee, T. Kimura, and W. Lovenberg, *Proc. Natl. Acad. Sci. U.S.A.*, **68**, 3015 (1971).

94. J. W. Boon and C. H. MacGillavry, *Rec. Trav. Chim.*, **61**, 910 (1942).

95. R. Mason and J. A. Zubieta, *Angew. Chem. Int. Ed.*, **12**, 390 (1973); W. M. Scovell and T. G. Spiro, *Inorg. Chem.*, **13**, 304 (1974); J. J. Mayerle, R. B. Frankel, R. H. Holm, J. A. Ibers, W. D. Phillips, and J. F. Weiher, *Proc. Natl. Acad. Sci. U.S.A.*, **70**, 2429 (1973); D. Coucouvanis, D. Swenson, P. Stremple, and N. C. Balzziger, *J. Amer. Chem. Soc.*, **101**, 3394 (1979).

96. R. H. Holm, *Acc. Chem. Res.*, **10**, 427 (1977).

97. J. J. Mayerle, S. E. Denmark, B. V. DePamphilis, J. A. Ibers, and R. H. Holm, *J. Amer. Chem. Soc.*, **97**, 1033 (1975).

98. J. G. Norman, B. J. Kalbacher, and S. C. Jackels, *J. C. S. Chem. Comm.*, **1978**, 1027.

99. J. Rawlings, S. Wherland, and H. B. Gray, *J. Amer. Chem. Soc.*, **99**, 1968 (1977).

100. S. W. Carter, J. Kraut, S. T. Freer, R. A. Alden, L. C. Sieker, E. Adman, and L. H. Jensen, *Proc. Natl. Acad. Sci. U.S.A.*, **69**, 3526 (1972).

101. L. C. Sieker, E. Adman, and L. H. Jensen, *Nature*, **235**, 40 (1972).

102. W. D. Phillips, M. Poe, C. C. McDonald, and R. G. Bartsch. *Proc. Natl. Acad. Sci. U.S.A.*, **67**, 682 (1970).

103. R. Cammack, *Biochem. Biophys. Res. Comm.*, **54**, 548 (1973).

104. W. V. Sweeney, A. J. Bearden, and J. C. Rabinowitz, *Biochem. Biophys. Res. Comm.*, **59**, 188 (1974).

105. C. W. Carter, *J. Biol. Chem.*, **252**, 7802, (1977).

106. T. Herskovitz, B. A. Averill, R. H. Holm, J. A. Ibers, W. D. Phillips and J. F. Weiher, *Proc. Natl. Acad. Sci. U.S.A.*, **69**, 2437 (1972).

107. L. Que, R. H. Holm, and L. E. Mortensen, *J. Amer. Chem. Soc.*, **97**, 463 (1975).

108. W. O. Gillum, L. E. Mortensen, J. S. Chen, and R. H. Holm, *J. Amer. Chem. Soc.*, **99**, 584 (1977).

109. M. A. Bobrik, K. O. Hodgson, and R. H. Holm, *Inorg. Chem.*, **16**, 1851 (1977); G. B. Wong, M. A. Bobrik, and R. H. Holm, *Inorg. Chem.*, **16**, 578 (1977).

110. R. H. Holm, B. A. Averill, T. Herskovitz, R. B. Frankel, H. B. Gray, O. Siiman, and F. J. Grunthaner, *J. Amer. Chem. Soc.*, **96**, 2644 (1974).

111. R. Maskiewicz and T. C. Bruice, *Biochemistry*, **16**, 3024 (1977).

112. M. W. W. Adams, S. G. Reeves, D. O. Hall, G. Christou, B. Ridge, and H. N. Ryden, *Biochem. Biophys. Res. Comm.*, **79**, 1184 (1977).

113. R. Maskiewicz and T. C. Bruice, *J. C. S. Chem. Comm.*, **1978**, 703.

114. P. J. Stephens, A. J. Thomson, T. A. Keiderling, J. Rawlings, K. K. Rao, and D. O. Hall, *Proc. Natl. Acad. Sci. U.S.A.*, **75**, 5273 (1978); *Biochemistry*, **17**, 4770 (1978).

115. S. P. J. Albracht and J. Subramanian, *Biochim. Biophys. Acta*, **462**, 36 (1977).

116. E. M. Meijer, R. Wever, and A. H. Stouthamer, *Eur. J. Biochem.*, **81**, 261 (1977).

185

117. F. J. Ruzicka and H. Beinert, *J. Biol. Chem.*, **252**, 8440 (1977).
118. R. Yoshida, K. Hori, M. Fujiwara, Y. Saeki, H. Kagamiyama, and M. Nozaki, *Biochemistry*, **15**, 4048 (1976).
119. Y. Tatsuno, Y. Saeki, M. Iwaki, T. Yagi, M. Nozaki, T. Kitagawa, and S. Otsuka, *J. Amer. Chem. Soc.*, **100**, 4614 (1978).
120. W. E. Blumberg and J. Peisach, *Ann. N. Y. Acad. Sci.*, **222**, 539 (1973); L. Que, J. D. Lipscomb, R. Zimmerman, E. Münck, N. R. Orme-Johnson, and W. H. Orme-Johnson, *Biochim. Biophys. Acta*, **452**, 320 (1976).
121. L. J. M. Spaapen, J. F. G. Vliegenthart, and J. Boldingh, *Biochim. Biophys. Acta*, **488**, 517 (1977); E. K. Pistorius, B. Axelrod, and G. Palmer, *J. Biol. Chem.*, **251**, 7144 (1976).
122. R. T. Ruettinger, G. R. Griffith, and M. J. Coon, *Arch. Biochem. Biophys.*, **183**, 528 (1977).
123. B. -M. Sjoberg, P. Reichard, A. Graslund, and A. Ehrenberg, *J. Biol. Chem.*, **252**, 536 (1977).
124. J. Peisach, P. Aisen, and W. E. Blumberg (Eds.), *The Biochemistry of Copper*, Academic Press, New York, 1966.
125. R. Österberg, *Coord. Chem. Revs.*, **12**, 309 (1974).
126. E. Ulrich and J. L. Markley, *Coord. Chem. Revs.*, **27**, 109 (1978).
127. J. A. Fee, *Structure and Bonding*, **23**, 1 (1975).
128. *Adv. Chem. Ser.*, **162**, 145, 173, 179, 263 (1977).
129. P. M. Colman, H. C. Freeman, J. M. Guss, M. Murata, V. A. Norris, J. A. M. Ramshaw, and M. P. Venkatappa, *Nature*, **272**, 319 (1978).
130. E. T. Adman, R. E. Stenkamp, L. C. Sieker, and L. H. Jensen, *J. Mol. Biol.*, **123**, 35 (1978).
131. T. D. Tullius, P. Frank, and K. O. Hodgson, *Proc. Natl. Acad. Sci. U.S.A.*, **75**, 4069 (1978).
132. C. Bergman, E. K. Gandvik, P. O. Nyman, and L. Strid, *Biochem. Biophys. Res. Comm.*, **78**, 1052 (1977). See *J. Biol. Chem.*, **254**, 432 (1979).
133. V. Sakaguchi and A. W. Addison, *J. Amer. Chem. Soc.*, **99**, 5189 (1977); H. Yokoi and A. W. Addison, *Inorg. Chem.*, **16**, 1341 (1977); A. S. Brill and G. F. Bryce, *J. Chem. Phys.*, **48**, 4398 (1968).
134. E. I. Solomon, J. W. Hare, and H. B. Gray, *Proc. Natl. Acad. Sci. U.S.A.*, **71**, 4760 (1976); E. I. Solomon, J. Rawlings, D. R. McMillin, P. J. Stephens, and H. B. Gray, *J. Amer. Chem. Soc.*, **98**, 8046 (1976).
135. G. S. Patterson and R. H. Holm, *Bioinorg. Chem.*, **4**, 257 (1975).
136. H. A. O. Hill and B. E. Smith, *Biochem. Biophys. Res. Comm.*, **81**, 1201 (1978).
137. J. K. Markley, E. E. Ulrich, S. P. Berg, and D. W. Krogmann, *Biochemistry*, **14**, 4428 (1975).
138. D. L. Tennent and D. R. McMillin, *J. Amer. Chem. Soc.*, **101**, 2307 (1979).
139. N. S. Ferris, W. H. Woodruff, D. B. Rorabacher, T. E. Jones, and L. A. Ochrymowycz, *J. Amer. Chem. Soc.*, **100**, 5939 (1978).
140. E. I. Solomon, P. J. Clendening, H. B. Gray, and F. J. Grunthaner, *J. Amer. Chem. Soc.*, **97**, 3878 (1975); S. Larsson, *J. Amer. Chem. Soc.*, **99**, 7708 (1977); M. Thompson, J. Whelan, D. J. Zemon, B. Bosnich, E. I. Solomon, and H. B. Gray, *J. Amer. Chem. Soc.*, **101**, 2483 (1979).
141. O. Farver, M. Goldberg, S. Wherland, and I. Pecht, *Proc. Natl. Acad. Sci. U.S.A.*, **75**, 5245 (1978).
142. L. E. Andréasson, R. Brändén, B. G. Malmström, and T. Vänngard, *FEBS Letters*, **32**, 187 (1973).
143. T. Manabe, N. Manabe, K. Hiromi, and H. Hatano, *FEBS Letters*, **23**, 268 (1972).
144. R. Brändén and J. Deinum, *FEBS Letters*, **73**, 144 (1977).
145. L. E. Andréasson, R. Brändén, and B. Reinhammar, *Biochim. Biophys. Acta*, **438**,

370 (1976).

146. B. Mondovi, M. T. Graziani, W. B. Mim, R. Oltzik, and J. Peisach, *Biochemistry*, **16**, 4198 (1977).

147. J. H. Dawson, D. M. Dooley, and H. B. Gray, *Proc. Natl. Acad. Sci. U.S.A.*, **75**, 4078 (1978); *J. Amer. Chem. Soc.*, **101**, 5046 (1979).

148. Y. Henry and J. Peisach, *J. Biol. Chem.*, **253**, 7751 (1978).

149. Y. Agnus, R. Louis, and R. Weiss, *J. Amer. Chem. Soc.*, **101**, 3381 (1979).

150. R. E. Weiner, M. J. Ettinger, and D. J. Kosman, *Biochemistry*, **16**, 1602 (1977).

151. D. J. Kosman, M. J. Ettinger, R. D. Bereman, and R. S. Giordano, *Biochemistry*, **16**, 1597 (1977).

152. J. Grant, I. Kelly, P. Knowles J. Olsson, and G. Pettersson, *Biochem. Biophys. Res. Comm.*, **83**, 1216 (1978).

153. M. Erecinska and D. F. Wilson, *Arch. Biochem. Biophys.*, **188**, 1 (1978).

154. D. F. Wilson, M. Erecinska, and E. S. Brocklehurst, *Arch. Biochem. Biophys.*, **151**, 180 (1972).

155. T. Brittain, J. Springall, C. Greenwood, and A. Thomson, *Biochem. J.*, **159**, 811 (1976); A. J. Thomson, T. Brittain, C. Greenwood, and J. P. Springall, *Biochem. J.*, **165**, 327 (1977); G. T. Babcock, L. E. Vickery, and G. Palmer, *J. Biol. Chem.*, **251**, 7907 (1976).

156. M. F. Tweedle, L. J. Wilson, L. Garcia-Iniguez, G. T. Babcock, and G. Palmer, *J. Biol. Chem.*, **253**, 8065 (1978) (and earlier papers in the series).

157. T. H. Moss, E. Shapiro, T. E. King, H. Beinert, and C. Hartzell, *J. Biol. Chem.*, **253**, 8072 (1978).

158. R. H. Petty and L. J. Wilson, *J. C. S. Chem. Comm.*, **1978**, 483.

159. H. L. Van Camp, Y. H. Wei, C. P. Scholes, and T. E. King, *Biochim. Biophys. Acta*, **537**, 238 (1978).

160. D. F. Wilson and Y. Miyata, *Biochim. Biophys. Acta*, **461**, 218 (1977); R. Wever, J. H. Van Drooge, A. O. Muijsers, E. A. Barber, and B. F. van Gelder, *Eur. J. Biochem.*, **73**, 149 (1977).

161. Y. Orii and T. E. King, *J. Biol. Chem.*, **251**, 7487 (1976).

162. B. Chance, C. Saronio, and J. S. Leigh, *J. Biol. Chem.*, **250**, 9226 (1975); B. Chance, C. Saronio, A. Waring, and J. S. Leigh, *Biochim. Biophys. Acta*, **503**, 37 (1978).

163. E. Antonini, M. Brunori, A. Colosimo, C. Greenwood, and M. T. Wilson, *Proc. Natl. Acad. Sci. U.S.A.*, **74**, 3128 (1977).

164. I. Fridovich, *Acc. Chem. Res.*, **10**, 321 (1972).

165. S. J. Lippard, A. R. Burger, K. Ugurbil, J. S. Valentine, and M. W. Pantoliano, *Adv. Chem. Ser.* **162**, 251 (1977).

165a. *Superoxide and Superoxide Dismutases* (Eds. A. M. Michelson, J. M. McCord, and I. Fridovich), Academic Press, London, 1977.

166. J. S. Richardson, K. A. Thomas, B. H. Rubin, and D. C. Richardson, *Proc. Natl. Acad. Sci. U.S.A.*, **72**, 1349 (1975).

167. A. E. G. Cass, H. A. O. Hill, B. E. Smith, J. V. Bannister, and W. H. Bannister, *Biochemistry*, **16**, 3061 (1977).

168. A. Desideri, M. Cerdonio, F. Mogno, S. Vitale, L. Calabrese, D. Cocco, and G. Rotilio, *FEBS Letters*, **89**, 83 (1978).

169. A. Rigo, P. Viglino, L. Calabrese, D. Cocco, and G. Rotilio, *Biochem. J.*, **161**, 27, 31 (1977); A. Rigo, P. Viglino, M. Bonori, D. Cocco, L. Calabrese, and G. Rotilio, *Biochem. J.*, **169**, 277 (1978); A. Rigo, F. Marmocchi, D. Cocco, P. Viglino, and G. Rotilio, *Biochemistry*, **17**, 534 (1978).

170. S. J. Lippard, A. Burger, K. Ugurbil, M. Pantoliano, and J. S. Valentine, *Biochemistry*, **16**, 1136 (1977).

171. G. Kolks, C. R. Frihart, H. N. Rabinowitz, and S. J. Lippard, *J. Amer. Chem. Soc.*, **98**, 5720 (1976); P. K. Coughlin, J. C. Dewan, S. J. Lippard, E. Watanabe, and

J. M. Lehn, *J. Amer. Chem. Soc.*, 101, 266 (1979).

172. E. I. Stiefel, *Progr. Inorg. Chem.*, 22, 1 (1977); R. C. Bray and J. C. Swann, *Structure and Bonding*, 11, 107 (1972); R. A. D. Wentworth, *Coord. Chem. Revs.*, 18, 1 (1976); E. I. Stiefel, W. E. Newton, G. D. Watt, K. L. Hadfield, and W. A. Bulen, *Adv. Chem. Ser.*, 162, 353 (1977).

173. P. C. H. Mitchell, *Coord. Chem. Revs.*, 1, 315 (1966); *Quart. Rev.*, 1966, 103; J. T. Spence, *Coord. Chem. Revs.*, 4, 475 (1969); B. Spivack and Z. Dori, *Coord. Chem. Revs.*, 17, 99 (1975).

174. R. C. Bray, in *The Enzymes* (Ed. P. Boyer), (3rd edn.), Vol. XII, Academic Press, New York, 1975, 299.

175. R. C. Bray and P. J. Knowles, *Proc. Roy. Soc.*, A302, 351 (1968); L. S. Meriwether, W. F. Marzluff, and W. G. Hodgson, *Nature*, 212, 465 (1966); S. Gutteridge, S. J. Tanner, and R. C. Bray, *Biochem. J.*, 175, 869 (1978).

176. R. C. Bray, P. F. Knowles, and L. S. Meriweather, in *Magnetic Resonance in Biological Systems* (Eds. A. Ehrenberg, B. G. Malmstrom, and F. Vanngard), Pergamon Press, Oxford, 1967, p. 249.

177. V. Massey, H. Komai, G. Palmer, and G. B. Elion, *J. Biol. Chem.*, 245, 2837 (1970).

178. S. Gutteridge, S. J. Tanner, and R. C. Bray, *Biochem. J.*, 175, 887 (1978).

179. N. Pariyadath, W. E. Newton, and E. I. Stiefel, *J. Amer. Chem. Soc.*, 98, 5388 (1976).

180. M. P. Coughlan, *FEBS Letters*, 81, 1 (1977); D. Edmondson, D. P. Ballou, G. Palmer, and V. Massey, *J. Biol. Chem.*, 248, 6135 (1973); J. S. Olson, D. P. Ballou, G. Palmer, and V. Massey, *J. Biol. Chem.*, 249, 4363 (1974).

181. V. Massey, in *Iron–Sulphur Proteins* (Ed. W. Lovenberg), Vol. I, Academic Press, 1973, p. 301.

182. H. I. Cohen, I. Fridovich, and K. V. Rajagopalan, *J. Biol. Chem.*, 246, 374 (1971).

183. A. E. Dennard and R. J. P. Williams, in *Transition Metal Chemistry* (Ed. R. L. Carlin), 2, 158 (1966).

184. J. L. Johnson, H. J. Cohen, and K. V. Rajagopalan, *J. Biol. Chem.*, 249, 5046 (1974).

185. J. A. Pateman, D. J. Cove, B. M. Rever, and D. B. Roberts, *Nature*, 201, 58 (1964); K.-Y. Lee, S.-S. Pan, R. Erickson, and A. Nason, *J. Biol. Chem.*, 249, 3941 (1974).

186. P. G. Lenhert and D. C. Hodgkin, *Nature*, 192, 937 (1961); D. C. Hodgkin, *Proc. Roy. Soc.*, A288, 294 (1965).

187. A. I. Scott, *Acc. Chem. Res.*, 11, 29 (1978); R. H. Abeles and D. Dolphin, *Acc. Chem. Res.*, 9, 114 (1976); B. M. Babior, *Acc. Chem. Res.*, 8, 376 (1975); D. G. Brown, *Progr. Inorg. Chem.*, 18, 177 (1973); J. M. Pratt, *Inorganic Chemistry of Vitamin B$_{12}$*, Academic Press, 1972; J. M. Wood and D. G. Brown, *Structure and Bonding*, 11, 47 (1972); G. N. Schrauzer, *Angew. Chem. Int. Ed.*, 16, 233 (1977); H. P. C. Hogenkamp and G. N. Sands, *Structure and Bonding*, 20, 23 (1974).

188. P. G. Lenhert, *Chem. Comm.*, 1967, 980.

189. L. G. Marzilli et al, *J. Amer. Chem. Soc.*, 101, 6754 (1979).

190. B. T. Golding and G. J. Leigh, *Inorg. Biochem.*, Specialist Periodical Report (Ed.H. A. O. Hill), 1, 77 (1979).

191. R. N. Katz, T. M. Vickrey, and G. N. Schrauzer, *Angew. Chem. Int. Ed.*, 15, 542 (1976).

Not added in proof

EXAFS data are now available on sulphite oxidase and xanthine oxidase (*J. Amer. Chem. Soc.*, 101, 2772, 2776 (1979)), iron–sulphur proteins and models (*ibid*, 101, 5624 (1979)), and cytochrome *c* (*Nature*, 277, 150, (1979)). The persulphide molybdenum cluster $[Mo_3S(S_2)_6]^{2-}$ undergoes sulphur abstraction by cyanide ion to give $[Mo_3S_4(CN)_9]$.$^{S-}$ (*Angew. Chem. Int. Ed.*, 19, 72 (1980)).

CHAPTER SIX

NITROGEN FIXATION AND THE NITROGEN CYCLE

The element nitrogen is a constituent of many naturally occurring compounds. These complex molecules are built up from smaller organic molecules, while the original source of the element lies in simple inorganic compounds. The 'Nitrogen Cycle' represents the transformation of inorganic nitrogen to organic nitrogen, together with the reverse degradation process. In this cycle, soil ammonia is converted to nitrate, *via* a number of intermediates, by the action of microorganisms in the process of nitrification. Nitrate is assimilated by plants and built up into organic molecules. Higher species feeding on these plants convert such molecules into more complex compounds. Ultimately, on the death and decay of living matter, these compounds are broken down, leading to the reformation of ammonia. There is also the process of denitrification in which nitrates are reduced by microorganisms; in this some dinitrogen is lost to the atmosphere, but in turn atmospheric dinitrogen is 'fixed', that is converted to ammonia, by various bacteria.

The nitrogen cycle is of vital importance in agriculture in supplying soluble nitrates. Unless nitrates are extensively added as fertilizers the important step is that of nitrogen fixation. Dinitrogen may be fixed industrially by a number of processes, of which the Haber process is most important, but it is obviously of great interest to investigate the mechanism of biological fixation of dinitrogen partly in the hope of establishing new commercially viable processes and also for an enhanced understanding of the fundamental chemical and biochemical processes involved. For at present man can only do with great difficulty under extreme conditions what Nature does apparently quite readily under mild aqueous conditions with no extremes of pH and temperature. The particular aspects of the problem of interest to the inorganic chemist lie not only in the general difficulties associated with the reactivity of what has always been traditionally regarded as an inactive molecule, but also in the fact that the dinitrogen reducing species in nature, the enzyme nitrogenase, requires the presence of the transition metals molybdenum and iron. This led investigators in recent years to postulate schemes for the reduction of dinitrogen to ammonia in which the nitrogen molecule is coordinated to a transition metal ion. This was stimulated by reports on the preparation of dinitrogen complexes of a number of transition metals, and, more recently by the demonstration of the reduction of coordinated dinitrogen.

While less dramatic than the area of nitrogen fixation, there are other problems in the nitrogen cycle that should be of interest to inorganic chemists. A number of intermediates must lie between ammonia and nitrate, in both nitrification and denitrification pathways. Hydroxylamine and nitrite (with nitrogen in formal oxidation states $-I$ and $+III$ respectively) are now well established as intermediates, but the nature of the species with nitrogen in the oxidation state $+I$ is less certain (assuming that changes of two in oxidation states are occurring). The chemistry of possible intermediate compounds is not very well established, particularly under the pH conditions of biological processes, and much remains to be done. Some of the biochemical processes involved in these oxidation and reduction reactions are now being investigated at the molecular level. Thus a hydroxylamine oxidase has been isolated from *Nitrosomonas*: this is an electron-transfer particle containing cytochromes. Again, in denitrification, the reduction of nitrate to nitrite is dependent on the presence of molybdenum, with the metal undergoing transition between the oxidation states VI and V. Here again is a potentially useful area of overlap between biochemistry and inorganic chemistry.

In this chapter, we will consider recent developments in the biochemistry and inorganic chemistry of both nitrogen fixation and, to a lesser extent, other parts of the nitrogen cycle. In each case a summary of the relevant biochemistry will be given first.

THE BIOCHEMISTRY OF NITROGEN FIXATION[1]

Nitrogen-fixing microorganisms

These fall into two main groups.

(a) Free bacteria (asymbiotic). A well-known example is *Azotobacter*, but this is atypical in one sense in that this is an aerobic microbe, one requiring oxygen. A very much larger number of anaerobic species fix dinitrogen in these absence of oxygen, including *Clostridium pasteurianum*, on which many of the earlier studies were carried out. A third type of free bacteria which fixes nitrogen is the facultative group, those that have the property of being able to grow aerobically or anaerobically. These can only fix nitrogen when growing in the absence of oxygen as their biochemistry will depend upon the conditions. This general dependence upon the absence of oxygen seems intuitively reasonable as the process concerned is a reductive one.

(b) Symbiotic microorganisms. These fix dinitrogen in association with plants, e.g. the bacterium *Rhizobium*, which is associated with the nodules on the roots of leguminous plants. In this case, the microorganism does not function until it has become a degenerate form inside the root nodule. Other examples include the lichens, where there is a combination of a fungus and nitrogen-fixing blue–green algae. In general much less progress has been made with the study of nitrogen fixation by the symbiotic bacteria. One common feature of these root nodule systems is the presence of a hemoglobin-like protein which has been isolated and purified. This has been named leghemoglobin and contains one atom of iron per molecule. There

TABLE 6.1 Some biological nitrogen-fixing systems

Asymbiotic	Symbiotic

(a) Anaerobic
   C. pasteurianum
   Chromatium
   Chlorobium (photosynthetic)
   Methanobacterium
   Desulphovibrio sp.
(b) Aerobic
   Azotobacter
   Anabena (photosynthetic)
   Azotomonas
   Beijerinckia
   Derxia
   Nocardia
   Pseudomonas
(c) Facultative
   B. polymyxa
   Klebsiella sp.
   Rhodospirillum (photosynthetic)
	(a) Root nodules
	   legumes, *Rhizobium* and
	   leguminous plant
	   non-legumes, microbe and tree
	   or shrub
	(b) Leaf nodules
	   bacteria and leaf of *Psychotria*
	   emetica
	(c) Lichens
	   Nostic, *Anabena*, *Tolypothrix*
	   and fungus
	(d) Mycorrhiza of pines
	   fungus and tree

appears to be a direct correlation between the presence of leghemoglobin and nitrogen-fixing properties.

Examples of both types are listed in Table 6.1.

Purification, analysis, and properties of nitrogenase

The six-electron reduction of dinitrogen to amonia is catalysed by the enzyme nitrogenase in an ATP-dependent reaction. It has long been postulated that reduction occurs in stages giving diimine (NH=NH) and hydrazine as intermediates. Only recently however has evidence been obtained for partially reduced nitrogen intermediates.[2] Hydrazine has been detected on treating the enzyme with acid or alkali while functioning under dinitrogen. This does not necessarily imply that hydrazine is an intermediate, as further chemical hydrolysis on treatment with acid or alkali may have contributed to its formation, but the results are of considerable interest.

ATP is hydrolysed during the reduction of dinitrogen and the enzyme also catalyses the reduction of H^+ to dihydrogen. This process always competes with the reduction of dinitrogen and in the absence of substrate all six electrons are utilized in this way. The physiological electron donors to nitrogenase are ferredoxin and flavodoxin. It is noteworthy that ferredoxin was first isolated during studies on dinitrogen fixation. Sodium dithionite is usually used as the electron donor in *in vitro* studies, while ATP is generated by creatine kinase, creatine phosphate, and ADP. Two moles of ATP are consumed for each electron transferred to the substrate. In blue–green algae dinitrogen fixation is linked to photosynthesis, the requirements for reducing power and ATP being met photochemically.

Nitrogenase has been isolated from some twenty sources, and at the present time purified from five of these: *Rhizobium lupini, R. japonicum, K. pneumoniae, C. pasteurianum*, and *A. vinelandii*. It is made up of two proteins, one of which contains iron (the Fe protein), while the other contains both molybdenum and iron (MoFe protein). Fe and MoFe proteins from different sources can be combined to give active preparations. Both proteins are sensitive to dioxygen, particularly the Fe protein. It is interesting that such sensitivity should be found in an aerobic microorganism. Postgate[3] has suggested that *Azotobacter* excludes dioxygen but not dinitrogen from nitrogenase as a result of its high energy requirements which utilize the dioxygen in other reactions.

At present, there is no agreement on the precise metal content of these proteins. Crystalline MoFe protein from *A. vinelandii* contains 2Mo, 34–38 Fe, and 26–28 S^2 per mole (Mol. wt. $\sim$ 270,000), while clostridial MoFe protein contains 2 Mo, 22–24 Fe, and 22–24 S^{2-} per mole (Mol. wt. 220,000). It appears reasonable that equal numbers of Fe and S^{2-} are present. This protein is tetrameric, containing two each of two types of subunit. The four subunits from *A. vinelandii* MoFe protein are arranged in the corners of a square 90 Å × 90 Å. The Fe protein is much smaller having a molecular weight in the range 55,000–70,000, made up of two equal subunits, and contains 4 Fe and 4 S^{2-} present as a Fe_4S_4 cluster.

Divalent cations are also necessary. ATP must bind Mg^{2+} before it can interact with nitrogenase. It is accepted that the Fe protein acts as a store of electrons and that the MoFe protein binds dinitrogen, probably at molybdenum. The reduced Fe protein binds MgATP, which induces a conformational change, after which it binds to the MoFe protein. The role of MgATP is more complex than this, as immunological studies suggest that it is also bound to the MoFe protein, during turnover only.[4]

Metal clusters in nitrogenase

It is clear that many of the metal atoms and sulphide groups in nitrogenase exist in clusters. All four Fe and S^{2-} in the Fe protein may be extruded as a Fe_4S_4 cluster. The MoFe protein probably contains several clusters, both Fe_4S_4 clusters and more complex ones in which molybdenum replaces iron. An iron- and molybdenum-containing compound (termed the iron–molybdenum cofactor) has been isolated from the MoFe protein from three sources,[5] and has a composition of Mo, Fe, and S^{2-} in the ratio 1:8:6. Mössbauer and EPR data suggest that both the native protein and the cofactor contain one $S = 3/2$ centre per Mo and that each centre contains about six iron atoms in a spin-coupled structure.[6] Furthermore the EPR spectrum of the MoFe protein in *N*-methylformamide can be interpreted as the sum of the spectra of the Fe–Mo cofactor and Fe_4S_4 clusters. This cofactor is distinct from the molybdenum cofactor which activates nitrate reductase in *Neurospora crassa*.

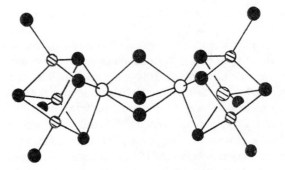

Figure 6.1 Arrangement of the central atoms in the [Fe$_6$Mo$_2$S$_8$(SR)$_9$]$^{3-}$ cluster. Mo = O, Fe = ◐, S = ●.

EXAFS studies[7] on the cofactor and the MoFe protein lead to the conclusion that the Mo in the MoFe protein is bound by three or four sulphur atoms at 2.35 Å and one or two sulphur atoms at 2.49 Å, similar to the situation found for the cofactor. This shows that the iron–molybdenum cofactor has been successfully extracted. This implies further that the MoFe protein contains up to eight Fe$_4$S$_4$ clusters in addition to the cofactor.

Some light is thrown on the possible arrangement of Mo and Fe in the cofactor by structural studies[8] on some Mo/Fe/S clusters which have been synthesized quite readily under anaerobic conditions by the reaction of (R$_4$N)$_2$MoS$_4$, FeCl$_3$, and a thiol in methanol. Each molybdenum atom is involved in a MoFe$_3$S$_4$ cluster and the two molybdenum atoms are bridged by three sulphur atoms (Figure 6.1). Stoichiometries of the type [Fe$_6$Mo$_2$S$_8$(SR)$_9$]$^{3-}$ and [Fe$_6$Mo$_2$S$_9$(SR)$_8$]$^{3-}$ have been established. The Mo—Fe, Mo—S(cube), and Mo—SR distances in these compounds compare favourably with the EXAFS data on nitrogenase. It appears likely, therefore, that the MoFe$_3$S$_4$ cube is present in nitrogenase, but at this stage it cannot be said that the whole sulphur-bridged double cube [Fe$_6$Mo$_2$S$_9$(SR)$_8$]$^{3-}$ is so present. Nevertheless the syntheses of these compounds represents an important advance in providing analogues of molybdenum sites in nitrogenase.

Binding of other substrates

In addition to catalysing the reduction of N$_2$ and H$_3$O$^+$, nitrogenase will catalyse the reduction of a range of other substrates, shown in Table 6.2. Many of these are triply bonded molecules. It is noteworthy that acetylene is reduced to ethylene and not ethane. If this is carried out in D$_2$O the product is always *cis*-1,2-deuteroethylene. Acetylene appears to be the most active substrate for the enzyme as it alone can compete 100% effectively with H$_3$O$^+$, as no dihydrogen is produced when acetylene is being reduced. The reduction of acetylene is commonly used as an assay for nitrogenase.

Azide undergoes a two-electron reduction to N$_2$ and NH$_3$. The dinitrogen thus

TABLE 6.2 The reduction of various substrates by nitrogenase

Substrate	Product
N_2	NH_3
N_2O	$N_2 + H_2O$
N_3^-	$NH_3 + N_2$
CN^-	$CH_4 + NH_3$
$CH{\equiv}CH$	$CH_2{=}CH_2$
$CH_3C{\equiv}CH$	$CH_3CH{=}CH_2$
CH_3NC	$CH_3NH_2 + CH_4$

produced is not reduced further, implying that it is not in the correct bonding position for reduction.

The result for the reduction of methyl isocyanide (CH_3NC) is interesting. It is usual in the chemical reduction of this species to obtain dimethylamine as product. This is not the product of enzymatic reduction. However, it has been shown that the reduction of isocyanide coordinated to a transition metal ion does lead to the formation of methane as a major product. This suggests that a transition metal may be directly involved in the biological reduction. Carbon monoxide inhibits the reduction of dinitrogen but not the evolution of dihydrogen, while chelating ligands inhibit both activities. This led to the suggestion that one metal is the centre of reducing power while the other binds and activates the nitrogen molecule.

These substrates can all bind side-on to a metal (*via* π bonding), and, with the exception of acetylene, can also bind end-on, *via* σ bonding. Side-on binding of acetylene would account for the *cis* addition of deuterium to acetylene. Nevertheless it appears unlikely that they all bind to the same site on nitrogenase, although their sites do appear to be on the MoFe protein. The presence of multiple sites is suggested by the non-competitive nature of the inhibition shown by some of these compounds and by the non-reduction of the dinitrogen produced from N_2O and N_3^- (Table 6.2). In summary it appears that CN^-, N_3^-, and CH_3NC bind at one site, which is different from the C_2H_2 site, the H_3O^+ site, and the N_2 site.

Redox properties and kinetics

Electron transfer in nitrogenase, particularly in the MoFe protein, is a complex process because of the number of redox-active metal clusters available to receive and transfer electrons. The relative redox potentials of these centres will determine the distribution of electrons in the overall system at equilibrium. Such redox potentials may vary with pH or temperature, and with the binding of various substrates and inhibitors. This will result in a redistribution of electrons. Some insight into these possibilities has come from EPR and Mössbauer spectroscopy, although the former technique has been of especial value in characterizing intermediates at low concentration occurring during enzyme turnover, and which are undetectable by Mössbauer techniques.

During nitrogenase activity, MgATP binds to the Fe protein resulting in a lowering of its potential from $-290\,mV$ to $-400\,mV$. This MgATP–Fe protein then binds to the MoFe protein with transfer of electrons to the MoFe protein and hydrolysis of the MgATP. It appears that this hydrolysis reaction is linked with electron transfer, as the pre-steady state burst of MgATP hydrolysis occurs with the same time constant as electron transfer between the Fe and MoFe proteins.[9] It is difficult to envisage a mechanism for this process, but it was noted earlier that the MgATP is also bound to the MoFe protein during turnover, and possibly it may serve to bridge the two proteins to allow electron transfer to take place.

The Fe proteins from different sources have very similar EPR spectra and show several unusual features. They have very anisotropic line widths and only 0.2–0.45 electrons in the integrated spectra, even though the protein is capable of accepting two electrons.[10] This has been explained by assuming that one of the two electrons accepted is located at a rapidly relaxing paramagnetic centre, that is unobservable by EPR but causes anisotropic broadening of the EPR signal of the other electron.[11] Alternatively some spin pairing could be postulated but this would not account for all features of the spectra.

Transfer of electrons to the MoFe protein of *K. pneumoniae* results in loss of its EPR spectrum, while changes in the Mössbauer parameters indicates that two Fe_4S_4 clusters have become reduced and that the other Fe atoms are distributed in at least two other environments.[12] This EPR inactive form of the MoFe protein is described as 'super-reduced', as it is reduced further than is possible with dithionite alone. On mild oxidation with dyes the EPR signal is also lost. Thus three stable oxidation levels for the MoFe protein have been characterized, although it is clear that the super-reduced form is crucial in enzyme turnover. These redox forms of the MoFe protein are continuing to receive much attention.[13]

The study[14] of nitrogenase from *K pneumoniae* at $10\,°C$ showed the presence of four new EPR signals, assignable to intermediates present in low concentrations in the steady state. The use of ^{57}Fe-substituted MoFe protein shows that they originate in Fe—S clusters in the MoFe protein of nitrogenase. These signals may then be used to monitor the effect of various substrates and inhibitors on the course of electron transfer in the MoFe protein. Such studies have been reported[14] for acetylene and ethylene, and schemes presented for the electron-transfer sequences involved in the reduction of H^+ and acetylene. Some of this material will be discussed later.

THE NITROGEN MOLECULE, DINITROGEN COMPLEXES AND THEIR REACTIVITY

A basic aspect of the problem of nitrogen fixation must be that of the structure of the molecule itself and the way in which it may be activated prior to chemical reaction. Until recently nitrogen has always been regarded as an inert molecule, its inertness being one of its characteristic features. The only reaction that it readily undergoes is the formation of nitrides, particularly with the most electropositive metals. Thus it reacts directly with lithium at room temperature and under certain conditions with magnesium. In addition there are the catalytic processes such as the Haber process which involve the use of high temperature and pressure and appropriate catalysts. From some points of view this inertness of dinitrogen is unexpected, for analogous triple-bonded molecules are quite reactive. Thus acetylene may be reduced readily, while the isoelectronic carbon monoxide forms a vast number of carbonyl complexes.

Coordination of dinitrogen offers one means of increasing its reactivity. Dinitrogen would be expected to be a very poor donor, but this does not exclude complex formation. Thus carbon monoxide has poor σ-donor properties, but stable carbonyl complexes are formed as a result of the well-known synergic effect in which the essentially weak σ-bonding and π-bonding effects reinforce each other to form strong metal–carbonyl bonds at the expense of the C—O bond.

The molecular orbital scheme for dinitrogen is shown in Figure 6.2. The highest filled orbital is a σ orbital formed by overlap of a $2s$ orbital of one nitrogen atom with the $2p$ orbital of the other. It has been shown spectroscopically that N_2^+ has its unpaired electron in a σ orbital. A different scheme holds in carbon monoxide in which the highest filled orbitals are the degenerate π orbitals. This constitutes an important difference, as the highest filled orbital in dinitrogen, a σ orbital, will lie deep in the molecule, shielded by the π-molecular orbitals, which are of low energy and strongly bonding. This means that the donor properties of the σ orbital will be limited and explains why the first ionization potential is high. It is then possible to write down a bonding scheme for the formation of the dinitrogen–transition metal complex, analogous to that for a carbonyl complex, in which the nitrogen molecule will be bonded end on to the metal with $N_2 \rightarrow$ metal σ bonding, and metal $\rightarrow N_2$ π bonding via overlap of the metal d orbitals and the appropriate π^* MO of the nitrogen molecule

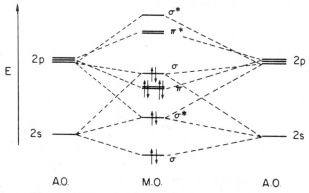

Figure 6.2 Energy levels in the nitrogen molecule.

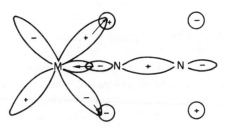

Figure 6.3 Bonding scheme for a transition metal–dinitrogen complex.

(Figure 6.3). The overall result of this bonding is that electron density is removed from the bonding σ orbital of nitrogen and transferred *via* the metal ion to the antibonding π orbital of nitrogen. This means that there is a doubly weakening effect on the nitrogen–nitrogen bond; indeed, the metal–nitrogen bond is produced at the expense of the nitrogen–nitrogen bond. The stronger the former then the weaker the latter bond, and the more likely that it may be involved in chemical reaction. In addition, the result of the electronic effects will be seen in an increasing polarization of the nitrogen molecule, again possibly making it more susceptible to attack. All this is confirmed by spectroscopic studies.

Normally the nitrogen molecule has a Raman-active nitrogen–nitrogen stretching frequency at $2331 \, \text{cm}^{-1}$. This vibration is not active in the infrared as there is no associated change in dipole moment. However, the infrared spectra of metal–dinitrogen complexes usually show a strong band at around $2100 \, \text{cm}^{-1}$ assigned to v_{N_2}, although in fact a fairly wide range of values has been noted. Two points are important. First, the activity of the vibration in the infrared confirms that the nitrogen molecule is now dipolar and secondly, the reduced frequency indicates a reduction in bond strength. It is also of interest to note that in the complexes $[Ru(NH_3)_5(N_2)]X_2$, v_{N_2} varies with the nature of the anion, ranging from 2105 (X = Cl) to $2167 \, \text{cm}^{-1}$ (X = PF_6). This phenomenon also confirms the dipolar character of coordinated nitrogen. Values of v_{N_2} are a useful indication of the bond strength of the nitrogen molecule.

The much greater ease of formation of carbonyl complexes can be understood in terms of the rather different molecular orbital scheme for carbon monoxide which results in a σ-donor and π-acceptor orbitals of more appropriate energy being available on the ligand.

Transition metal complexes of dinitrogen

The first characterized dinitrogen complex, $[Ru(NH_3)_5(N_2)]^{2+}$, was prepared in 1965 by Allen and Senoff.[15] Since then a large number of other complexes have been prepared.[16,17] Some of the early examples were prepared accidentally but there are now some clear-cut preparative routes. These include (a) direct reaction with dinitrogen either through replacement of a labile ligand or by reduction of a transition metal complex in the presence of dinitrogen, (b) the reaction of certain coordinated ligands, for example, the oxidation of hydrazine, and

TABLE 6.3 Some dinitrogen complexes*

Complex	N_2 (Cm^{-1})	Reference/comments
$[\{Ti(\eta^5\text{-}C_5H_5)_2\}_2(N_2)]$	1280	27
$[\{Zr(\eta^5\text{-}C_5Me_5)_2\}_2(N_2)]$	2041, 2006, 1556	28
$cis\text{-}[Cr(N_2)_2(PMe_3)_4]$	1919	16
$trans\text{-}[Mo(N_2)_2(diphos)_2]$	2020(w), 1970	Diphos $= Ph_2PCH_2CH_2PPh_2$; 16
$[\{Mo(C_6H_6)(PPh_3)_2\}_2(N_2)]$	1910	Raman
$trans\text{-}[W(N_2)_2(diphos)_2]$	1953	16
$trans\text{-}[W(N_2)_2(PPh_2Me)_4]$	1910, 1975	24
$[ReCl(N_2)(PR_3)_4]$	2075–1810	Range of phosphines
$[MCl_4\{(N_2)ReCl(PMe_2Ph)_4\}_2]$	$\rightarrow 1800$	Polynuclear, M $=$ Ti, Mo
$[ReCl(PMe_2Ph)_4(N_2)MCl_4\text{-}(PEtPh_2)]$		
$[FeH_2(N_2)(PR_3)_3]$	$\rightarrow 1680$	Various M
$[Fe(N_2)(diphos)_2]$	2065–1989	PR$_3$ = PEtPh$_2$, PMePh$_2$, PBuPh$_2$
$[FeH(Pr^i)(PPh_3)_2(N_2)Fe\text{-}(Pr^i)(PPh_3)_2]$	2068	16
$[Ru(NH_3)_5(N_2)]X_2$	1761	16
$[Ru(NH_3)_5(N_2)Ru(NH_3)_5]\text{-}(BF_4)_4$	2105–2167	X $=$ Cl, Br, I, BF$_4$, PF$_6$; 15
$[RuH_2(N_2)(PPh_3)_3]$	2100	Raman
$[Os(NH_3)_5(N_2)]X_2$	2147	
$cis\text{-}[Os(NH_3)_4(N_2)]X_2$	2033	X $=$ Cl, Br, I
$[Os(N_2)Cl(PBu_2Ph)_3]$	2147	
$[CoH(N_2)(PPh_3)_3]$	2088	
$Na[Co(N_2)(PR_3)_3]$	1840	PR $=$ PEt$_2$Ph, PPh$_3$
$[RhCl(N_2)(PPh_3)_2]$		
$K_2[\{Rh(NO_2)_3(NH_3)(OH)\}_2(N_2)]$	2070	Raman
$[IrCl(N_2)(PPh_3)_2]$	2095	
$[Ni(N_2)(PR_3)_3]$	2070–2060	PR$_3$ = PEt$_3$, PEt$_2$Ph
$Ni(N_2)$	2088	Matrix isolation*
$Ni(N_2)_2$	2104	Matrix isolation*

*See reference 16 (p. 606) for a comprehensive list of dinitrogen complexes obtained by matrix isolation techniques.

(c) metathetical reactions. In addition some interesting complexes, particularly of Ni, have been prepared by matrix isolation techniques in which metal atoms and dinitrogen are cocondensed at 6–25 K.

Dinitrogen complexes have now been prepared for almost every transition element. Mono, bis (cis and trans), and bridging dinitrogen complexes are known, including an example where bridging and terminal groups are bound to the same metal. Some examples are listed in Table 6.3, with values of v_{N_2}, the nitrogen–nitrogen stretching frequency.

The structures of many of these complexes have been determined. Dinitrogen is usually bound end-on to transition metals, as shown in Figure 6.4 for terminal and bridging examples. The N—N bond lengths in such compounds (1.123 and 1.124 Å in the examples in Figure 6.4) usually do not differ much from the value found for the free molecule (1.0976 Å), except in situations where dinitrogen bridges complexes without a closed shell of electrons. In this case the σ N_2 bonding MO is unoccupied and the N—N bond length increases, as in trans-$[MoCl_4\{(N_2)ReCl(PMe_2Ph)_4\}_2]$, where its value is 1.28 Å.[18]

A few compounds are known in which the dinitrogen is bound side-on, for both terminal and bridging modes. The structure of the side-on terminal complex, $[RhCl(N_2)(PPr^i_3)_2]$ shows that the N—N bond length differs only slightly from that of dinitrogen.[19]

The extent of electron release to dinitrogen from the metal $d\pi$ orbitals is indicated by the v_{N_2} frequency and also by the intensity of the absorption. Strong $d\pi$–$p\pi^*$ interaction is shown by low v_{N_2} and high intensity. The polarity of such compounds has been demonstrated by X-ray photoelectron spectroscopy, for example in trans-$[ReCl(N_2)(PMe_2Ph)_4]$, but it has not been possible to determine which nitrogen atom in the ligand is the more negatively charged.

The donor properties of dinitrogen have been compared[16] with other ligands. It is suggested, mainly on Mössbauer evidence, that dinitrogen is a very poor σ donor, but a moderately good π acceptor.

The reactivity of coordinated dinitrogen

The location of additional electron density on the coordinated dinitrogen means that the dinitrogen will be subject to electrophilic attack. This has been amply

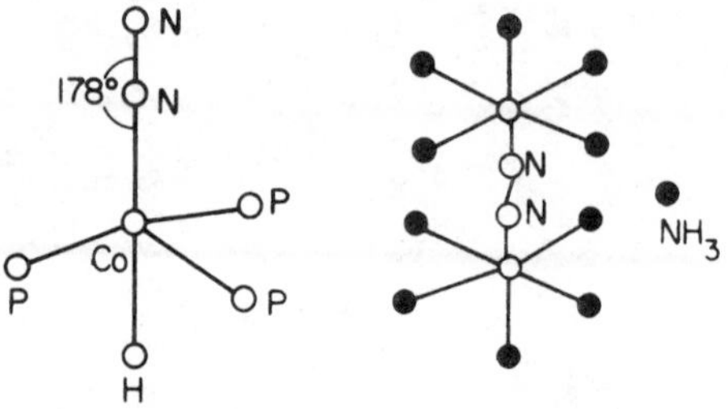

Figure 6.4 Structures of $Co(N_2)H(PPh_3)_3$ and $[(NH_3)_5Ru(N_2)Ru(NH_3)_5](BF_4)_4$.

demonstrated in the formation of various hydride species of dinitrogen through protonation, and in the formation of nitrogen–carbon bonds by, for example, alkylation and acylation at the dinitrogen. This latter topic is of considerable industrial potential in the formation of organonitrogen compounds but will not be considered in this account. References to some recent work are given.[16,20]

Attempts to reduce dinitrogen met with little success for some time—indeed many reactions just resulted in the solvolysis of dinitrogen. The first demonstration of partial reduction of dinitrogen involved the treatment of the chelating phosphine complex $[M(N_2)_2(Ph_2PCH_2CH_2PPh_2)_2]$ (M = Mo or W) with HBr. This led to the formation of diazine N_2H_2.[21] Treatment of the analogous complex involving tertiary phosphines instead of the chelating phosphine with methanol/sulphuric acid[22] gave almost complete conversion of one mole of N_2 to NH_3, with a little hydrazine. A 90% yield of NH_3 was obtained

$$[Mo(N_2)_2(diphos)_2] \xrightarrow{HBr} [MoBr_2(N_2H_2)(diphos)_2] + N_2$$

$$[M(N_2)_2(PR_3)_4] \xrightarrow[CH_3OH]{H_2SO_4} 2NH_3 + N_2 + \text{other products}$$

with the complex cis-$[W(N_2)_2(PMe_2Ph)_4]$. Chatt has accounted for the further reduction of dinitrogen in this second reaction by assuming that the complex with unidentate phosphines is able to lose a phosphine ligand, so allowing attack of sulphate (a π donor) at the metal with increase in electron density, which eventually ends on the diazine ligand, so triggering its further protonation and reduction to hydrazine and then to ammonia. Later it was appreciated[23] that by use of a suitable solvent such as propylene carbonate the chelating phosphine could also be displaced with resulting reduction of dinitrogen. This work was of considerable interest as the first reported instance of reduction of dinitrogen in an aqueous environment. It is interesting that one metal centre appears to be involved as it is difficult to conceive of a six-electron transfer process taking place. Following the observation that in biological systems clusters of metals always seem to be associated with multielectron transfer it appears possible that some sort of Mo or W cluster may be formed as intermediates in this system. Such a view is consistent with the chemistry of these elements, but is only speculative.

Reactions of this type have been explored further for a range of molybdenum and tungsten complexes with varying phosphine ligands.[24] In the case of cis-$[W(N_2)_2(PMe_2Ph)_4]$ sulphuric acid is not necessary for ammonia production as methanol is acidic enough to provide the necessary protons. Of particular interest is the fact that for the complexes of the chelating phosphine 'diphos' discussed earlier, stepwise protonation and reduction of the dinitrogen species has been observed, to give complexes containing N_2H (diazenido-), N_2H_2 (diazine), and NNH_2 (hydrazido(2-)) ligands, while the complex $[WCl_3(N_2H_3)(PMePh_2)_2]$ has also been prepared. There is some doubt as to whether hydrazine is formed as conflicting reports have appeared.[25] On nitrogenase, hydrazine is a poor substrate for reduction and would accumulate.

Therefore the following scheme may be put forward[24] in which the fourth proton attacks the NH_2 entity of a —NNH_2 group with subsequent loss of ammonia and further protonation and reduction of the remaining NH fragment,

$$M^0 {\cdots} N{\equiv}N \xrightarrow{H^+} M{\cdots}N{\equiv\equiv}NH \xrightarrow{H^+} M{\equiv}N{-}NH_2 \xrightarrow{H^+} M{\cdots}NH{-}NH_2$$

$$\xrightarrow{H^+} M{\cdots}NH{-}NH_3^+ \longrightarrow M{\cdots}NH + NH_3 \xrightarrow{H^+} M{\cdots}NH_2 \xrightarrow{H^+} M^{VI} + NH_3$$

This scheme holds for the tungsten complex $[W(N_2)P_4]$, but the mechanism does appear to diverge at the (N—NH_2) stage in the case of molybdenum, possibly because the ligand is released from this metal and subsequently disproportionates into dinitrogen and ammonia. Such conclusions are drawn from stoichiometry studies.

The scheme is clearly an attractive one. Chatt has pointed out that all species with the exception of(—$NHNH_3^+$) represent known compounds. The mechanism may be applied to nitrogenase-catalysed reduction of dinitrogen, where electrons would be supplied *via* the iron–sulphur centres to the dinitrogen site, so that the molybdenum centre would not undergo such a wide change in oxidation state. The ready availability of electrons (and perhaps binding constrictions) may be postulated to overcome the problem described above in which the (—N—NH_2) group appeared to be lost readily from molybdenum.

Other work in this area has concentrated on titanium, zirconium, and iron. This began in 1964 with the report of Vol'pin and Shur on the reduction of transition metal halides under dinitrogen and the hydrolysis of the product to give ammonia.[26] Since that time it has been shown that the reaction of $[TiCl_2(\eta^5\text{-}C_5H_5)_2]$ with the Grignard reagent Pr^iMgBr under dinitrogen gives a binuclear complex of dinitrogen having an extremely low ν_{N_2} frequency of $1255\,cm^{-1}$. This reacts with hydrogen chloride at $-60\,°C$ to give hydrazine or ammonia depending on the solvent, together with dinitrogen.[27] More recent examples of studies on binuclear complexes are typified by reaction of the complex $[(C_5(CH_3)_5)_2Zr]_2(N_2)_3$ (formed by reaction of dinitrogen with bis(pentamethylcyclopentadienyl)zirconium(II)) with $10\,mol\,dm^{-3}$ HCl in toluene at $-80\,°C$, which leads to the release of two moles of dinitrogen and the formation of hydrazine.[28] The hydrolysis of $Co(N_2)(PR_3)_3Mg(THF)_2$ to give hydrazine and a little ammonia[29] provides an interesting example of the reduction of one of the first dinitrogen complexes prepared, and is one of the few cases known in which a late transition metal is involved.

Fixation via nitride formation

Several reactions are known in which the reduction of transition metal complexes (usually of titanium and vanadium) by strong reducing agents in aprotic solvents in the presence of dinitrogen led to the formation of nitrides. These nitrides could then react with water or alcohols to give ammonia.

Thus the reduced species formed by the reaction of metal halides with lithium naphthalide in THF reacts with dinitrogen to give a nitride which may be converted to ammonia on hydrolysis.[30] The best yield of ammonia (2.0 moles per mole of metal) was obtained on the hydrolysis of the product from the reduction of vanadium(III) chloride under a dinitrogen pressure of 120 atmospheres for 30 minutes at 20 °C.

This type of reaction was developed by van Tamelen and his coworkers.[31] They reduced dichlorodiisobutoxytitanium(IV) with potassium in diglyme or THF. On passing dinitrogen through the solution ammonia was evolved. If, on the cessation of ammonia production, more potassium was added then the flow of ammonia recommenced. In this case the solvent is the source of hydrogen. Later it was shown that a titanium(II) dinitrogen complex is formed under certain conditions. A dioxygen-free benzene solution of biscyclopentadienyl-titanium(II) absorbs dinitrogen at room temperature. Such solutions show an infrared absorption band at about $1960\,cm^{-1}$ attributable to ν_{N_2}. On flushing argon through the solution dinitrogen is liberated, while the band at $1960\,cm^{-1}$ disappears. Molecular weight studies on the benzene solution indicate that the complex is dimeric. Treatment of a solution of this titanium(II) dinitrogen complex with sodium naphthalide and subsequent hydrolysis results in the production of ammonia.

One interesting aspect of this system is reflected in the development[32] of cyclic processes for the reduction of dinitrogen to ammonia in which the reducing agent is regenerated electrolytically. Figure 6.5 represents one example, and involves a cell fitted with an aluminium anode and a nichrome cathode. A typical electrolyte solution would contain the following: 1.68 mmol titanium tetraisopropoxide, 7.6 mmol naphthalene, 42 mmol aluminium isopropoxide, 8.6 mmol tetrabutyl-ammonium chloride in 60 ml 1.2 dimethoxyethane. After electrolysis for eleven days at 40 V, treatment with 40 ml of $8\,mol\,dm^{-3}$ aqueous alkali and heating resulted in the formation of 10.2 mmol ammonia. It appears that naphthalene functions as an electron carrier, being reduced to naphthalide at the electrode and oxidized by the titanium nitrogen species. The aluminium isopropoxide, in addition to acting as an electrolyte, removes nitride from the titanium species, so allowing further interaction with dinitrogen.

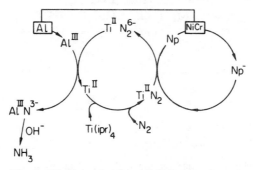

Figure 6.5 The reduction of dinitrogen to ammonia *via* nitride formation.

NITROGEN FIXATION IN AQUEOUS SOLUTION

The basic approach in this work has involved attempts to reduce dinitrogen by using 'soups' containing a number of compounds in aqueous solution. Initially the composition of these 'soups' reflected the chemical components of natural fixing systems, and so contained, for example, iron and molybdenum salts, sulphur ligands such as cysteine or mercaptoethanol, plus reducing agents such as borohydride. Subsequent modifications have led to successful systems rather dissimilar from the 'soup' described above. Over the years this work has aroused much controversy, and claims for successful reduction of dinitrogen were not substantiated by later workers. Early difficulties probably centred around the problems associated with the analysis of small amounts of ammonia and the presence of catalytic impurities in reagents. It should also be stressed that the mechanisms of these reactions are ill defined and that their relation to nitrogenase is suspect. Nevertheless they involve some interesting chemistry. Some illustrative examples will be discussed for systems based on molybdenum, but it should be noted that V(II) systems are also effective.

G. N. Schrauzer and his colleagues[33] have published a substantial series of papers on nitrogenase model systems using molybdate, a thiol, and $NaBH_4$ or $Na_2S_2O_4$ at pH 9, and have shown that such systems reduce various nitrogenase substrates in an ATP-stimulated reaction. Dinitrogen has been reduced, although the amounts of ammonia produced are very small relative to the molybdenum. Schrauzer has stressed the role of diazene as an intermediate; this has been established by several approaches. Diazene is not reduced further, but disproportionates to give dinitrogen, dihydrogen, and hydrazine. Hydrazine is then reduced to ammonia, a reaction which has been verified directly. More recently[33] some success has been achieved with cyano complexes of molybdenum, $[MoO(CN)_4(H_2O)]^{2-}$.

In 1971 Shilov and his colleagues[34] reported the catalytic reduction of dinitrogen to hydrazine in protic solutions. Dinitrogen was reduced by Ti^{3+} or Cr^{2+} in the presence of magnesium salts and a molybdenum compound, such as MoO_4^{2-}, $MoCl_5$, and $MoOCl_3$. High pH (≥ 10) was necessary, and optimum yields of ammonia were obtained at about 100 °C, 100–150 atmospheres pressure of dinitrogen and a Mg/Ti ratio of 0.5. Up to 100 molecules of hydrazine could be formed per molybdenum site.[35] The following scheme has been suggested.

The role of Mg^{2+} is partly structural, but it is suggested also to control the redox potential of the Ti(III) to prevent the reduction of water to dihydrogen, which does not occur in this system. Molybdenum has been replaced by V(II), in which case Mg^{2+} and a high pH are still necessary. Dinitrogen is reduced to hydrazine at 20 °C and ammonia at 70 °C. Shilov has suggested a different scheme in which the dinitrogen is bound to V(II) and reduced by it, the V(II) being oxidized to V(III). To accommodate the stoichiometry, the oxidation of a tetranuclear cluster of V(II) is postulated, with transfer of four electrons to dinitrogen, subsequent protonation leading to hydrazine. At higher temperatures the hydrazine reacts further with vanadium(II).[36]

GENERAL CONCLUSIONS ON NITROGEN FIXATION

The biochemical evidence summarized in this chapter may be accommodated by the following scheme for nitrogenase-catalysed fixation of dinitrogen. In this scheme the Fe and MoFe proteins are represented by (Fe) and (MoFe)

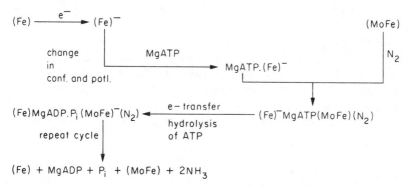

respectively. Earlier models postulated a dimetal site with bridging dinitrogen. Chemical evidence suggests that this feature is not necessary and that dinitrogen can be reduced at a single metal site. Reduction of dinitrogen in model systems may be triggered off by substitution at the metal centre by a hard ligand. In view of this it is tempting to speculate that ATP interaction at the Mo site in nitrogenase could trigger electron transfer to dinitrogen. Stepwise reduction of dinitrogen will then take place with concomitant protonation, although it is unlikely that hydrazine lies on the reaction pathway.

At present, work on nitrogenase using EPR and Mössbauer techniques is leading to the characterization of the detailed course of electron transfer between the redox-active iron−sulphur centres. Some twelve EPR signals have been characterized under different conditions. These may be associated with, for example, low- and high-pH forms of the MoFe protein, the Fe protein, and its MgATP and MgADP complexes, and the ethylene- and CO-bound forms of the MoFe protein. Schemes have been suggested[13] for the detailed mechanism of nitrogenase, in particular for the evolution of dihydrogen, the reduction of acetylene and its inhibition by carbon monoxide.

Some of the inorganic studies carried out on dinitrogen fixation may well offer feasible new routes for industrial processes. At present though, they do not compete with the Haber process on economic grounds, although this situation

204

could well change. In any case it is now realized that synthesis of more ammonia does not provide an easy answer to the problem of nitrogen fertilizers. Indiscriminate and excessive use of nitrates has caused considerable problems. The ultimate answer to the problem of nitrogen fixation may well lie in the cultivation of new strains of cereals which are able to fix their own dinitrogen as required. The first steps in this direction have already been taken, in that the genes for dinitrogen fixation have been transferred successfully to the bacterium *Escherichia coli*.

OTHER ASPECTS OF THE NITROGEN CYCLE

The ammonia formed by fixation of molecular nitrogen together with any ammonia resulting from decomposition of living matter is utilized by microorganisms which use the energy derived from the oxidation of reduced nitrogen compounds to reduce carbon dioxide, in order to obtain carbon for synthesis of new cellular material. These are autotrophic species. These oxidative reactions result in the eventual conversion of ammonia to nitrate in the process of nitrification. There are also reductive pathways for nitrate which lead to the production of ammonia, which may be incorporated into the cell, and also other gaseous products including dinitrogen. These reduction pathways are more diverse than the oxidative reactions. These various processes are summarized in the nitrogen cycle depicted in Figure 6.6.

Nitrification

This is brought about mainly by *Nitrosomonas* species and *Nitrobacter* species that oxidize ammonia to nitrite and nitrite to nitrate respectively. It is now generally accepted that the oxidation to nitrite proceeds *via* two-electron changes, although the species with nitrogen in the formal oxidation state $+$ I has not been satisfactorily identified. Hydroxylamine appears to be the N^{-1} intermediate.

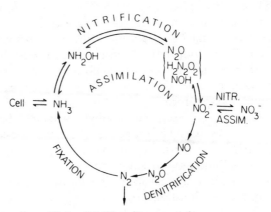

Figure 6.6 The nitrogen cycle.

$$NH_3 \rightarrow NH_2OH \rightarrow ? \rightarrow NO_2^-$$
$$-III \quad\quad -I \quad\quad +I \quad +III$$

It is not to be expected that the reaction will involve appreciable build up of intermediate species as the rate of appearance of nitrite practically equals the rate of loss of ammonia. Whole cells of *Nitrosomas* will oxidize hydroxylamine nearly as rapidly as ammonia.

It has been shown that the use of different inhibitors affected the oxidation of ammonia to hydroxylamine and hydroxylamine to nitrite in different ways. In the presence of certain inhibitors an accumulation of hydroxylamine has been observed indicating that the two processes are independent. However, the presence of oxygen is essential if hydroxylamine is to be oxidized right through to nitrite. In the absence of oxygen, treatment of hydroxylamine with *Nitrosomonas* leads to the production of nitrous and nitric oxides.

Trace amounts of the *trans*-hyponitrite ion $(N_2O_2^{2-})$ are present in nitrification systems, leading to the suggestion that this is the intermediate species with nitrogen in the formal oxidation state $+I$. However added sodium hyponitrite is not oxidized under these conditions, and so it is most likely that hyponitrite is formed as a result of a side reaction and does not lie on the main reaction pathway. An alternative N^{+I} intermediate is the nitroxyl ion NO^-, dimerization of which would lead to hyponitrite and some N_2O. Nitroxyl ion is now commonly accepted to be an intermediate in nitrification, although detailed subsequent pathways to nitrite have not been established. One suggestion postulates the reaction of NO^- with nitrite to give what is often referred to as nitrohydroxylaminate but which is more properly known as the trioxodinitrate ion, $(N_2O_3^{2-}$, Angeli's salt). Some nitric oxide has also been identified as a product of oxidation of hydroxylamine by *Nitrosomonas*.

The oxidation of nitrite to nitrate by *Nitrobacter* is a much simpler reaction. The process is an aerobic one, although the role of dioxygen is that of an electron acceptor, as it may be shown that the additional oxygen atom in nitrate is generated from water and not from dioxygen.

Studies with cell-free extracts of N. europaea

While whole cells of *N. europaea* quantitatively oxidize both ammonia and hydroxylamine to nitrite, cell-free extracts only oxidize hydroxylamine, and require an artificial terminal electron-accepting compound (such as phenazine methosulphate) for this to take place. The enzymes responsible for the initial oxidation of ammonia are thought to be near the cell membranes and are lost on disruption of the cell. The naturally occurring terminal electron acceptor for the oxidation of hydroxylamine is not physically associated with the hydroxylamine oxidoreductase and so is separated from it in the purification procedure.

The hydroxylamine oxidoreductase-containing extracts of *N. europaea* oxidize hydroxylamine to nitrite and nitrate in the presence of dioxygen. This provides a notable difference from the intact cell which only oxidizes the hydroxylamine to nitrite. Furthermore, the nitrate is not produced from the nitrite, but from an earlier intermediate.[37] The production of nitrate (but not nitrite) was inhibited by diethyldithiocarbamate which was oxidized to bis(diethyldithiocarbamoyl)-disulphide. The total reaction system may thus be summarized as follows:

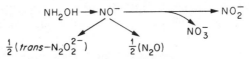

The enzyme hydroxylamine oxidoreductase has been purified. It has a molecular weight of about 196,000[38] and has a notably high content of cytochromes, including[39] the unique cytochrome P460, which has been suggested to be involved in the dioxygen metabolizing portion of the reaction.[40]

The chemistry of nitrification

The compounds postulated as intermediates are often unstable, with the result that their chemistry has only been characterized fairly recently. There is now very good evidence for the existence of nitroxyl ion in aqueous solution. Its pK_a value of ~ 4 shows that at biological pH it will exist as NO^- rather than the parent acid, while a survey of reactions in which it occurs as an intermediate shows that there appears to be two distinct forms, which are suggested to be triplet and singlet.[41] NO^- dimerizes to give a mixture of cis- and trans-hyponitrite. The former species is very unstable and will give N_2O, while the latter species is much more stable and will decompose at pH 7 and room temperature to N_2O with a half life of about 60 minutes.[42] Thus the observation that in the absence of dioxygen, hydroxylamine is oxidized by Nitrosomonas to N_2O and NO is in accord with the chemistry of NO^-, provided the NO may be obtained from it by a one-electron oxidation.

Of particular interest is the reaction of NO^- with dioxygen, as this reaction is essential to give nitrite in nitrification. The product of this reaction is the species peroxonitrite, which has received very little attention as a possible intermediate in nitrification,[43] but whose chemical properties are quite appropriate for this role. This reaction to give peroxonitrite takes place readily when dioxygen is bubbled through reaction solutions in which NO^- occurs as an intermediate.[44] One useful model system is the chemical autoxidation of hydroxylamine in alkaline solution. This leads to the formation of nitrite in a metal ion catalysed reaction.[45] At pH 11 with a trace of metal ion, up 80% of the hydroxylamine is converted to nitrite. It is suggested that the nitroxyl ion, NO^-, is the first product of interaction of hydroxylamine and oxygen. In the absence of oxygen this would dimerize giving nitrous oxide, but in the presence of more oxygen, further reaction occurs giving the species peroxonitrite ONO_2^-. Peroxonitrite then reacts with hydroxylamine giving nitrite and regenerating nitroxyl ion, from which peroxonitrite ion may once more be formed. This means that a trace of peroxonitrite will catalyse the autoxidation of hydroxylamine directly to nitrite. Furthermore the peroxonitrite may isomerize to nitrate.[46]

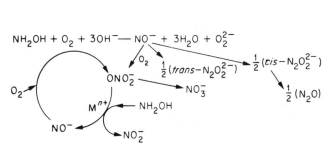

Alternatively, to avoid postulating this cyclic process, it should be noted that peroxonitrite reacts[47] very readily with a range of compounds with the transfer of an oxygen atom and the formation of NO_2^-

$$\text{i.e. } NH_2OH \longrightarrow NO^- \xrightarrow{O_2} ONO_2^- \xrightarrow{.S} NO_2^- + SO$$
$$\searrow NO_3^-$$

The parallel with hydroxylamine oxidoreductase is marked. The products of oxidation of hydroxylamine under different conditions correspond well in both cases, while it should be noted that the production of nitrate from peroxonitrite is in accord with the result that it is not derived by oxidation of nitrite. The key feature in both cases is the formation of NO^-, by a two-electron oxidation of hydroxylamine, and its subsequent reactions. These reactions are chemical rather than biochemical in character, but the involvement of peroxonitrite is not excluded by this, as these are facile reactions.

The suggestion that NO^- may react with NO_2^- to give 'nitrohydroxylaminate' is in fact correct, although the relevance of this compound to nitrification still seems rather doubtful. The decomposition of trioxodinitrate is retarded by added nitrite, due it is suggested to the existence of an equilibrium between the products of $N{=}N$ cleavage, NO^- and NO_2^-.[48]

$$^-ON{=}NO_2^- \rightleftharpoons NO^- + NO_2^-$$

Reduction of nitrate

Several pathways are available for the reduction of nitrate. The molybdenum-containing enzyme nitrate reductase is widely distributed in plants and microorganisms. In some cases the nitrate acts as a terminal electron acceptor in electron transport (instead of dioxygen), and the nitrogen of the nitrate is not incorporated into the cell. This dissimilatory reduction of nitrate is usually found under anaerobic conditions, for example in *Halobacterium*.[49] In other cases the enzyme nitrite reductase is present and catalyses the reduction of nitrite to ammonia in the process of assimilation. There is also a denitrification pathway that leads to gaseous products.

The assimilatory NADPH nitrate reductase of the mould *Neurospora crassa* has been much studied. This 230,000 mol. wt. multimeric protein contains flavin, molybdenum, and cytochrome b_{557} and the following electron-transfer pathway has been postulated:

$$NADPH \rightarrow FAD \rightarrow cyt\ b_{557} \rightarrow Mo \rightarrow NO_3^-$$

Several other enzymatic activities are associated with the NADPH nitrate reductase, including NADPH cytochrome c reductase. Extracts of *nit-1*, a mutant form of *N. crassa*, cannot reduce nitrate but still possesses the other reductase activities. However the nitrate reducing properties may be restored by treating *nit-1* with the neutralized products of the acid hydrolysis of several molybdoenzymes such as xanthine oxidase. It is suggested that these preparations donate a molybdenum-containing polypeptide cofactor[50] which is common to several enzymes, but which is absent in the mutant *nit-1*. As noted earlier the molybdenum cofactor in nitrogenase differs from this cofactor.

There is some doubt over the oxidation states of Mo found in nitrate reductase during activity. It was always thought that the molybdenum cycled between oxidation states V and VI, in which case a two-electron reduction of nitrate is coupled to a one-electron change in molybdenum (Mo(V)$\rightarrow$Mo(VI)). EPR studies will help to resolve this problem, as Mo(V) and Mo(III) are both observable by this technique. Such studies have led to the suggestion that reduction of molybdenum to Mo(III) occurs, but more recent work has failed to find evidence for this species.[51] Detailed results on the nitrate reductase from *E. coli* show at least five Mo(V) species, four of which are mechanistically significant. These are high and low pH forms of the enzyme (linked by $pK_a = 8.26$, probably due to a proton on a nitrogen ligand to Mo), and complexes of the low pH form of the enzyme with nitrate and nitrite. No other anion formed such complexes, in accord with the function of this enzyme. The low pH form of the Mo(V) enzyme probably involves loss of one ligand (by protonation), so providing a vacant coordination site for the nitrate and nitrite anions. This EPR work also showed the molybdenum centres in nitrate reductase and sulphite oxidase to be similar. Redox potential measurements on molybdenum in nitrate reductase show that Mo(IV) and Mo(V) are both capable of reducing nitrate. It is not yet clear whether this process involves a one-electron or two-electron pathway. The need for molybdenum in this enzyme and the importance of oxo groups in its chemistry leads to the intuitive conclusion that reduction of nitrate should involve oxo group transfer to the molybdenum centre, but evidence is not available as yet for this pathway.

Nitrite reductases

These are found in a number of nitrate-assimilating species and catalyse the six-electron reduction of nitrite to ammonia. This reaction sequence is the reverse of the steps involved in nitrification. The pathway is not completely defined but hyponitrous acid and hydroxylamine have been isolated. Indirect evidence that hydroxylamine is an intermediate comes from the observation that nitrite reductase catalyses the reduction of hydroxylamine by NADH.[52] It is probable that nitrite reductase is made up of two similar subunits.

The nitrite reductase of spinach contains a specific heme prosthetic group called 'siroheme'. This is an iron–tetrahydroporphyrin of the isobacteriochlorin type, having eight carboxylate side chains. This is also present in sulphite reductase but has not yet been identified elsewhere. While it is present in all sulphite reductases it is not yet certain that it is present in all nitrite reductases. It is an interesting comparison, however, particularly as both reductases are involved in six-electron oxidations. Indeed sulphite reductases also show nitrite reductase activity, but there is a high K_m for nitrite binding.[53] The nitrite reductases from the leaves of higher plants utilize reduced ferredoxin as electron donor. EPR spectra from both siroheme and iron–sulphur centres have been measured.[54] Evidence has been put forward for a heme–NO complex as an intermediate, but little is really known about the detailed course of electron transfer in these reactions. Models are available for siroheme.[55]

Reduction pathways in denitrification are no better defined. Several denitrifying species have enzymes that are able to catalyse the reduction of nitrite to nitrogen monooxide. The use of whole cells leads to the production of dinitrogen oxide and dinitrogen.

REFERENCES

1. For general reviews on the biochemistry of N_2-fixation see: W. Newton, J. R. Postgate, and C. Rodriguez–Barrueco (Eds.), *Recent Developments in Nitrogen Fixation*, Academic Press, New York, 1977; W. G. Zumft, *Structure and Bonding*, **29**, 1 (1976); R. R. Eady and J. R. Postgate, *Nature*, **249**, 805 (1974); R. W. F. Hardy (Ed), *Dinitrogen Fixation*, Wiley–Interscience, 1978.
2. R. N. F. Thorneley, R. R. Eady, and D. J. Lowe, *Nature*, **272**, 557 (1978).
3. J. R. Postgate, *Proc. Roy. Soc.* **B172**, 355 (1969).
4. R. J. Rennie, A. Funnell, and B. E. Smith, *FEBS Letters*, **91**, 158 (1978).
5. V. K. Shah and W. J. Brill, *Proc. Natl. Acad. Sci. U.S.A.* **74**, 3249 (1977).
6. J. Rawlings, V. K. Shah, J. R. Chesnell, W. J. Brill, R. Zimmerman, E. Munck, and W. H. Orme-Johnson, *J. Biol. Chem.*, **253**, 1001 (1978).
7. S. P. Cramer, K. O. Hodgson, W. O. Gillum, and L. E. Mortensen, *J. Amer. Chem. Soc.*, **100**, 3398 (1978); S. P. Cramer, W. O. Gillum, K. O. Hodgson, L. E. Mortensen, E. I. Steifel, J. R. Chisnell, W. J. Brill, and V. K. Shah, *J. Amer. Chem. Soc.*, **100**, 3814 (1978).
8. T. E. Wolff, J. M. Berg, C. Warrick, K. O. Hodgson, R. H. Holm, and R. B. Frankel, *J. Amer. Chem. Soc.*, **100**, 4630 (1978); S. R. Acott, G. Christou, C. D. Garner, T. J. King, F. E. Mabbs, and R. M. Miller, *Inorg. Chim. Acta*, **35**, L337 (1979); G. Christou, C. D. Garner, F. E. Mabbs, and J. J. King, *J.C.S. Chem. Comm.*, **1978**, 740; **1979**, 91.
9. R. R. Eady, D. J. Lowe, and R. N. F. Thorneley, *FEBS Letters*, **95**, 211 (1978).
10. R. N. F. Thorneley, M. G. Yates, and D. J. Lowe, *Biochem. J.*, **155**, 137 (1976).
11. D. J. Lowe, *Biochem. J.*, **175**, 955 (1978).
12. E. Munck, H. Rhodes, W. H. Orme–Johnson, L. C. Davis, W. J. Brill, and V. K. Shah, *Biochim. Biophys. Acta*, **400**, 32 (1975); B. E. Smith and G. Lang, *Biochem. J.*, **137**, 169 (1974).
13. D. J. Lowe, R. R. Eady, and R. N. F. Thorneley, *Biochem. J.*, **175**, 277 (1978); B. K. Burgess, E. I. Stiefel and W. E. Newton, *J. Biol. Chem.*, **255**, 353 (1980).
14. M. J. O'Donnell and B. E. Smith, *Biochem. J.*, **173**, 831 (1978).
15. A. D. Allen and C. V. Senoff, *Chem. Comm.* **1965**, 621.

16. For a recent review see J. Chatt, J. R. Dilworth and R. L. Richards, *Chem. Rev.*, **78**, 589 (1978).
17. D. Sellman, *Angew. Chem. Int. Ed.*, **13**, 639 (1974).
18. P. D. Cradwick, J. Chatt, R. H. Crabtree, and R. L. Richards, *J.C.S. Chem. Comm.*, **1975**, 351.
19. C. Busetto, A. D. D'Alfonso, E. Maspero, G. Perego, and A. Zazzetta, *J. C. S. Dalton Trans.*, **1977**, 1828.
20. J. Chatt, R. A. Head, G. J. Leigh, and C. J. Pickett, *J. C. S. Dalton Trans.* **1978**, 1638; V. W. Day, T. A. Geage, and S. D. A. Iske, *J. Amer. Chem. Soc.*, **97**, 4127 (1975); M. Hidai, Y. Mizobe, M. Sato, T. Kodama, and Y. Uchida, *J. Amer. Chem. Soc.*, **100**, 5740 (1978); H. M. Colqhoun, *J. Chem. Res. (s)*, 325 (1979).
21. J. Chatt, G. A. Heath, and R. L. Richards, *J. C. S. Chem. Comm.*, **1972**, 1011.
22. J. Chatt, A. J. Pearman, and R. L. Richards, *Nature*, **253**, 40 (1975).
23. C. R. Brulet and E. E. Van Tamelen, *J. Amer. Chem. Soc.*, **97**, 911 (1975).
24. J. Chatt, G. A. Heath, and R. L. Richards, *J. C. S. Dalton Trans.*, **1974**, 2074; J. Chatt, A. J. Pearman, and R. L. Richards, *J. C. S. Dalton Trans.*, **1977**, 1852, 2139; 1976, 1520; *J. Organometallic Chem.*, **101**, C45 (1975).
25. T. Takahashi, Y. Mizobe, M. Sato, Y. Uchida and M. Hidai, *J. Amer. Chem. Soc.*, **101**, 3405 (1979); *Chem. Letters*, **1978**, 1187.
26. M. E. Vol'pin and V. B. Shur, *Nature*, **209**, 1236 (1966).
27. Yu.G. Borodko, I. N. Ivleva, L. M. Kachapina, S. J. Salienko, A. K. Shilova, and A. E. Shilov, *J.C.S. Chem. Comm.*, **1972**, 1178.
28. J. Manriquez and J. E. Bercaw, *J. Amer. Chem. Soc.*, **96**, 6229 (1974).
29. Y. Muira and A. Yamamoto, *Chem. Letters*, **1978**, 937.
30. G. Henrici-Olive and S. Olive, *Angew. Chem. Int. Ed.*, **6**, 873 (1967).
31. E. E. van Tamelen, *Acc. Chem. Res.*, **3**, 362 (1970).
32. E. E. van Tamelen and J. D. A. Seeley, *J. Amer. Chem. Soc.*, **91**, 5194 (1969).
33. E. L. Moorehead, B. J. Weathers, E. A. Ufkes, P. R. Robinson, and G. N. Schrauzer, *J. Amer. Chem. Soc.*, **99**, 6089 (1977) and earlier papers in this series.
34. A. Shilov, N. Denisov, O. Efimov, N. Shuvalov, N. Shuvalova, and A. Shilova, *Nature*, **231**, 460 (1971).
35. N. T. Denisov, A. E. Shilov, N. I. Shuvalova, and T. P. Panova, *React. Kin. Cat. Letters*, **2**, 237 (1975).
36. N. T. Denisov, *Kinet. Catal.*, **17**, 906 (1976).
37. J. H. Anderson, *Biochem. J.*, **91**, 8 (1964); A. B. Hooper, K. R. Terry and P. C. Maxwell, *Biochim. Biophys. Acta*, **462**, 141 (1977).
38. M. K. Rees, *Biochemistry*, **7**, 353, 366 (1968).
39. A. B. Hooper, K. R. Terry, and P. C. Maxwell, *Biochemistry*, **17**, 2984 (1978).
40. A. B. Hooper and K. R. Terry, *Biochemistry*, **16**, 455 (1977).
41. F. T. Bonner, L. S. Dzelzkalns, and J. A. Bonucci, *Inorg. Chem.* **17**, 2487 (1978).
42. M. N. Hughes, *Quart. Rev. Chem. Soc.*, **1969**, 1.
43. M. N. Hughes and H. G. Nicklin, *Biochim. Biophys. Acta*, **222**, 650 (1970).
44. G. Yagil and M. Anbar, *J. Inorg. Nucl. Chem.*, **26**, 453 (1964).
45. M. N. Hughes and H. G. Nicklin, *J. Chem. Soc. (A)*, **1971**, 164.
46. M. N. Hughes and H. G. Nicklin, *J. Chem. Soc. (A)*, **1968**, 450.
47. M. N. Hughes, H. G. Nicklin, and W. A. C. Sackrule, *J. Chem. Soc. (A)*, **1971**, 3722.
48. M. N. Hughes and P. E. Wimbledon, *J. Chem. Soc. (Dalton)*, **1977**, 1975.
49. M. M. Werber and M. Mevarech, *Arch. Biochem. Biophys.*, **186**, 60 (1978).
50. J. A. Paleman, D. J. Cove, B. M. Rever, and D. B. Roberts, *Nature*, **201**, 58 (1964); A. Nason, A. D. Antoine, P. A. Ketchum, W. A. Frazier, and D. K. Lee, *Proc. Natl. Acad. Sci. U.S.A.*, **65**, 137 (1970); P. A. Ketchum and R. S. Swarin, *Biochem. Biophys. Res. Comm.*, **52**, 1450 (1973); P. A. Ketchum, H. Y. Cambier, W. A. Frazier, C. H. Madansky, and A. Nason, *Proc. Natl. Acad. Sci. U.S.A.*, **66**, 1016 (1970).

51. S. P. Vincent and R. C. Bray, *Biochem. J.*, **171**, 639 (1978).
52. K. J. Coleman, A. Cornish-Bowden, and J. A. Cole, *Biochem. J.*, **175**, 483 (1978).
53. M. J. Murphy, L. M. Siegel, S. R. Tove, and H. Kamin, *Proc. Natl. Acad. Sci. U.S.A.*, **71**, 612 (1974).
54. R. E. Cammack, D. P. Hucklesby, and E. J. Hewitt, *Biochem. J.*, **171**, 519 (1978); J. R. Lancaster, J. M. Vega, H. Kamin, N. R. Orme-Johnson, W. H. Orme-Johnson, R. J. Krueger and L. M. Siegel, *J. Biol. Chem.*, **254**, 1268 (1979).
55. P. F. Richardson, C. K. Chang, L. D. Spaulding and J. Fajer, *J. Amer. Chem. Soc*, **101**, 7738 (1979).

Note added in proof

Further models for the Fe-Mo cluster in nitrogenase have been prepared: R. H. Holm et al., *J. Amer. Chem. Soc.*, **101**, 4140 (1979); *Inorg. Chem.*, **19**, 430 (1980); D. Coucouvanis et al., *J. C. S. Chem. Comm.*, 360 (1979); B. K. Teo and B. A. Averill, *Biochem. Biophys. Res. Comm.*, **88**, 1454 (1979).

Core extrusion techniques have been applied to the iron–sulphur centres in the MoFe protein of nitrogenase from several sources (*A. vinelandii* and *C. pasteurianum.*) R. H. Holm et al., *Proc. Natl. Acad. Sci. U.S.A.*, **76**, 4986 (1979).

CHAPTER SEVEN

OXYGEN CARRIERS

The transport and storage of molecular oxygen* is an essential physiological function. It is carried out by a number of well-known iron and copper species, which occur in the blood of animals. These are listed in Table 7.1, where it may be seen that both heme and non-heme iron proteins are involved. The heme oxygen carriers, responsible for red colour of human blood, absorb one mole of dioxygen per mole of iron(II), while the blue pigment of crab blood, hemocyanin, absorbs one mole of dioxygen* for every two moles of metal ion. Certain marine worms (e.g. *Golfingia elongata*) have a violet colour which is due to the presence of the non-heme iron protein hemerythrin. Vanadium is present in the blood cells of certain ascidians, which are also of interest in that they involve a pH of *ca.* 1 due to the presence of sulphuric acid. The vanadium is present as V(III) and was once thought to be an oxygen carrier. This is now known to be incorrect.

These carriers all provide sites at the transition metal centre for the reversible attachment of dioxygen (Equation 7.1). In simple compounds these metal ions (e.g. Fe(II)) would be readily and irreversibly oxidized by molecular oxygen, but in hemoglobin iron(II) will bind dioxygen and not be oxidized by it.

$$\text{Carrier} + O_2 \overset{K_{O_2}}{\rightleftharpoons} \text{Carrier}\,(O_2) \tag{7.1}$$

For myoglobin the binding of dioxygen is represented straightforwardly by Equation (7.1). Myoglobin has a greater affinity for dioxygen than does hemoglobin, in accord with the fact that transfer of dioxygen takes place from hemoglobin to myoglobin for use in respiration in muscle cells. The uptake of dioxygen by hemoglobin is more complex due to interactions between its four subunits. Each subunit contains a heme group and protein and binds dioxygen reversibly. However, uptake of dioxygen by one unit enhances the ability of the other groups to take up dioxygen, so that the ratio of successive binding constants is 1:4:24:9. This 'cooperative effect' explains why hemoglobin carries dioxygen so efficiently. For the more dioxygen bound to hemoglobin the more likely it is that further dioxygen will be bound. Conversely the less dioxygen bound the more easily will it be released. This ensures that hemoglobin's binding capacity is fully utilized, and that all the dioxygen is released under conditions of low dioxygen concentration. Thus hemoglobin will be fully saturated in the lungs and completely deoxygenated in the capillaries.

---

*The term *dioxygen* is used to describe any form of O_2, in a free or complexed state. When used of complexes, this term has no significance as far as defining the extent of reduction of dioxygen is concerned. This will be specifically referred to by the use of 'superoxo' or 'peroxo'. *Molecular oxygen* refers to uncombined oxygen in the ground state, and *oxygenation* to the process of addition of dioxygen to a complex.

212

TABLE 7.1 Oxygen carrier and storage proteins

	Metal	Function	M/O$_2$	Ligands	Colour oxy	deoxy
Hemoglobin	Fe	Carrier	1 : 1	Porphyrin	Red	Red–blue
Myoglobin	Fe	Storage	1 : 1	Porphyrin	Red	Red–blue
Hemerythrin	Fe	Storage	2 : 1	Protein	Violet	Colourless
Hemocyanin	Cu	Carrier	2 : 1	Protein	Blue	Colourless

The extent of cooperativity is measured by the Hill coefficient, n, in the expression for the fraction of sites oxygenated (Y) (Equation 7.2). For myoglobin and isolated subunits of hemoglobin $n = 1$; for hemoglobin $n = 2.8$; for hemerythrin $n = ca.$ 1.2, suggesting that it is not a carrier but an oxygen storage protein like myoglobin, a view confirmed by its high affinity for dioxygen.

$$Y = K_{O_2} P_{O_2} / (1 + K_{O_2} P_{O_2})^n \qquad (7.2)$$

Reference has been made already to aspects of these oxygen carriers; for example, the five coordination of Fe(II) in hemoglobin and myoglobin. This provides an open site for the addition of dioxygen and implies a decrease in the activation energy for this addition reaction compared to that of a substitution reaction. The addition of dioxygen to hemoglobin is associated with a change in the magnetic properties of hemoglobin from paramagnetism to diamagnetism, while much interest has centred on X-ray studies which appear to show a change in the position of the iron atom with respect to the porphyrin plane. This has been linked to the question of the cooperative binding of dioxygen to the four subunits of the hemoglobin molecule.

A large number of transition metal complexes, particularly those of cobalt(II), carry dioxygen reversibly. Much useful information has come from a study of these model compounds particularly with respect to the structure of the M(O$_2$) group and the extent of electron transfer from the metal to the dioxygen ligand. An important group of model compounds has been the metal–porphyrin systems, particularly those with modified porphyrins such as the 'capped' and 'picket-fence' porphyrins, all of which are discussed in this chapter.

In theory, dioxygen may be bound in one of several ways in the metal complex. First of all we will discuss the electronic properties of the oxygen molecule and the structures taken up by M(O$_2$) systems. We will then examine a range of model compounds and attempt to assess the factors that determine whether or not these model systems will carry oxygen reversibly. The chapter will be concluded with a survey of current views on the transport of dioxygen by natural carriers, with a particular emphasis on hemoglobin. Various aspects of this topic are covered in recent reviews.[1-6]

THE OXYGEN MOLECULE AND ITS REDUCTION PRODUCTS

In the ground state the oxygen molecule has two unpaired electrons. This is explained readily by simple molecular orbital theory with its degenerate π orbitals. Molecular orbital designations for dioxygen and its possible reduction products, superoxide (O_2^-) and peroxide (O_2^{2-}) are given in Figure 7.1, together with the excited-state configurations of O_2, $^1\Delta$ and $^1\Sigma$ which are 95.3 and 154.5 kJ mol^{-1} above the ground state respectively.

As the oxygen molecule is reduced, electrons are fed into the antibonding π molecular orbitals with a decrease in bond order from 2 to 1.5 and 1 as O_2 is reduced to O_2^- and O_2^{2-}. Four electrons in all may be added, in which case all the antibonding orbitals are filled, the bond order is zero and oxide is formed. The reduction in bond order is associated with an increase in O—O bond length and a decrease in the stretching frequency, v_{O-O}, all of which are useful guides to the assessment of the extent of electron transfer from metal to dioxygen. These are summarized in Table 7.2. The overall geometry of the $M(O_2)$ or $M(O_2)M$ group is also a useful check on this matter. Peroxo complexes $M(O_2)M$ are frequently non-planar.

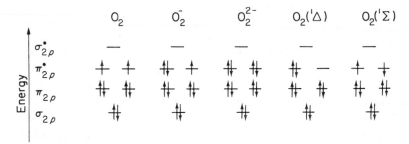

Figure 7.1 MO schemes for O_2, O_2^-, O_2^{2-} and excited states of O_2.

TABLE 7.2 Characterization of O_2, O_2^-, and O_2^{2-}

	O_2	O_2^-	O_2^{2-}
Bond order	2	1.5	1
Bond length (Å)	1.21	1.28	1.49
v_{O-O} (cm^{-1})	1560	1150–1100	850–740

THE REACTIONS OF DIOXYGEN

Many reactions of ground-state dioxygen are complex chain reactions, reflecting its diradical character. Molecular mechanisms are also observed but usually for excited-state molecular oxygen. Reactions with transition metal ions are often much simpler. It is instructive to compare redox potentials for the various reduction reactions of oxygen with other redox potentials[7] (Table 7.3). The

TABLE 7.3 Some redox potentials

Couple	E^0 (V)	Equation no.
$O_2 + e \rightleftharpoons O_2^-$	-0.4	(7.3)
$O_2 + 2H^+ + 2e \rightleftharpoons H_2O_2$	$+0.68$	(7.4)
$O_2 + 4H^+ + 4e \rightleftharpoons 2H_2O$	$+1.23$	(7.5)
$O_2 + H^+ + e \rightleftharpoons HO_2$	-0.1	(7.6)
$Fe^{3+} + e \rightleftharpoons Fe^{2+}$	$+0.77$	(7.7)

potential for reaction (7.5) indicates that molecular oxygen is a strong oxidizing agent. The full redox potential will only be realized when the oxidation carried out involves an essentially synchronous four-electron reaction. Such reactions have been suggested for biological oxidations, but if this cannot occur, then reduction will take place either by one- or two-electron steps (Equations 7.3 and 7.4). The geometry of the π antibonding molecular orbitals of the oxygen molecule will, of course, impose limitations on the feasibility of different redox reactions occurring. Reduction will involve the transfer of electrons from the valence shell of the metal ion to the oxygen molecule and there will therefore be a symmetry requirement on the orbital of the metal ion. The implications of the Franck–Condon principle, discussed in Chapter 2, will also hold here.

In a number of reactions of oxygen with transition metal species in acid solution (such as iron(II)) it has been shown that the rate law involves a square dependence on metal ion concentration. This can be understood in terms of the redox potentials in Table 7.3. This data shows that the reduction of oxygen to superoxide (O_2^-) by Fe(II) is not favoured. However, as the stoichiometry involves two moles of Fe(II) it may be seen that under these conditions the oxygen will now be reduced to peroxide O_2^{2-}, a more favoured reaction, particularly as the peroxo product is probably stabilized by coordination to the Fe(III) produced. The presence of ligands other than water in the coordination shell of the ferrous ion has a marked effect on the rate of reaction. Oxygen donor ligands will particularly affect the rate of reaction as these bases have π-donor properties and hence will enhance the transfer of electron density from the appropriate metal d orbital into the oxygen molecule. In contrast σ donors, such as nitrogen ligands, *trans* to the oxygen molecule will have little influence.

While some oxidations of transition metal complexes by dioxygen proceed by an outer-sphere reaction, others involve reversible formation of a metal dioxygen complex as a first step, followed by electron transfer to form coordinated superoxide ion. In many cases this is followed by coordination of the superoxide ion to another complex cation with further reduction to peroxide, as illustrated in Equation (7.8) for Co(II) compounds. Mononuclear superoxocobalt(III) complexes may be stabilized in several ways, for example by the use of sterically hindered ligands, non-aqueous solvents, low temperature, or by the use of ligands with considerable π-acceptor character. While in this chapter it is convenient to refer to dioxygen complexes in terms of superoxo or peroxo

$$L_5Co^{II} \xrightarrow{O_2} [L_5Co^{III}-O^{\diagup O}] \xrightarrow{L_5Co^{II}} [L_5Co^{III}-O^{\diagup O-Co^{III}L_5}] \quad (7.8)$$

species, it must be emphasized that the transfer of electron density onto the dioxygen may not correspond exactly to that required to form such superoxo or peroxo complexes.

MODEL COMPOUNDS FOR OXYGEN CARRIERS

In this section we will consider various model systems in which the oxygen molecule is present as O_2^- or O_2^{2-} and may or may not be carried as such. There is no evidence for further reduction of bound oxygen in model compounds, except in the case of cobalt complexes with dipeptides. Discussion of porphyrin complexes of iron and cobalt is reserved until the consideration of natural carriers.

The air oxidation of cobalt(II) complexes with ammine and amine ligands

It has been known since the time of Werner that cobalt(III) complexes may be prepared by assembling the ligands and cobalt(II) (which is labile and therefore the complex is rapidly formed), and then air oxidizing to give the cobalt(III) complex, which is inert. In particular it was known that ammoniacal solutions of cobalt(II) complexes were weak oxygen carriers but that they were ultimately air oxidized to give brown salts containing the diamagnetic cation $[(H_3N)_5CoO_2Co(NH_3)_5]^{4+}$. Treatment with hydrogen peroxide gives the green paramagnetic $[(H_3N)_5CoO_2Co(NH_3)_5]^{5+}$. Much work has been done on these diamagnetic and paramagnetic species, which are fairly common. Thus air oxidation of cobalt(II) cyanide leads to the formation of brown, diamagnetic $[(NC)_5CoO_2Co(CN)_5]^{6-}$, while further treatment with bromine in alkali gives the red paramagnetic species $[(NC)_5CoO_2Co(CN)_5]^{5-}$. The diamagnetic species involve two low-spin Co(III) ions bridged by a peroxo group, $Co^{III}-O_2^{2-}-Co^{III}$, formed as described above (Equation 7.8). Further oxidation gives the paramagnetic complexes which Werner originally formulated as peroxo-bridged Co(IV) and Co(III). EPR studies show however that both atoms are equivalent, a conclusion that is intuitively reasonable as the two cobalt atoms have identical environments. It appears that both cobalt atoms are cobalt(III) with bridging superoxide, a view confirmed by the electronic spectra of the complexes.[3] Molecular orbital theory predicts that the odd electron is in the orbital which extends over both cobalt atoms and both oxygen atoms. Spectroscopic measurements confirm that the electron is concentrated on the oxygen atoms in accord with the presence of an O_2^- bridging group. The oxygen–oxygen bond length in the decaammine is 1.31 Å, a value consistent with a superoxide species (Figure 7.2)[8]. The structure of $[(NH_3)_5CoO_2Co(NH_3)_5]^{4+}$ has the longer oxygen–oxygen bond (1.47 Å) expected for a peroxo group.[9]

Mononuclear superoxo complexes of Co(III) have been studied but they are usually unstable species reacting rapidly to give the bridged peroxo complexes.

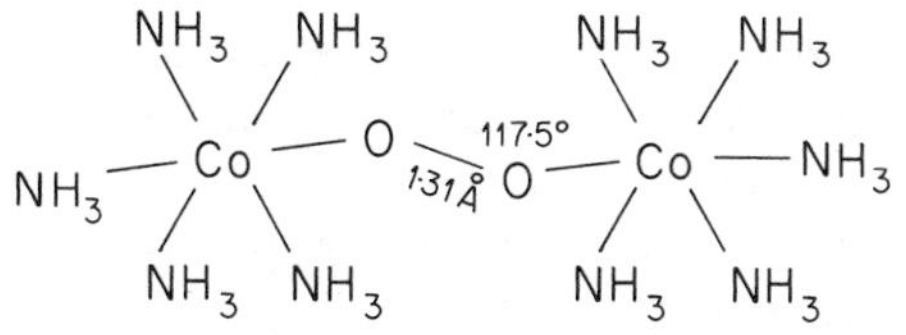

Figure 7.2 The structure of $[(NH_3)_5CoO_2Co(NH_3)_5]^{5+}$.

Particularly stable species may be prepared[10] by using N,N'-disubstituted ethylenediamines as in the case of $[Co(Me_2en)_2XO_2]^{n+}$, where $X = Cl^-$ or solvent. These are stable at room temperature in non-aqueous solvents in contrast to the lower temperatures required to stabilize Schiff-base systems. The ligand-to-metal charge-transfer transitions $O_2^- \rightarrow Co(III)$ and $Cl^- \rightarrow Co(III)$ have been observed at 340 and 236 nm respectively (a measure of the simplicity of this ligand system) providing confirmation of the cobalt oxidation state.[11]

A wide range of μ-dioxygencobalt complexes with various amine ligands have also been characterized. These will also contain hydroxo bridges if a vacant coordination site is available on each cobalt atom in the binuclear pair. Thus the complexes with ethylenediamine, diethylenetriamine, and triethylenetetramine have μ-hydroxo bridges and are formulated $[CoL_2(O_2)(OH)CoL_2]^{3+}$ for ethylenediamine and $[CoL(O_2)(OH)CoL]^{3+}$ for the latter two. The hydroxo bridge is absent when the polyamine ligand is tetraethylenepentamine and the complex $[CoL(O_2)CoL]^{4+}$ is formed. These complexes all undergo irreversible oxidation to Co(III) products.

Cobalt(II) complexes with Schiff-base ligands

Certain of these chelates, which provide a set of four good coplanar ligand atoms, are very effective oxygen carriers, both in the solid state and in solution in certain solvents. Complexes **I** and **II** in Figure 7.3 are particularly well known.[12] Carriers related to cobalt(II) salen (complex **I**) have been most well studied. Salen is formed by the Schiff-base condensation of two moles of salicylaldehyde with ethylenediamine, while complex **II** involves a Schiff-base ligand formed from salicylaldehyde and a triamine.

Both types of complex exist in a number of crystalline modifications, only some of which are active. Complexes of salen and related ligands have a square-

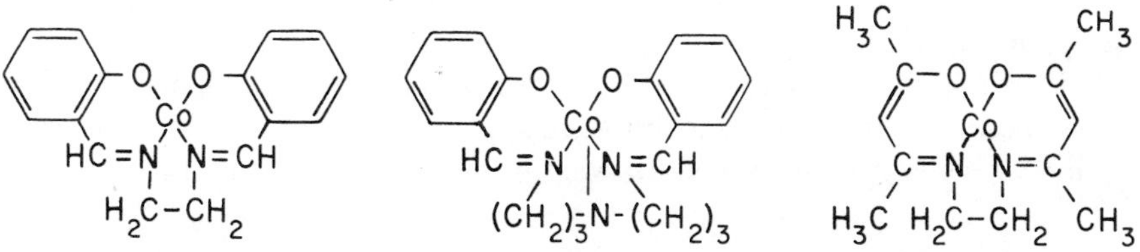

Figure 7.3 Synthetic dioxygen carriers with Schiff-base ligands.

planar structure and are packed parallel in the crystal. They absorb oxygen with a stoichiometry of one mole of dioxygen per two moles of complex, the dioxygen bridging two cobalt centres. The crystal structure of the inactive modifications shows that they exist as dimers, which presumably are unable to pick up dioxygen. Type II chelates have a tetragonal pyramidal structure and absorb one mole of oxygen per mole of complex. Magnetic measurements show that chelates of type I and II have one and three unpaired electrons respectively. After oxygenation type I chelates become diamagnetic and type II chelates have one unpaired electron suggesting the formation of a peroxo species in each case.

These chelates undergo a number of cycles of oxygenation and deoxygenation, but their efficiency gradually deteriorates. This is at least partly due to the oxidation of the ligands. An alternative view that loss of activity resulted from the gradual loss of a coordinated water molecule is no longer accepted.

The four-coordinate cobalt(II) complexes have a much lower affinity for dioxygen than the five-coordinate complexes.[1] Similarly the dioxygen affinity of the four-coordinate species is enhanced by the binding of a fifth ligand (such as pyridine) in one of the axial positions to give a square-pyramidal structure. Studies in solution[13] have led to the characterization of a range of substituted salen complexes of Co(III) of the form $(B)(salen)Co^{III}(O_2^{2-})$—$Co^{III}$ (salen)(B) where B is a base such as pyridine. The pyridine adduct is inactive in the solid state. With 3-methoxysalen in pyridine the monomeric form was isolated, with superoxide bound to Co(III). At temperatures below $0°$, the complex Co(acacen) (III in Figure 7.3) acted as an oxygen carrier, and solid $Co(acacen)B(O_2)$ could be isolated. This was shown to involve $Co(III)$—(O_2^-) with a structure $Co^{O\diagdown O}$. This type of superoxide structure has been confirmed by accurate crystal structure determination on several compounds,[14] which give oxygen–oxygen bond lengths in the range 1.27–1.30 Å and Co—O—O angles of about $117°$. The non-equivalence of the oxygen atoms and the location of nearly 100% of the spin density on dioxygen is shown by EPR techniques.[15] All this confirms beyond doubt the formulation of these compounds as superoxocobalt(III) species, a view supported unambiguously by X-ray photoelectron spectra.[16] The formation of bridged peroxo complexes from these mononuclear superoxo complexes is a facile process, and may be avoided by the use of low temperatures, dilute solutions, steric hindrance in the ligand, or by the selection of an axial ligand having appropriate electronic properties. As we shall see later, this type of mononuclear superoxo $M(O_2)$ unit is especially relevant to the understanding of oxyhemoglobin and oxymyoglobin, for which Pauling proposed end-on dioxygen–metal bonding in the 1930's.

Evidence has also been presented for the formation of a 1:1 mononuclear adduct between V(III) salen and dioxygen in pyridine, an observation of interest in view of the supposed oxygen-carrying properties of V(III) in ascidians.

Cobalt(II) complexes with amino acids and peptides

Many complexes of cobalt(II) with these ligands display oxygen-carrying properties. Most studies on them have been carried out in solution, but some oxygenated solids have been isolated in which one mole of dioxygen is added to two moles of complex. Complexes of histidine have been well studied, and it is noteworthy that while Fe(II) complexes are irreversibly oxidized to Fe(III), both Cu(I) and Ni(II) species are unaffected by oxygen. Only the cobalt complex acts as an oxygen carrier. Clearly the redox potential of the Co(II)/Co(III) couple lies in an intermediate state in terms of the equilibrium between M^{n+}, O_2, and $M^{(n+1)+}$, i.e. Fe < Co < Ni < Cu. The ease of oxidation of a metal in a given complex is related to the ease of uptake of dioxygen by that complex.

It should be noted, however, that the nickel complex of the tetrapeptide tetraglycine is readily oxidized by molecular oxygen under mild conditions. It appears that in this case nickel(II) is activated by peptide coordination so that it reacts with molecular oxygen and catalyses the oxidation of the peptide to a number of products (i.e. probably through the formation of nickel(III)).

Gillard and Spencer[17] have measured the uptake of oxygen by alkaline solutions of cobalt(II) with nineteen dipeptides and have observed three different types of stoichiometric behaviour: (1) O_2:4Co:8 dipeptide: (2) O_2:2Co:4 dipeptide, and (3) continuous oxygen uptake with catalytic oxygenation of the dipeptide. Oxygen uptake is very pH dependent. The 2:1 complex of glycylglycine with cobalt(II) at pH 10 reacts rapidly with oxygen giving yellow (or at higher concentrations, brown) solutions of binuclear, diamagnetic oxygenated species. Depending upon the ligand, these complexes decompose at a variety of rates to yield red mononuclear bis(dipeptide) cobalt(III) chelates. The authors conclude that a minimum of three N donors is necessary for formation of an oxygenated complex.

The irreversible formation of these Co(III) complexes is usually the limiting factor in determining the number of reversible oxygenation cycles through which a cobalt(II) complex will function. With dipeptides this oxidation is often rapid enough to compete with the oxygenation reaction. A recent report[18] on the irreversible redox rearrangement of cobalt oxygen complexes of dipeptides to give these red products leads to the conclusion that the reaction occurs in two steps. The first (and rapid one) involves the oxidation of one ligand molecule for each pair of cobalt atoms and the conversion of the bridging dioxygen group to water, an interesting reaction. The second step involves the replacement of the oxidized ligand by the excess dipeptide present in solution.

Vaska's iridium complex

The complex Ir(PPh$_3$)$_2$COCl behaves[19] as an oxygen carrier. The oxygenated form is diamagnetic and has a trigonal bipyramidal structure, containing π- bonded oxygen (Figure 7.4). This compound has proved a useful model for more complex systems. The parent complex only reacts with oxygen in benzene solution, taking up one mole of oxygen per mole of complex, with a colour change from yellow to red. Molecular weight studies show it to be monomeric in benzene. It yields hydrogen peroxide on treatment with acid, and is diamagnetic. The infrared spectrum of the carrier shows a band at 860 cm^{-1} characteristic of a coordinated peroxo group. It is now accepted however that this does not

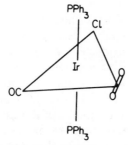

Figure 7.4 Vaska's iridium complex.

correspond to a pure 'O—O' stretching frequency and that a symmetrical ν_{M-O} mode is mixed with it. Vaska's complex will also reversibly absorb other molecules such as hydrogen, chlorine, and the hydrochloric acid molecule. It appears that the oxygen molecule is π bonded only and it has been suggested that the carrier properties of the complex may be associated with this. Certainly Vaska's complex will reversibly bind a number of π-bonding ligands. The reversible addition of molecular oxygen may then be represented by Equation (7.9).

$$Ir(PPh_3)_2(CO)Cl + O_2 \underset{k_{-1}}{\overset{k_2}{\rightleftharpoons}} Ir(PPh_3)_2(CO)Cl(O_2) \qquad (7.9)$$
$$K = k_2/k_{-1}$$

X-ray structural studies[20] show that both oxygen atoms are equidistant from the iridium atom. The structure may then be regarded formally as being either five or six coordinate depending on whether the interaction with the metal is through both oxygen atoms or the double bond of the oxygen molecule. The use of matrix isolation techniques[21] on the products of reaction between dioxygen and Ni, Pd, or Pt in pure oxygen and oxygen/argon mixtures has led to the conclusion in this case that the symmetrical triangular structure for the $M(O_2)$ complex is correct. In Vaska's complex the oxygen–oxygen bond length (1.30 Å) is close to that of a superoxo species but this is not a true representation of the complex as this implies iridium in an oxidation state of II. Iridium(II) species (d^7) are paramagnetic. However, a peroxide structure implies iridium in the III oxidation state (d^6), consistent with a diamagnetic complex. It is probably best, at this stage, to regard this complex as a dioxygen complex of iridium(I) and not to attempt to interpret the O—O bond length, but it is probably a peroxo complex of Ir(III).

The function of the phosphine ligand may be readily understood in broad terms. This is a group that stabilizes low oxidation states, and will therefore help to prevent the irreversible reduction of oxygen by electron transfer from the iridium ion. The role of this ligand has been examined in more detail.[22] It has been shown that the rate constants for oxygen addition (k_2) to *trans*-[IrCl(CO)(R$_3$P)$_2$] (Equation 7.9) and the stability of the dioxygen adduct (K) increase with the basicity of the group R in the phosphine ligand, provided the groups being considered are similar in geometry. The geometry of group R can have a considerable effect. Thus oxygen addition does not occur when $R = o$-$CH_3C_6H_4$, due to the steric effect of the methyl group, although the electronic properties of the group are favourable. No reaction occurs, in addition, when $R = C_6F_5$ due to the lack of electron density on the metal. Vaska and Chen also noted a correlation between constants k_2 or K and the carbonyl stretching frequency (ν_{CO}) in the complexes when steric effects are absent. Thus for $R = C_6F_5, \nu_{CO} = 1994 \, cm^{-1}$ while values lie between 1966 and $1934 \, cm^{-1}$ for complexes with other phosphines which do absorb oxygen. These results are important in helping to define the conditions under which a transition metal ion will bind oxygen reversibly, even though dioxygen complexes of this type do not appear to be relevant to biological systems.

Complexes with dimethylglyoxime and related ligands

It has been seen that a number of oxygen-carrying cobalt(II) complexes involve ligands that provide four in-plane donor atoms, such as the 'salen' complexes discussed earlier. Thus corrin, porphyrin, and dimethylglyoxime[23] complexes carry oxygen, while the use of EPR techniques has shown[24] that phthalocyanine complexes also form 1:1 complexes with dioxygen. Both the cobalt(II) corrinoids and phthalocyanines do this in the absence of a nitrogen donor ligand in the axial position. A dioxygen complex of manganese phthalocyanine has been isolated.[25]

Bis(dimethylglyoximato)iron(II) readily reacts with oxygen in the presence of ligands such as pyridine, ammonia, histidine, or imidazole. The reaction may be reversed by bubbling nitrogen through the solutions. Strongly alkaline solutions of bis(dimethylglyoximato)nickel(II) also reversibly absorb oxygen, while the corresponding cobalt(II) complexes have been studied[26] as models for the oxygen carrier vitamin B_{12r}. Bis(dimethylglyoximato)cobalt(II), CoD_2H_2 (where DH_2 = dimethylglyoxime) is a paramagnetic solid ($\mu = 1.70$) which will react with certain bases to give 1 : 1 and 1 : 2 adducts which are soluble in solvents such as acteone, benzene, and dichloromethane. Such solutions absorb 0.5 moles of oxygen per mole of cobalt, giving intensely coloured products in which the oxygen is held only weakly, being readily removed by a stream of nitrogen or by the action of heat. These species are easily oxidized to cobalt(III), particularly in the presence of traces of water. The initial oxygen adducts are diamagnetic, μ-peroxo complexes which may be isolated as solids. The solutions gradually trun brown with the evolution of oxygen, and the formation of μ-superoxo species. (Equation 7.10)

$$2BCo(D_2H_2)\text{—}O_2\text{—}Co(D_2H_2)B \rightarrow BCo(D_2H_2)\text{—}O_2\text{—}Co(D_2H_2)B^+$$
$$+ 2Co(D_2H)B + \tfrac{1}{2}O_2 + H_2O \qquad (B = Base) \qquad (7.10)$$

On standing, the solution of the superoxo species (B = pyridine) undergoes further change, as is evidenced by the change in the EPR spectrum, apparently to the mononuclear peroxo complex $py\text{-}Co(D_2H_2)O_2^{\cdot}$. Vitamin B_{12r}, on reaction with oxygen, gives a peroxo radical having an EPR spectrum very similar to that of $py\text{-}Co(D_2H_2)O_2^{\cdot}$, suggesting that the electronic environment of the $Co\text{—}O_2^{\cdot}$

group must be very similar in both cases. It is probable that the cobalt(II) chelates react with oxygen in two stages (Equation 7.8) to give the μ-peroxo complexes, and that steric reasons prevent the formation of the bridging peroxide in the case of vitamin B_{12r}.

Model complexes in which singlet oxygen is postulated

Singlet dioxygen has been postulated for certain of the $Co(O_2)$ and $Ir(O_2)$ complexes discussed so far, but has later been recognized to be incorrect as there appears now to be no doubt that cobalt(II) is oxidized in the oxygenation process. However two recent reports on other systems suggest that excited-state oxygen is bound in dioxygen complexes.

Oxygenation of the iron(II) complex cis-[Fe(bt)$_2$(NCS)$_2$], where bt = 2,2'-bi-2-thiazoline, in non-aqueous solvents gives a product which on the basis of magnetic and Mössbauer evidence is suggested[27] to involve a binuclear structure with seven-coordinate Fe(II), in the triplet ground state, bridged by singlet oxygen. The alternative explanation invoking antiferromagnetism is eliminated by the temperature dependence of the magnetic moments.

Mn(II) complexes $MnLX_2$, where L is a tertiary phosphine and X an anion, carry dioxygen reversibly through thousands of cycles without apparent deterioration.[28] Of great interest is the observation that uptake of oxygen is cooperative, the second stage of the uptake being faster than the first and being subject to photochemical catalysis. It is suggested that the uptake of the first 50 mol % of dioxygen involves ground-state, triplet O_2 giving a total of seven unpaired electrons when complexed to Mn(II). The second 50 mol % is suggested to be singlet oxygen, giving a total of five unpaired electrons in the Mn(II) complex, so accounting for the apparent presence of six unpaired electrons in the totally oxygenated system. The structures of these compounds have not yet been determined, but are probably polymeric in solution. They offer a unique duplication of essential features in the behaviour of hemoglobin.

SOME COMMENTS ON DIOXYGEN COMPLEXES

The important metal ions are iron(II) and cobalt(II), while manganese(II) is also significant. Four structural types have been distinguished corresponding to mononuclear and bridged superoxo and peroxo complexes. These are listed in Table 7.4, with illustrative examples from cobalt(II) chemistry. These structures are formulated on the basis of magnetic measurements, oxygen–oxygen bond lengths, and stretching frequencies and in a few cases the planarity or otherwise of the $M(O_2)M$ group. Solid-state packing problems sometimes cause confusion by giving planar peroxo groups. These formulations may be over-simplified as in the case of Vaska's complex, or in the dioxygen adduct of Co(II) salen, which contains a twisted $Co(O_2)$ Co group[29] and yet has a bond length much shorter than expected for a peroxide, suggesting that complete transfer of electrons to oxygen from the cobalt has not occurred. Such phenomena may well be related to reversibility.

TABLE 7.4 Structural types of dioxygen complexes of cobalt

Complex	O—O (Å)	∠Co—O—O (deg)	Reversible?
Superoxo Co—O⋯O			
Co(acacen) (py) (O$_2$)			Yes
Co(bzacen) (py) (O$_2$)	1.26	126	Yes
(Et$_4$N)$_3$[Co(CN)$_5$)(O$_2$)].5H$_2$O	1.24	153	No
Co—O⋯O—Co			
[Co$_2$(NH$_3$)$_{10}$(O$_2$)](NO$_3$)$_5$	1.317	117.3	
[Co$_2$en$_4$(NH$_2$)(O$_2$)](NO$_3$)$_4$	1.353	119.2	
Peroxo			
[Co(2 = phos)(O$_2$)](BF$_4$).2C$_6$H$_6$	1.420		No
[Co$_2$(CN)$_4$(PMe$_2$Ph)$_5$(O$_2$)]	1.44		
Co—O—O—Co			
[Co$_2$(NH$_3$)$_{10}$(O$_2$)](SO$_4$)$_2$.4H$_2$O	1.473	113	No
[Co$_2$(NH$_3$)$_{10}$(O$_2$)](SCN)$_4$	1.469	110.8	No
Co$_2$(salen)$_2$(DMF)$_2$(O$_2$)	1.339	120.3	Yes
Co$_2$(salen)$_2$(piperidine)$_2$ (O$_2$)	1.383	120.0	Yes

Range of O—O bond lengths known at present:
superoxo, monomeric 1.1–1.302 Å
superoxo, dimeric 1.243–1.353 Å
peroxo, dimeric 1.308–1.488 Å

The superoxo bridged structure is confined to cobalt(II) and arises from a one-electron oxidation of the peroxo bridged complex. The side-on, π-bonded peroxo complex, while well known in low-valent complexes of the precious metals, is not important in first row transition metals. The only exceptions for Co(II) are $[Co(2 = phos)_2(O_2)]$ BF_4 and $Co_2(CN)_4(PMe_2Ph)_5(O_2)$. The latter complex is binuclear;[30] but the cobalt atoms are bridged by cyanide ion rather than the O_2 molecule, the oxygen molecule being coordinated as a bidentate ligand to one of the cobalt atoms in a manner similar to the $Ir(O_2)$ system in Vaska's complex. Both superoxo-bridged and side-on π-bonded dioxygen do not appear to be important from a biological point of view. In contrast the mononuclear end-on superoxo complex is almost certainly relevant to oxyhemoglobin, while the peroxo-bridged species is probably present in hemerythrin and hemocyanin.

The bonding of dioxygen to the metal ions may be readily understood. In Figure 7.5(a) is shown the 'side-on' π bonding of oxygen to Vaska's complex. Both π and π^* orbitals of dioxygen can interact with the bonding set of s, p, and d orbitals on the metal, giving a σ bond by electron transfer from the filled dioxygen π orbital to appropriate orbitals on the metal, and π bonds by electron transfer from filled metal d orbitals into unfilled π^* orbitals of dioxygen.

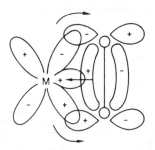

(a) π-bonding of dioxygen oxygen to the metal ion.

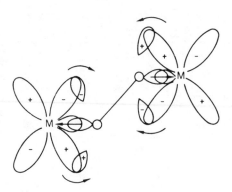

(b) σ-bonding of dioxygen oxygen in the cation $[(NH_3)_5Co—O—O—Co(NH_3)_5]^{4+}$

Figure 7.5 Dioxygen oxygen binding by a transition metal ion.

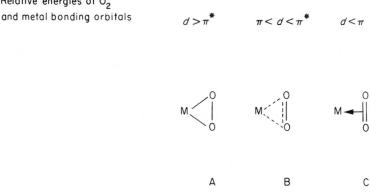

Figure 7.6 Schematic representation of side-on metal–dioxygen bonding.

According to the extent of electron transfer *via* back bonding, so reduction of dioxygen will occur. If the metal orbitals are of similar or higher energy than the π^* orbitals of dioxygen then the bonding orbitals formed from metal orbitals and dioxygen π orbitals will be largely dioxygen in character, so there will be little dioxygen–metal σ bonding. The π orbitals however will have both metal and oxygen character, and so will form strong bonds. If back donation is considerable, then the O—O bond order will be reduced to one, with lengthening of the bond.

On the other hand, if the energy of the d orbital is less than that of the dioxygen π orbital, then the σ bond between dioxygen and metal ion will be strong and the π bonding will be weak. In this case the O—O bond length will be little affected. Both situations are represented in Figure 7.6 together with a third possibility (B), intermediate between these two (A and C).

For Vaska's complex (and others involving iridium, rhodium, and platinum) the bonding situation is A or B. These forms as noted above are less likely for Co(II) and Fe(II) as the metal d orbitals will be less able to overlap well with the π^* dioxygen orbitals. However the alternative 'side-on' scheme, C, does not allow metal → dioxygen back donation which is necessary to lower the formal charge on the metal. Accordingly the 'end-on' bonding situation is favoured which allows back donation. This is shown in Figure 7.5(b) (and will hold for the mononuclear structure).

Figure 7.7 recapitulates the relationships between the various types of

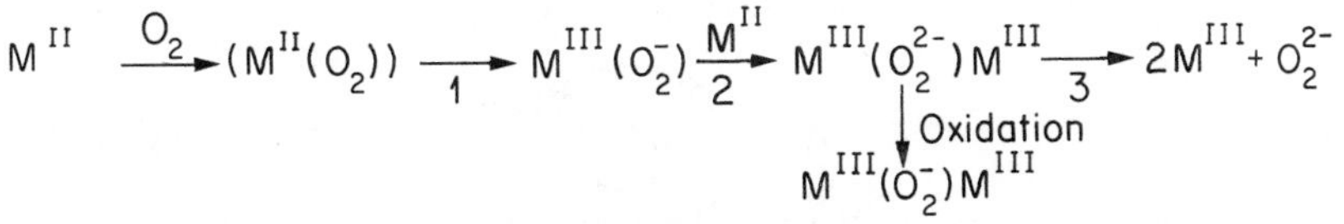

$$M^{II} \xrightarrow{O_2} (M^{II}(O_2)) \xrightarrow{1} M^{III}(O_2^-) \xrightarrow{M^{II}} M^{III}(O_2^{2-})M^{III} \xrightarrow{3} 2M^{III} + O_2^{2-}$$

$$\downarrow \text{Oxidation}$$

$$M^{III}(O_2^-)M^{III}$$

Figure 7.7 Relationships between various dioxygen complexes (ligands are omitted).

dioxygen compound. Step (3) is important because once this has occurred the reaction cannot be reversed. The peroxo-bridged species is unstable with respect to the products of dissociation and so this reaction will occur readily for Fe(II). In cobalt systems this step is slow because Co(III) is inert. Reversible reactions with Fe(II) can only take place if bridge formation is prevented. It will be recalled that for Co(II) this could be achieved by using dilute solutions, or sterically hindered ligands, or low temperature or by the use of ligands with considerable π-acceptor character. This approach has also been used with Fe(II) porphyrins (to be discussed later) and with the Fe(II) complex of octaaza (14) annulenes with 9,10-bridged 9,10-dihydroanthracenes. Here it was not possible for the complex molecules to approach each other closely enough to form a bridge, so this species readily gave reversible formation of a mononuclear dioxygen complex.[31] This complex is shown in Figure 7.8.

As noted in the discussion of metal–peptide complexes, the redox potential of the M(II)–M(III) couple is important. This has been demonstrated for a range of Schiff-base cobalt(II) complexes with varying axial ligands by the establishment of a linear correlation between the equilibrium constants for dioxygen adduct formation and the ease of oxidation of Co(II) to Co(III), as measured by cyclic voltammetry.[32] This correlation exists because the redox potential of the cobalt chelate is a measure of the electron density on the cobalt, which is a key factor in determining the oxygen affinity of the complexes. A similar result has been obtained with a series of cobalt porphyrins. This will account too for the requirement of three nitrogen donor atoms for the oxygenation of cobalt(II)

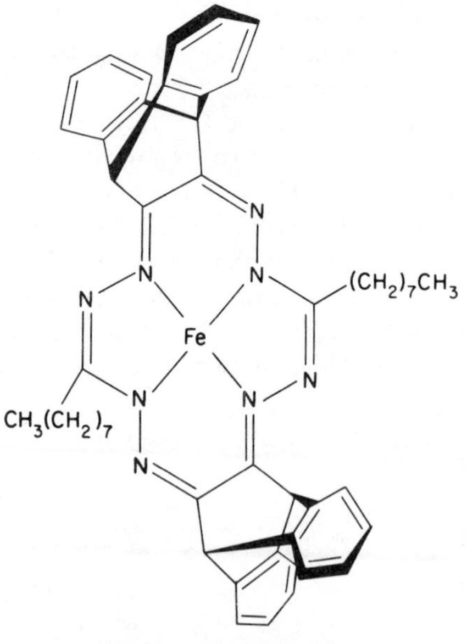

Figure 7.8 Fe (II) complex of octaazaannulene with 9,10-bridged 9,10-dihydroanthracene[31].

peptide complexes, as nitrogen will donate electron density more effectively than oxygen.

Loss of π-electron density from cobalt to a ligand will cause a decrease in dioxygen affinity as metal-to-dioxygen donation is decreased. Examples include the case of the sulphur analogue of salen[33] (where the sulphur atoms accept electron density from cobalt) and the lower tendency of Co(II) porphyrins to bind dioxygen compared to the Co(II) Schiff-base complexes. Ligands *trans* to dioxygen in a complex with a macrocyclic planar ligand are particularly important in controlling the extent of interaction with dioxygen. In general carboxylate and heterocyclic ligands reduce oxygen-carrier properties.

Variation in one ligand may have a considerable effect on these properties as there is a delicate balance between partial electron transfer to oxygen, irreversible reduction of oxygen, and failure to show oxygen-carrying properties. Thus the replacement of the chloro group in Vaska's salt by an iodo group gives a complex which is irreversibly oxidized by oxygen. The greater charge-releasing properties of the iodide result in the metal having greater electron density and hence a more facile reduction of oxygen. Appropriately, the oxygen–oxygen bond length in this compound is increased to 1.51 Å.

NATURAL OXYGEN CARRIERS AND PORPHYRIN MODEL COMPOUNDS

The hemoproteins

As noted previously, heme is a square-planar iron(II) complex of protoporphyrin (Figure 5.4). The four-coordinate iron may accept two additional ligands. In hemoproteins these are usually histidine groups, but the sixth position is often filled by weakly bound ligands, so explaining why they may be readily replaced by π-bonding ligands such as cyanide and carbon monoxide. This usually results in an electronic rearrangement to a low-spin complex. In hemoglobin and myoglobin, the iron(II) is bound by one axial group, an imidazole of a histidine residue, giving a five-coordinate species.

Iron(III) protoporphyrins, the hemins, are formed by irreversible oxidation of the hemoproteins, particularly in acid solutions or in media of high dielectric constant. The question of the ease of oxidation of iron(II) porphyrins is an important one, and depends upon a number of factors, such as the nature of the axial ligands, the porphyrin substituents, and the configuration of the protein.

The structure of hemoglobin and myoglobin

Perutz and Kendrew and their coworkers have published X-ray diffraction data of high resolution on a number of hemoglobins and myoglobins respectively, and have laid the foundation for the investigation of the molecular basis of their physiological function. Hemoglobin is a tetramer of molecular weight 64,450. It is made up of two identical pairs of units (the α and β units) arranged roughly in a tetrahedron.

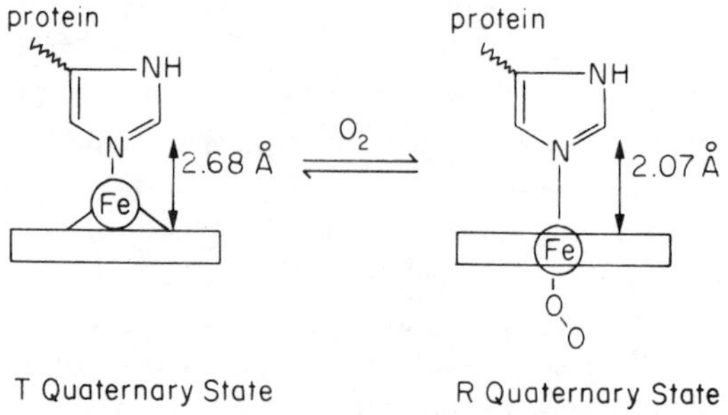

Figure 7.9 Trigger for the cooperative effect in hemoglobin, after Hoard and Perutz.

An explanation of the cooperative effect is central to the understanding of the role of hemoglobin. The distance between heme groups in the tetramer is considerable and so the effect must be transmitted through the protein structure. Uptake of O_2 is associated with substantial conformational changes, explaining why crystals of hemoglobin crumble on conversion to the oxy form. These conformational changes are associated with the reversible interconversion of a tensed (T), low-affinity form of hemoglobin into a relaxed (R), high-affinity form. It is usually thought that this process is triggered by the movement of the iron atom towards the plane of the heme group in response to the binding of dioxygen. This mechanism (Figure 7.9) was put forward by Perutz[34] and followed an earlier suggestion by Hoard[35] which was based on the structures of iron(II) porphyrins. The deoxy form of hemoglobin is high-spin, paramagnetic with the iron(II) lying above the plane of the heme, due it is suggested to its large size, which prevents it moving into the 'hole' in the porphyrin. Binding of dioxygen then results in the formation of low-spin iron(II), which is smaller and able to move into the plane of the porphyrin group. The resulting change in Fe—N(imidazole) bond length is magnified through the interactions in hemoglobin so that substantial conformational changes occur. The papers of Perutz[34,36] should be consulted for full details, but one result is a change in quaternary structure, so that the distance between heme groups is decreased on oxygenation. The interactions between subunits must therefore be considered; these are assumed to be salt bridges, as dissociation of the subunits is enhanced by neutral electrolytes in high concentration, presumably by a weakening of these polar interactions. Myoglobin shows no conformational change on deoxygenation, but here there is only one heme group and a heme–heme cooperative effect is not possible.

This view of the trigger effect is now thought to be at least partly incorrect. Thus it seems unlikely that high-spin iron(II) is too large to fit into the porphyrin hole, as the tetraphenylporphyrin complexes Sn^{IV} (TPP)Cl_2 and Mo^{VI}(TPP) $(O_2)_2$ have in-plane metal ions. Indeed five-coordinate complexes usually have the metal ion displaced out of the basal plane towards the axial ligand. Recent calculations[37] have shown that high-spin iron(II) is able to fit into the porphyrin ring, so any movement of the iron(II) in hemoglobin cannot be attributed directly to the high-spin–low-spin electronic rearrangement. In any case it is suggested[37] that iron in oxyhemoglobin is not low-spin Fe(II) but Fe(III). Other complications arise from several measurements which indicate that the movement of the iron with respect to the porphyrin plane is much smaller than the 0.7 Å suggested originally. Thus structural studies[38] on myoglobin and oxymyoglobin at 2 Å resolution show that on oxygenation the iron moves from 0.55 to 0.33 Å above the plane, while extended X-ray absorption fine-structure (EXAFS) measurements on hemoglobin[39] lead to the conclusion that there are at the most only slight structural differences around the iron atom in the high- and low-affinity forms of hemoglobin. Thus it seems that alternative trigger effects for the cooperative effect in hemoglobin should be sought, although it is still possible that the movement of iron(II) could be implicated in this.

Myogolobin is a simpler molecule, rather similar to the monomeric units of hemoglobin, having a molecular weight of 17,500. The structure of deoxymyoglobin has been determined at 1.4 Å resolution. The protein contains 153 amino acid residues, with no disulphide or SH groups. Like hemoglobin, a large fraction has the α-helix structure. Another common feature revealed by the X-ray studies, is that all the polar groups of the protein are on the outside of the molecule, with a resulting hydrophobic interior. This is almost certainly a very important aspect of the structure of these two proteins and may be correlated with the known fact that iron(II) porphyrins are much more stable with respect to irreversible oxidation to iron(III) porphyrins in media of low dielectric constants, such as could be provided by the non-polar aromatic amino acid residues found inside the hemoglobin and myoglobin.

The binding of the heme group to the protein is through the binding of the imidazole of a protein histidine group to the metal ion, but considerable extra stabilization occurs through a number of other heme–globin interactions. Thus high-resolution X-ray studies on oxyhemoglobin have indicated the complexity of these interactions. Each heme group of oxyhemoglobin is in van der Waals contact with about sixty atoms of the protein. Almost all these atoms belong to amino acid residues common to the globins of all species of mammal hemoglobin examined so far, suggesting that they are all necessary for function. Further specific stabilization of the heme–globin interaction may result from the interaction of the carboxyl groups of the porphyrin with basic groups of the protein. Thus the X-ray model of myoglobin suggests that one such carboxylic acid grouping interacts with an arginine residue of the protein. Again the X-ray model shows that the vinyl groups of the heme are directed towards the hydrophobic interior of the peptide chains and hence a stabilizing interaction with these aromatic amino acids residues will result.

The Bohr effect

The oxygen affinity of hemoglobin varies with the pH of the medium. This is the Bohr effect and it may be attributed, in general terms, to the effect of pH on the interaction between the heme group and ionizable groups of the protein. The effect is complex, and the study of this phenomenon in hemoglobin is very difficult, due to the interactions between the four heme groups. It appears that a further imidazole group (the distal imidazole group) is associated with oxygen uptake. Thus the pH range in which the Bohr effect is operative is in the range in which imidazole dissociates, while pK_a measurements over a temperature range show that the heat of ionization of this 'oxygen-linked' group is characteristic of the heat of ionization of imidazole.

The Fe(O$_2$) group in hemoglobin and myoglobin

It is now certain that Griffith's suggestion[40] that dioxygen is bound side-on to iron with a triangular structure is incorrect, and that the end-on bonding suggested by Pauling is favoured,[41] as shown for example by the structural studies on oxymyoglobin,[38] where dioxygen is bound end-on with an Fe—O—O angle of 121°. The Pauling model postulates low-spin Fe(II) with neutral dioxygen, which must be in a singlet state to account for the overall diamagnetism of the system. In contrast Weiss[42] has put forward a scheme in which electron transfer to dioxygen from iron has taken place, giving $Fe^{3+}(O_2^-)$. The diamagnetism then results from strong antiferromagnetic interaction between low-spin iron(III) and superoxide ion. Indeed temperature-dependent magnetic susceptibility data[43] have been interpreted in terms of two antiferromagnetically coupled $S = \frac{1}{2}$ systems. If this is correct then the matter will be resolved conclusively in favour of $Fe^{3+}(O_2^-)$. Other lines of argument in favour of this formulation include (a) the assignment[44] of dioxygen stretching frequencies for both oxymyoglobin and oxyhemoglobin at 1103 and 1107 cm^{-1} respectively, clearly in the superoxide range, (b) the displacement of superoxide from oxyhemoglobin by weak nucleophiles such as chloride ion,[44] and (c) spectral similarities between oxyhemoglobin and iron(III) heme systems.[1] It must be accepted that many of these arguments are not conclusive and several calculations have been published[37,45,46] which support low-spin iron(II) with singlet dioxygen, with some transfer of paired electron density to the dioxygen molecule. It has also been suggested that the geometry of the Fe(O$_2$) group is dependent upon the protein environment.[46]

Models for hemoglobin

Iron(II) porphyrins

Simple iron(II) porphyrins are readily oxidized to give μ-peroxo dimers which then react further to give μ-oxo dimers. (Equation 7.11).

$$
\text{L}-\text{Fe}-\text{O}^{\nearrow\text{O}} \longrightarrow \text{L}-\text{Fe}-\text{O}^{\nearrow}\text{O}-\text{Fe}-\text{L} \longrightarrow \text{L}-\text{Fe}-\text{O}-\text{Fe}-\text{L} \tag{7.11}
$$

As noted earlier for cobalt(II) carriers, this bridge formation must be prevented if carrier properties are to be observed. This has been achieved by the use of low temperature and sterically hindered or immobilized iron(II) porphyrins. Irreversible oxidation by a different pathway is prevented by the use of hydrophobic environments, while model systems should also be five coordinate. In this section it is hoped to examine these factors and see their relevance to natural carriers, and further, to investigate cooperative interactions by examining the results of the introduction of different degrees of strain into the model porphyrin complexes.

The use of solid supports. The first synthesis of an Fe(II) porphyrin having oxygen-carrying properties was due to Wang.[47] He had suggested earlier that the stability of oxyhemoglobin could be attributed to the hydrophobic environment produced by the protein and this was incorporated into his model. Another important feature was the immobilization of the heme group by incorporation into a polymer.

Wang treated a benzene solution of the diethyl ester of hemin and an excess of 1-(2 phenylethyl)imidazole with aqueous alkaline sodium dithionite, under carbon monoxide. After centrifugation, the bright red benzene solution was mixed with a 10% solution of polystyrene in benzene and dried in a warm stream of carbon monoxide. A transparent film was obtained having the 1-(2-phenylethylimidazolecarbonmonoxyheme diethyl ester embedded in a matrix of polystyrene and the substituted imidazole. The bound carbon monoxide could be removed by treatment with nitrogen gas for several hours. On exposure to air, the resulting material rapidly combined with oxygen giving a product having an oxyhemoglobin-type spectrum. The oxygenation is reversible and the cycle may be repeated a number of times. The film may also be kept in air-saturated water for days without oxidation to iron(III).

This approach has been continued, and reversible oxygen-carrying properties demonstrated for various Fe(II) porphyrin systems attached to, for example, water-soluble polymers[48] and a rigid silica gel support[49] (Figure 7.10), although in the latter case only low dioxygen absorption is seen.

The use of porphyrins with special structural features. Much effort has been put into the synthesis of iron(II) porphyrin complexes with special structural features designed to prevent bridge formation, to prevent or hinder binding in the sixth

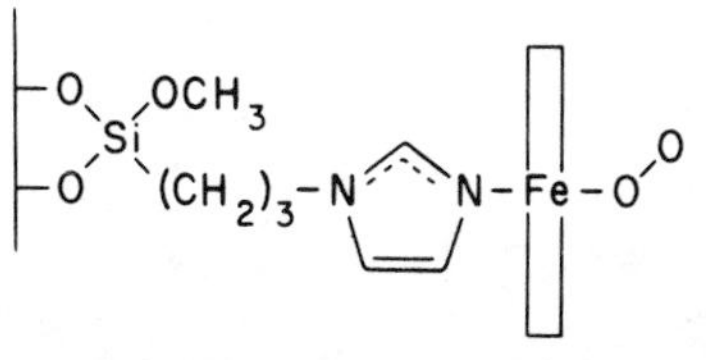

Figure 7.10 An oxygen carrier attached to a silica gel support.

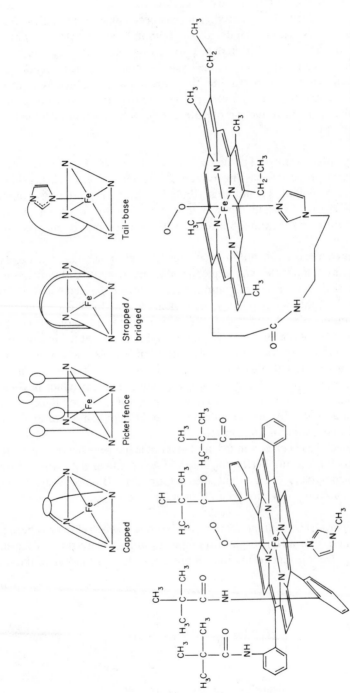

Figure 7.11 Schematic representation of iron(II) complexes of some synthetic porphyrins, and examples of picket fence and tail-base porphyrins

coordination position (simple hemes of coordination number five cannot be prepared), and also to provide a protected environment for the binding of dioxygen. These are shown schematically in Figure 7.11, and include the capped porphyrin of Baldwin,[50] the picket fence porphyrin of Collman,[4] the strapped or bridged porphyrin of Baldwin[51] and Battersby,[52] in which opposite *meso* positions are linked by a chain of atoms, and the tail-base porphyrins of Traylor.[53] In the last-named example the axial imidazole group is bridged to the porphyrin molecule by an amide link. The length and configuration of the tail can be varied, and so much work has been done on these. Specific examples will carry dioxygen reversibly in polystyrene films and in solution.

The capped iron(II) porphyrin reversibly oxygenates in solution at room temperature (in contrast to the species in Figure 7.8, which does this only at low temperature). The original picket fence porphyrin had four pivaloylamide groups above one plane of the porphyrin ring, but different pendant groups have now been incorporated. In the presence of excess of N-methylimidazole in benzene, these compounds reversibly add O_2 at room temperature. The pivaloyl groups do not prevent the coordination of a sixth axial ligand, but this process is hindered substantially. Dioxygen binding occurs within the cage of the pivalamido group, as is evidenced by the failure to obtain μ-oxo dimers with the unhindered faces of the porphyrin ring facing each other. Several solid iron–dioxygen complexes with picket fence porphyrins have been isolated. The structure has been reported[54] for Fe(TpivPP)(NMeimid)(O_2) (TpivPP = $\alpha,\alpha,\alpha,\alpha$-tetrapivaloylamidophenylporphyrin). This shows end-on binding of O_2 with an Fe—O—O angle of about 130°. Unfortunately accurate structural parameters for the Fe(O_2) group could not be obtained. However v_{O-O} has been assigned[55] at $1159 \, cm^{-1}$, conforming the $Fe^{3+}O_2^-$ formalism, while resonance Raman spectra[56] show the Fe—O_2 stretching mode at $568 \, cm^{-1}$, close to the value in oxyhemoglobin ($567 \, cm^{-1}$).

A further interesting feature of the picket fence porphyrins is their high affinity for carbon monoxide compared to that of hemoglobin. The Fe—CO group in hemoglobin is bent with an angle of 145°. This structure is probably forced on it by the steric constraints of the heme cleft, resulting in a weakened Fe—CO bond. In contrast, the Fe—O—O group is intrinsically bent and so is able to fit well in the cleft and is probably stabilized by the resulting interactions. For the picket fence porphyrin, the linear Fe—CO group can be accommodated without complication and so carbon monoxide is bound strongly.

Coboglobins and other metal-substituted heme and porphyrins

The incorporation of other metals into these systems allows the introduction of paramagnetic probes, and gives insight into structural aspects of dioxygen binding and cooperativity in terms of their reaction to variation in the M(O_2) entity. Much has been done on the reconstitution of globins from hemoglobin and myoglobin with cobalt(II) protoporphyrin. The coboglobins are similar to hemoglobin and take up O_2 reversibly,[1,57] showing Bohr and cooperative effects. EPR studies show the similarity of oxycobalt hemoglobin with oxygenated cobalt porphyrin and suggest therefore that the protein does not measurably influence the electronic structure of the heme group (although it does control the extent of oxygen binding). Such a conclusion could not be reached using iron(II) systems. Coboglobin is low-spin in the deoxy state and in the oxy form is low-spin Co(III) (with O_2^-). Thus cooperativity does not depend upon the spin state of the metal. Furthermore, it is estimated that the movement of the cobalt in the two forms is at the most 0.3 Å. All this emphasizes the need for fresh proposals on the trigger mechanism, but it should be noted that the cooperativity in cobalt hemoglobin is less than that in hemoglobin and it is possible that this may reflect the smaller movement. Mixed iron–cobalt hemoglobins have been prepared,[58] so allowing the study of subunit interactions through paramagnetically shifted proton NMR. The thermodynamics and kinetics of dioxygen binding to cobalt(II) porphyrins have been reviewed.[1] The π-bonding properties of the axial ligand have been emphasized, while a comparison of thermochemical cycles for binding of dioxygen to the five-coordinate cobalt porphyrin complexes and cobalt myoglobin shows that the protein provides a favourable entropy contribution.

The porphyrin complex, *meso*-tetraphenylporphyrinatopyridine-manganese(II), Mn(TPP)(py), carries dioxygen in toluene at $-79°C$.[1] The dramatic changes in the visible spectrum on oxygenation suggest the formation of Mn(III) but EPR studies[59] suggest a quartet ground state for a t_2^3 ion and a $Mn^{IV}(TPP)(O_2^{2-})$ formulation, possibly with a Vaska-type structure. The complexes Mn(T(p-X)PP)B exist in equilibrium[60] with the dioxygen complex Mn(T(p-X)PP)(O$_2$), in which the fifth, axial ligand is displaced. This is reflected in the low extent of dioxygen binding in these complexes.

The oxygen-carrying properties of tetraphenylporphinato- and octaethylporphinatoruthenium(II) at room temperature have been studied;[61] this is dependent on the nature of the axial base. The dioxygen complexes have been formulated as $Ru^{III}(O_2^-)$. A ruthenium myoglobin has been prepared[62] but is irreversibly oxidised by dioxygen.

Models for cooperative interaction in hemoglobin

The difference between the T and R states of hemoglobin is usually attributed to strain in the T state, either to the sum of a number of small strains or to a specific effect in a localized part of the molecule. Most recent work has been concerned with the effects of strain introduced through the use of axial ligands with sterically active substituents. The position of the axial ligand relative to the plane of the porphyrin group is clearly central to the Hoard–Perutz view of cooperativity in hemoglobin, and this will be altered by the use of appropriately substituted axial ligands. This strain cannot be manifested in gross structural effects as these have not been observed in hemoglobin itself despite the use of many physical techniques. Nevertheless it now seems clear that the use of hindered and non-hindered bases mimic well the reactivity of the T and R states of hemoglobin.[63,64] Thus Collman and his coworkers[63] have shown that oxygen binding to iron(II) and cobalt(II) picket fence porphyrins with N-methylimidazole as an axial ligand is similar to that with myoglobin and subunits of hemoglobin and their cobalt analogues. But the use of 1,2-dimethylimidazole as the axial ligand causes a decrease in affinity for dioxygen, down to the level found in T state hemoglobin and coboglobin. The decrease in the latter case is smaller, in accord with the lower cooperativity shown by cobalt hemoglobin. The ratio of oxygen affinities for N-methyl- and 1,2-dimethylimidazole as axial ligands is 80:1, rather similar to the ratio of 100:1 observed for R and T hemoglobin.

Certain of these iron(II) picket fence porphyrins with the sterically restrained base 2-methylimidazole actually show cooperativity[4] in the uptake of dioxygen in the solid state. No detailed information is yet available on this remarkable observation but it appears probable that oxygenation of some iron(II) sites induces some change in the crystal lattice.

The reversible addition of dioxygen to hemoglobin

The studies on model compounds described so far indicate that in addition to providing a five-coordinate deoxy state, the protein in hemoglobin influences the chemistry of the heme group by providing a hydrophobic binding site and isolated heme groups for the stabilization of the $Fe(O_2)$ system, and is also the source of the cooperative effect through its restraining of the proximal histidine group. This axial imidazole ligand provides direct heme–globin interaction and also helps to influence electron distribution in the heme–dioxygen adduct through its π-acceptor and strong σ-donor properties.

The second imidazole group, the distal group, has a less well-defined role. The steric constraints it imposes on the dioxygen-binding site may favour the binding of dioxygen, as the bent $Fe(O_2)$ group can be accommodated by it. In contrast the binding of small molecules such as carbon monoxide which prefer a linear configuration will not be favoured. Thus the distal imidazole may contribute to the specificity that hemoglobin shows towards dioxygen. It is noteworthy that myoglobin from *Aplysia*, a Mediterranean mollusc, has a very different amino acid composition from mammalian myoglobins, and in particular has no oxygen-linked imidazole group. This species therefore shows no Bohr effect, and its affinity for dioxygen is much lower than that of other myoglobins.

The salt bridges in and between the subunits in hemoglobin have been much studied. The Bohr effect results from the effect on the equilibrium between T and R forms of breaking hydrogen bonds through proton dissociation as the pH is raised. As noted earlier the papers of Perutz should be consulted for detailed information on the role of these bridges in controlling the conformation of hemoglobin.[34, 36]

HEMERYTHRIN

This non-heme iron protein[65] is found in species of the following four phyla:brachiopods, polychaetes, priapulids, and sipunculids. It binds dioxygen with little if any cooperativity as shown by the Hill coefficient values of 1.2–1.4, in accord with its role in oxygen storage. Hemerythrin from erythrocytes of the worm *Golfingia gouldii* has a molecular weight of 107,000 to 108,000, and consists of eight identical subunits, each containing two iron atoms. Each pair of iron atoms binds one molecule of oxygen, possibly in a similar fashion to that seen for certain of the cobalt(II) model compounds, with bridging dioxygen. The octameric species is in equilibrium with monomeric species, and the treatment of cysteine residues with mercurials causes dissociation of the octamer. A monomeric form (myohemerythrin) has been isolated, but this appears to have rather different properties from those produced by dissociation of the octamer.[66]

Deoxyhemerythrin is a paramagnetic ($S = 2$) iron(II) species. Oxyhemerythrin may be formulated in several ways depending upon the extent to which electron transfer to the dioxygen has occurred. The oxy- and deoxyhemerythrins may be oxidized to the iron(III) (met) species. Oxyhemerythrin is slowly converted to the deoxygenated met form in the presence of oxygen, and it is difficult to prepare oxyhemerythrin which is free from the iron(III) (met) species. However the iron(III) (met) form may be reduced to the deoxy iron(II) form by dithionite, a reaction which has recently been studied kinetically.[66]

The electronic spectrum of oxyhemerythrin is similar to that observed for the iron(III) (met) species suggesting that the two iron centres are Fe(III), with a peroxo group. This view is confirmed by resonance Raman studies, in which a band at 848 cm^{-1} has been assigned to ν_{O-O}, typical of a peroxo structure.[67] The ν_{Fe-O} mode was observed at 500 cm^{-1}, and both assignments were confirmed by ^{18}O isotope shifts. Further studies with $^{16}O-^{18}O$ have shown the two atoms of the bound O_2 to have different environments.[68] Mössbauer spectroscopy[69] of the oxy form of hemerythrin from *G. gouldii* suggests that the two iron atoms also have different environments but that there is only one environment for iron in both deoxy- and methemerythrins. Magnetic susceptibility measurements on both oxy and met forms have shown the two iron(III) species to be antiferromagnetically coupled.[70] In the met form this probably takes place through an oxo bridge.

Structural studies and metal-binding groups

This topic has been reviewed.[65] Only recently have X-ray crystallographic data been available. The arrangement of subunits is shown schematically in Figure 7.12, while the tertiary structures of myohemerythrin[71] (from *T. zostericola*) and hemerythrin[72] (from *T. dyscritum*) have been determined at low resolution. The two iron atoms in each subunit are in close proximity to each other and are coordinated by eight amino acid residues, namely His-25, His-54, Glu-58, His-73, His-77, His-101, Asp-106, and Tyr-109. Both Glu-58 and Asp-106 have a carboxyl oxygen atom suitable for bridging the two iron atoms, so it appears that in methemerythrin the two iron atoms have approximately octahedral environments, with the octahedra sharing one triangular face with Asp-106, Glu-58, and a water molecule as bridging ligands. One iron atom will have three histidine

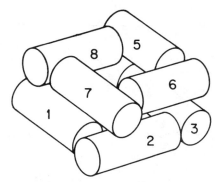

Figure 7.12 Schematic representation of subunit arrangement in hemerythrin.

(imidazole) ligands to complete its coordination shell, while the other will have two histidine and one tyrosine residues as ligands. The hydrogen bonding and other ionic interactions contributing to the quaternary structure have been characterized. The only cysteine residue in the amino acid sequence is Cys-50, modification to which is thought to produce dissociation of the octamer.

Previous chemical studies[73] had implicated at least four of the seven histidine residues per subunit as ligands, together with Tyr-109 and a glutamic acid residue, in good agreement with crystallographic results.

The binuclear structure of the subunit has been modelled[74] by several systems in which two metal ions are held close together. These have included modified picked fence porphyrins, and a crowned porphyrin, in which a crown ether occupies a position above the porphyrin plane.

Dioxygen binding

The Mössbauer measurements suggested that oxyhemerythrin had different environments for the two iron atoms in the subunit. This could be explained either in terms of the different ligand arrangements around the two iron atoms or in terms of the binding mode of the dioxygen. The latter suggestion seems most reasonable as Mössbauer spectroscopy does not distinguish two types of iron in the met and deoxyhemerythrins. Possible structures for the $Fe_2(O_2)$ group are shown in Figure 7.13; structures 2 and 5 give different iron environments due to the mode of dioxygen binding, but only structure 2 also gives different environments for the oxygen atoms of the peroxo grouping, as required by resonance Raman studies using $^{16}O-^{18}O$.

A further approach to this problem has involved the use of other unsymmetrically labelled ligands such as $^{15}N-^{14}N-^{14}N$ azide ion and ligands such as the thiocyanate ion. Resonance Raman spectra of the azidohemerythrin again support structure 2 with a peroxo group bound to only one iron(III). However structure 4 is also consistent with these results, while it is difficult to rule out *distorted* structures of types 1 and 5. It should also be stressed that the structure

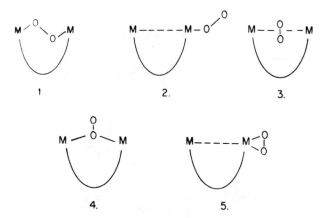

Figure 7.13 Possible structures for the $M_2(O_2)$ group in oxyhemerythrin and oxyhemocyanin, where $M = Fe(III)$ or $Cu(II)$, and (O_2)
is carried as a peroxo-like group.

of oxyhemerythrin may well differ from that of azidomethemerythrin. Some further comments on this problem will be reserved until the discussion on hemocyanin in view of the similarities between these two systems, but, clearly, detailed speculation at this stage is not profitable.

HEMOCYANIN

These oxygen-carrying copper non-heme proteins are found in the hemolymphs of various molluscs and arthropods. They have very high molecular weights but break down into subunits, depending upon the pH and ionic strength. The hemocyanin of *Loligo pealei* has a molecular weight of 3,750,000, with subunits of between 25,000 and 75,000. Aspects of the physiology and biochemistry of the hemocyanins have been reviewed.[75]

The deoxyhemocyanins are colourless, suggestive of copper(I), while the oxy form is blue, its d–d spectrum confirming the presence of copper(II). In addition to d–d bands at 440, 570, and 700 nm, an intense band at 347 nm ($\varepsilon \sim 8900$) is observed which is assigned to charge transfer between the two copper species or between copper and the dioxygen entity. One mole of dioxygen is carried for each pair of copper ions, and resonance Raman studies[76] on several hemocyanins show ν_{O-O} at around 745 cm^{-1} in accord with a peroxo-like structure. The use of $^{16}O-^{18}O$ has led to the suggestion[77] that the $Cu_2(O_2)$ group involves a non-planar bridging structure. Antiferromagnetic interaction between the two copper ions has been confirmed.[78] Clearly the overall situation resembles that of hemerythrin and the same alternative structures must be considered (Figure 7.13).

Metal-binding groups

The apoprotein may be prepared by removing the copper with cyanide. Analysis for —SH groups on native hemocyanins and their apoproteins has led to the suggestion that, in some cases only, the copper is bound by —SH groups. One or both of the copper ions in hemocyanin can be oxidized with nitrite or nitric oxide. EPR studies on the singly oxidized species indicate that the copper(II) is bound to at least two nitrogen atoms of amino acid residues.[79] The suggestion that at least two imidazole ligands per copper are involved in metal binding in hemocyanin is in accord with Raman data[76] and a recent X-ray absorption study[80] of both Cu(I) hemocyanin and Cu(II) oxyhemocyanin. The latter report excludes the possibility of a sulphur ligand.

Model compounds

Only recently have model compounds been prepared which exhibit reversible oxygenation of copper(I) complexes. The 2,6-bis[1-(2-imidazol-4-ylethylimino)-ethyl]pyridinecopper(I) cation (Figure 7.14) in dimethylsulphoxide solution changes from red to green on exposure to dioxygen, absorbing one mole of O_2 per two moles of copper.[81] The reaction is readily reversed and can be repeated with about 20% decrease in binding capacity per cycle.

A second example involves the ligand 1,4-bis(1-oxo-4,10-dithio-7-azacyclodo-decan-7-ylmethyl)benzene, which binds two copper(I) ions and reversibly absorbs dioxygen in the solid state only.[82]

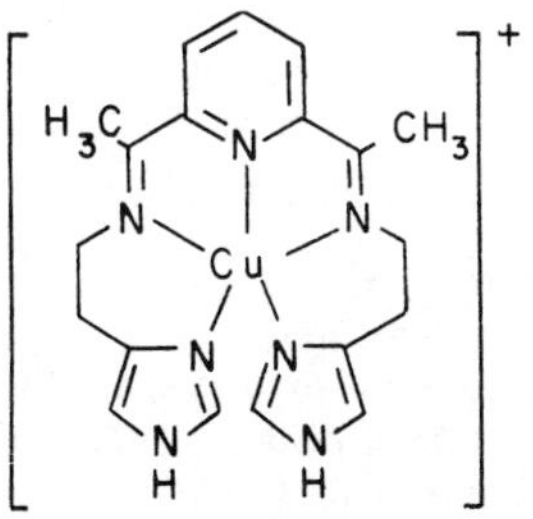

Figure 7.14 A model compound for hemocyanin.

Dioxygen binding—some comments on hemocyanin and hemerythrin

Infrared studies[83] on carbon monoxide bound to hemocyanin, show ν_{CO} at $2063\,cm^{-1}$, typical of mononuclear groups and higher by $100-200\,cm^{-1}$ than that expected for a bridging carbonyl group. Provided carbon monoxide binds in a similar fashion as dioxygen, this leads to the conclusion that dioxygen is bound to only one copper, and that structure 2 in Figure 7.13 is correct, one supported for hemerythrin by several lines of argument. It is interesting to note that if this is correct then the geometry of the binding of dioxygen by both these dinuclear sites is similar to that in hemoglobin, but the second Fe(II) (or Cu(I)) is then able to provide an additional electron via some metal–metal bridge to reduce further the dioxygen to peroxide.

REFERENCES

1. F. Basolo, B. M. Hoffman, and J. A. Ibers, *Acc. Chem. Res.*, **8**, 384 (1975); R. D. Jones, D. A. Summerville, and F. Basolo, *Chem. Rev.*, **79**, 139 (1979).
2. L. Vaska, *Acc. Chem. Res.*, **9**, 175 (1976).
3. A. P. B. Lever and H. B. Gray, *Acc. Chem. Res.*, **11**, 349 (1978).
4. J. P. Collman, *Acc. Chem. Res.*, **10**, 265 (1977); J. P. Collman and K. S. Suslick, *Pure Appl. Chem.*, **50**, 951 (1978).
5. J. S. Valentine, *Chem. Rev.*, **73**, 235 (1973).
6. J. A. Connor and E. A. V. Ebsworth, *Adv. Inorg. Radiochem.*, **6**, 279 (1964); G. McLendon and A. E. Martell, *Coord. Chem. Revs.*, **19**, 1 (1976).
7. W. M. Latimer, *Oxidation Potentials*, (2nd edn.), Prentice-Hall Inc., New Jersey, 1952.
8. W. P. Schaefer and R. E. Marsh, *J. Amer. Chem. Soc.*, **88**, 178 (1966); *Acta Cryst.*, **21**, 735 (1966).
9. F. R. Fronczek, W. P. Schaefer, and R. E. Marsh, *Acta Cryst. B*, **30**, 117 (1974).
10. G. P. Khare, E. L. Ruff, and A. P. B. Lever, *Can. J. Chem.*, **54**, 3424 (1976).
11. S. R. Pickens, A. E. Martell, G. McLendon, A. P. B. Lever, and H. B. Gray, *Inorg. Chem.*, **17**, 2190 (1978).
12. A. E. Martell and M. Calvin, *The Chemistry of the Metal Chelate Compounds*, Prentice-Hall Inc., New Jersey, 1952.
13. C. Floriani and T. Calderazzo, *J. Chem. Soc. (A)*, **1969**, 945.
14. A. Avdeef and W. P. Schaefer, *J. Amer. Chem. Soc.*, **98**, 5153 (1976). R. S. Gall, J. F. Rogers, W. P. Schaefer and G. C. Christoph, *J. Amer. Chem. Soc.*, **98**, 5136 (1976); G. B. Jameson, W. T. Robinson, and G. A. Rodley, *J. Chem. Soc. (Dalton)*, **1977**, 191; R. S. Gall and W. P. Schaefer, *Inorg. Chem.*, **15**, 2758 (1976).
15. E. Melamud, B. L. Silver, and Z. Dori, *J. Amer. Chem. Soc.*, **96**, 4689 (1974).
16. J. H. Burness, J. G. Dillard, and L. T. Taylor, *J. Amer. Chem. Soc.*, **97**, 6080 (1975).
17. R. D. Gillard and A. Spencer, *J. Chem. Soc. (A).*, **1969**, 2718.
18. W. R. Harris, R. C. Bess, A. E. Martell, and T. M. Ridgeway, *J. Amer. Chem. Soc.*, **99**, 2958 (1977).
19. L. Vaska, *Science*, **140**, 809 (1963).
20. J. A. McGinnety, R. J. Doedens, and J. A. Ibers, *Inorg. Chem.*, **6**, 2243 (1967).
21. H. Huber and W. Klotzbucher, *Can. J. Chem.*, **51**, 2722 (1973).
22. L. Vaska and L. S. Chen, *Chem. Comm.*, **1971**, 1080.
23. J. M. Pratt, *Inorganic Chemistry of Vitamin B_{12}*, Academic Press, London, 1972.
24. E. W. Abel, J. M. Pratt, and R. Whelan, *Chem. Comm.*, **1971**, 449.
25. K. Uchida, S. Naito, M. Soma, T. Onishi, and K. Tamaru, *J.C.S. Chem. Comm.*, **1978**, 217.
26. G. N. Schrauzer and L. P. Lee, *J. Amer. Chem. Soc.*, **92**, 1551 (1970).
27. V. McKee, S. M. Nelson, and J. Nelson, *J.C.S. Chem. Comm.*, **1976**, 225.
28. C. A. McAuliffe et al, *J. C. S. Chem. Comm.*, **1979**, 737.
29. M. Calligaris, G. Nardin, and L. Randaccio, *Chem. Comm.*, **1969**, 763.
30. J. Halpern, B. L. Goodall, G. P. Khare, H. S. Lim, and J. J. Pluth, *J. Amer. Chem. Soc.*, **97**, 2301 (1975).
31. J. E. Baldwin and J. Huff, *J. Amer. Chem. Soc.*, **95**, 5757 (1973).
32. M. J. Carter, D. P. Rillema, and F. Basolo, *J. Amer. Chem. Soc.*, **96**, 392 (1974).
33. M. F. Corrigan, K. S. Murray, B. O. West, P. R. Hicks, and J. R. Pilbrow, *J.C.S. (Dalton)*, **1977**, 1478.
34. M. F. Perutz, *Nature*, **228**, 726 (1970); **237**, 495 (1972); *Br. Med. Bull.*, **32**, 195 (1976); *New Scientist*, **1971**, 676.
35. J. L. Hoard, M. J. Hamour, T. A. Hamour, and W. S. Caughey, *J. Amer. Chem. Soc.*, **87**, 2311 (1965).
36. H. Muirhead, J. M. Cox, L. Mazzarella, and M. F. Perutz, *J. Mol. Biol.*, **28**, 117 (1967); M. F. Perutz, H. Muirhead, J. M. Cox, L. G. Goaman, F. S. Matthews, E. L.

McGandy, and L. E. Webb, *Nature*, **219**, 29 (1968); M. F. Perutz, H. Muirhead, J. M. Cox, and L. G. Goaman, *Nature*, **219**, 131 (1968); M. F. Perutz and H. Lehmann, *Nature*, **219**, 902 (1968).

37. B. D. Olafson and W. A. Goddard, *Proc. Natl. Acad. Sci. U.S.A.*, **74**, 1315 (1977).

38. S. E. V. Phillips, *Nature*, **273**, 247 (1978).

39. P. Eisenberger, R. G. Shulman, G. S. Brown, and S. Ogawa, *Proc. Natl. Acad. Sci. U.S.A.*, **73**, 491 (1976).

40. J. S. Griffith, *Proc. Roy. Soc.*, **A235**, 23 (1956).

41. L. Pauling, *Nature*, **203**, 182 (1964).

42. J. J. Weiss, *Nature*, **202**, 83 (1964).

43. M. Cerdonia, A. Congui-Castellano, F. Mogno, B. Pispisa, G. L. Romane, and S. Vitale, *Proc. Natl. Acad. Sci. U.S.A.*, **74**, 398 (1977). M. Cerdonio, A. Congui-Castellano, L. Calabrese, S. Morante, B. Pispisa, and S. Vitale, *Proc. Natl. Acad. Sci. U.S.A*, **75**, 4916 (1978).

44. C. H. Barlow, J. C. Maxwell, W. J. Wallace, and W. S. Caughey, *Biochem. Biophys. Res. Comm.*, **55**, 91 (1973); J. C. Maxwell, J. A. Volpe, C. H. Barlow, and W. S. Caughey, *Biochem. Biophys. Res. Comm.*, **58**, 166 (1974); W. J. Wallace, J. C. Maxwell, and W. S. Caughey, *Biochem. Biophys. Res. Comm.*, **57**, 1104 (1974).

45. B. H. Huynh, D. A. Case, and M. Karplus, *J. Amer. Chem. Soc.*, **101**, 4433 (1979).

46. M. W. Makinen, A. K. Churg, and H. A. Glick, *Proc. Natl. Acad. Sci. U.S.A.*, **75**, 2291 (1978).

47. J. H. Wang, A. Nakahara, and E. B. Fleischer, *J. Amer. Chem. Soc.*, **80**, 1109 (1958); J. H. Wang, *J. Amer. Chem. Soc.*, **80**, 3168 (1958).

48. E. Bayer and G. Holzbach, *Angew. Chem. Int. Ed.*, **16**, 117 (1977).

49. O. Leal, D. L. Anderson, R. G. Bowman, F. Basolo, and R. L. Burwell, *J. Amer. Chem. Soc.*, **97**, 5125 (1975).

50. J. Almog, J. E. Baldwin, and J. Huff, *J. Amer. Chem. Soc.*, **97**, 227 (1975); J. Almog, J. E. Baldwin, R. L. Dyer, and M. Peters, *J. Amer. Chem. Soc.*, **97**, 226 (1975).

51. J. E. Baldwin, T. Klose, and M. Peters, *J.C.S. Chem. Comm.*, **1976**, 881.

52. A. R. Battersby, D. G. Buckley, S. G. Hartley, and M. D. Turnbull, *J.C.S. Chem. Comm.*, **1976**, 829.

53. C. K. Chang and T. G. Traylor, *Proc. Natl. Acad. Sci. U.S.A.*, **70**, 2647 (1973); J. Geibel, C. K. Chang, and T. G. Traylor, *J. Amer. Chem. Soc.*, **97**, 5924 (1975).

54. J. P. Collman, R. R. Gagne, C. A. Reed, W. T. Robinson, and G. A. Rodley, *Proc. Natl. Acad. Sci. U.S.A.*, **71**, 1326 (1974).

55. J. P. Collman, J. I. Brauman, T. R. Halbert, and K. S. Suslick, *Proc. Natl. Acad. Sci. U.S.A.*, **73**, 3333 (1976).

56. J. M. Burke, J. R. Kincaid, S. Peters, R. R. Gagne, J. P. Collman, and T. G. Spiro, *J. Amer. Chem. Soc.*, **100**, 6083 (1978).

57. W. Snyder and J. C. W. Chien, *Eur. J. Biochem.*, **91**, 83 (1978).

58. M. Ikeda-Saito, T. Inubushi, G. C. McDonald, and T. Yonetani, *J. Biol. Chem.*, **253**, 7134 (1978).

59. C. J. Weschler, B. M. Hoffman, and F. Basolo, *J. Amer. Chem. Soc.*, **97**, 5278, 9976 (1975); **98**, 5473 (1975).

60. R. D. Jones, D. A. Summerville, and F. Basolo, *J. Amer. Chem. Soc.*, **100**, 4416 (1978).

61. N. Farrell, D. H. Dolphin, and B. R. James, *J. Amer. Chem. Soc.*, **100**, 324 (1978).

62. D. R. Paulson, A. W. Addison, D. Dolphin and B. R. James, *J. Biol. Chem.*, **254**, 7002 (1979).

63. J. P. Collman, J. I. Brauman, K. M. Doxsee, T. R. Halbert, and K. S. Suslick, *Proc. Natl. Acad. Sci. U.S.A.*, **75**, 564 (1978).

64. D. K. White, J. B. Cannon, and T. G. Traylor, *J. Amer. Chem. Soc.*, **101**, 2443 (1979); J. Geibel, C. K. Chang, and T. G. Traylor, *J. Amer. Chem. Soc.*, **97**, 5294 (1975); T. G. Traylor, Y. Tatsuno, D. W. Powell, and J. B. Cannon, *J.C.S. Chem. Comm.*, **1977**, 732.

242

65. D. M. Kurtz, D. F. Shriver, and I. M. Klotz, *Coord. Chem. Revs.*, **24**, 145 (1977).
66. P. C. Harrington, D. J. A. deWaal, and R. G. Wilkins, *Arch. Biochem. Biophys.*, **191**, 444 (1978).
67. J. B. R. Dunn, D. F. Shriver, and I. M. Klotz, *Proc. Natl. Acad. Sci. U.S.A.*, **70**, 2582 (1973).
68. D. M. Kurtz, D. F. Shriver, and I. M. Klotz, *J. Amer. Chem. Soc.*, **98**, 5033 (1976).
69. K. Garbett, C. E. Johnson, I. M. Klotz, M. Y. Okamura, and R. J. P. Williams, *Arch. Biochem. Biophys.*, **142**, 574 (1971).
70. J. W. Dawson, H. B. Gray, H. E. Hoening, G. R. Rossman, J. M. Shredder, and R. H. Wang, *Biochemistry*, **11**, 461 (1972).
71. W. A. Hendrickson and K. B. Ward, *J. Biol. Chem.*, **252**, 3012 (1977).
72. R. E. Stenkamp, L. C. Seiker, L. H. Jensen, and J. E. McQueen, *Biochemistry*, **17**, 2499 (1978); *J. Mol. Biol.*, **126**, 457 (1978).
73. J. L. York and G. C. Fan, *Biochemistry*, **10**, 1659 (1971). E. M. Gormley J. S. Loehr, B. Brimhall, and M. A. Hermodson, *Biochem. Biophys. Res. Comm.*, **85**, 1360 (1978).
74. C. M. Elliot, *J.C.S. Chem. Comm.*, **1978**, 399; D. A. Buckingham, M. J. Gunter, and L. N. Mander, *J. Amer. Chem. Soc.*, **100**, 2899 (1978); C. K. Chang, *J. Amer. Chem. Soc.*, **99**, 2819 (1977).
75. F. Ghiretti (Ed.), *Physiology and Biochemistry of Hemocyanins*, Academic Press, New York, 1968.
76. T. B. Freedman, J. S. Loehr, and T. M. Loehr, *J. Amer. Chem. Soc.*, **98**, 2809 (1976).
77. T. J. Thamann, J. S. Loehr, and T. M. Loehr, *J. Amer. Chem. Soc.*, **99**, 4187 (1977).
78. E. I. Solomon, D. M. Dooley, R. H. Wang, H. B. Gray, M. Cerdonio, F. Mogno, and R. L. Romani, *J. Amer. Chem. Soc.*, **98**, 1029 (1976).
79. H. van der Deen and H. Hoving, *Biochemistry*, **16**, 3519 (1977).
80. T. K. Eccles, Stanford Synchroton Radiation Laboratory Report 78/01, Stanford University, 1978.
81. M. G. Simmons and L. J. Wilson, *J.C.S. Chem. Comm.*, **1978**, 634.
82. J. E. Bulkowski, P. L. Burke, M. F. Ludmann, and J. A. Osborn, *J. C. S. Chem. Comm.*, **1978**, 498.
83. J. O. Alben, L. Yen, and N. J. Farrier, *J. Amer. Chem. Soc.*, **92**, 4475 (1970).

Note added in proof

The biochemistry of vanadium in *Ascidian* blood cells has been described by K. Kustin, G. C. McLeod et al, *J. Cell. Physiol.*, **93**, 309 (1977); *Biochem. J.*, **181**, 457 (1979).

The high-spin Fe(II) porphyrin, *meso*-tetraphenylporphinato)-bis(tetrahydro-furan)iron(II), has the Fe(II) lying in the *plane* of the porphyrin ring (C. A. Reed et al, *J. Amer. Chem. Soc.*, **102**, 2302 (1980).

Additional examples of protected porphyrins have been synthesised: tailed picket fence (*J. Amer. Chem. Soc.*, **102**, 4182 (1980); basket handle (*Nou. J. de Chimie*, **3**, 77 (1979); cyclophane and homologous capped (*J. Amer. Chem. Soc.*, **101**, 4749, 4761, 4762 (1979).

EXAFS and resonance Raman data on hemocyanin have led to a model in which two Cu are bound to the protein by three histidine ligands each, while in the oxy form the Cu(II) ions are also bridged by the bound O_2^{2-} and a protein group, probably tyrosine. (*J. Amer. Chem. Soc.*, **102**, 4210, 4217 (1980).

CHAPTER EIGHT

THE STORAGE AND TRANSPORT OF IRON

The transport and storage of iron has been much studied for both microbial and mammalian systems. Microbial iron transport is the better characterized of these, but involves a relatively simple process utilizing hydroxamate and phenolate ligands which are secreted by various organisms in order to solubilize and mobilize inorganic iron and transport it back into the cell. In higher organisms, there is clearly a need for a much more complex regulatory system to store iron and to deliver it to appropriate cells for differing synthetic purposes. This system also has to deal with the breakdown products of these iron-containing macromolecules, partly to conserve the correct concentration levels. In an adult human some 65% of the total iron is found in hemoglobin and myoglobin, and the bulk of the remainder is found in storage proteins (ferritin and hemosiderin), while only a very small amount is actually present in various iron enzymes and transport systems. In this chapter, it is hoped to discuss the specific transport and storage proteins, transferrin and ferritin respectively, and physiological aspects of the transfer of iron from one to the other. The chapter will be concluded with a survey of microbial iron-transport systems. A number of reviews and books are available on the topic of iron transport.[1, 2]

TRANSFERRIN

The transferrins [1b] are a class of iron-binding molecules that include lactoferrin (from milk), conalbumin or ovotransferrin (from egg white), and serum transferrin. Serum transferrin, the most studied of these proteins, is the transport protein that carries iron from the breakdown sites of hemoglobin in the reticuloendothelial cells of spleen and liver back to the sites in bone marrow for the synthesis of hemoglobin. Problems associated with the selective binding of iron(III) and its ultimate release thus have to be considered. Serum transferrin accounts for about 4 mg of body iron in humans, but actually delivers some 40 mg of iron daily to the marrow—a testimony to its efficiency as a transport protein. Humans with a genetic inability to synthesize transferrin thus suffer from both iron-deficiency anaemia and the toxic effects of an iron overload! It is probable that transferrin is also involved in the biosynthesis of other iron-containing enzymes.

Transferrin is a glycoprotein of molecular weight about 80,000. It consists of a single polypeptide chain, which is folded into two compact regions, each of which is able to bind one iron(III) ion. However iron binding is only possible when an anion is also bound. In the absence of an appropriate anion, iron is not bound at all. The naturally occurring anion is usually carbonate but other anions, such as oxalate, malonate, citrate, and EDTA, will also activate the metal-binding site.

Much controversy has centred on the question of whether or not the two metal sites are identical in terms of structure and function. The use of several instrumental techniques suggests that the two sites are very similar in a structural sense, but it is still not certain whether or not they are identical in terms of function. This matter will be discussed in detail later, but current views seem to suggest that they are non-equivalent.

Metal-binding sites

A remarkably wide range of spectroscopic properties have been applied to the characterization of the metal-binding sites in transferrin, particularly through the use of metal probes.[3]

Tyrosyl residues were suggested to be ligands for iron on the basis of pH studies, which also implicated histidyl residues. Thus the effect on pH of adding Fe(III) to the apoprotein over a range of pH values had indicated that the pK_a of metal-binding groups was near 11.2, suggestive of tyrosyl residues (the phenolic OH groups). Titration studies on the iron and apotransferrin also showed that above pH 5.5 some ten new groups were titratable in the apoprotein. It was proposed that six of these were phenolic OH groups, while the fact that the pK_a of the remaining four groups was near 7 suggested that these were histidyl residues. It appeared therefore that each iron(III) group was bound by three tyrosyl residues and two histidyl residues. However work on UV difference spectra indicated that only two tyrosyl residues were utilized. It is probably significant too, that two homologous regions in the primary structure of transferrin, which may correspond to components of the two active sites, actually only contain two tyrosyl residues each.[4]

Further support for the involvement of tyrosyl residues in metal binding comes from chemical modification of tyrosyl residues, which leads to the decrease or loss of metal-binding properties in transferrin, and from a range of spectroscopic techniques. Thus a comparison of proton magnetic resonance spectra of the iron and gallium complexes of transferrin showed differences in the aromatic regions which were attributed to the broadening of tyrosyl proton resonances in the proximity of paramagnetic Fe(III).[3]

Binding of Tb(III) to transferrin resulted in an increase in the intensity of fluorescence of Tb(III) by a factor of 10.[5] This was attributed to energy transfer from a tyrosyl ligand,[5] but it was suggested that each site contains two tyrosyl residues, not three. In addition, the resonance Raman spectra of the Cu(II) and Fe(III) complexes of transferrin closely resembled spectra of model Fe(III) phenolate compounds.[6, 7]

The involvement of histidyl residues in metal binding is also supported by resonance Raman spectra,[6] while the ligand hyperfine structure on the EPR spectra of Cu(II) transferrin is consistent with a nitrogen donor group. Chemical modification of histidine residues confirms their role in the active site, as an average of two histidyl residues per site were protected in the Fe(III) transferrin while all reacted in the apoprotein.[8] Analysis of the ^{1}H NMR spectra of transferrin and the apoenzyme also leads to the suggestion that two histidyl residues are linked to the metal ion.[9]

Distance measurements based on water proton relaxation rates suggest that the aquo group is also a ligand.[10] Thus it appears that each iron(III) in transferrin is bound by at least two and possibly three tyrosyl residues, probably two histidyl residues, and a water molecule. A minimum of five coordination positions around Fe(III) are satisfied. In addition the need for the anionic group as a prerequisite for iron binding suggests that this too should be considered as a ligand.

The anionic group could bind in three possible ways,[11] (a) to the metal only, (b) to the protein only, with some conformational change or an alosteric effect, or (c) to metal and protein. The third possibility seems reasonable. Attempts have been made[12] to explore these possibilities, using distance measurements with ^{13}C-labelled CO_3^{2-}. While a sharp ^{13}C NMR could be seen in the Ga(III) transferrin, no signal could be seen with the Fe(III) species probably as a result of line broadening from the paramagnetic ion. The use of spin-labelled malonate failed to give a detectable EPR signal, again due to broadening effects. The maximum distances that must separate the magnetic centres in two cases, so that the appropriate signals are not observed, are 9 Å and 11 Å respectively. Accordingly it has been postulated that the anion is a ligand to the metal, and is probably a bridging ligand. The protein is suggested to bind the anion by an ionized ε-amino group of lysine[13] or by a histidine group.[9]

Equivalence or non-equivalence of the two binding sites

Early work suggested that iron(III) bound randomly at the two sites, but it is becoming clear now that a distinction may be made between them. At pH 6.0 only one site is occupied, and Fe(III) and Cr(III) bind to different sites, designated A and B respectively.[14] The sequential filling of these sites by Fe(III) on increasing the pH was confirmed by labelling with ^{59}Fe and ^{55}Fe, and releasing one of them by lowering the pH to ~ 6. However it appears that the nature of the binding sequence is dependent upon the anion. The experiments described above were carried out using the nitrilotriacetate anion, but Fe(III) citrate occupies site B preferentially, while random binding apparently occurs with Fe(III) ascorbate.[15]

The situation has been clarified partly as a result of new techniques for separating the two different monoiron(III) transferrin species. Thus polyacrylamide gel electrophoresis in the presence of 6 mol dm^{-3} urea allows the separation of iron-free human serum transferrin, the two monoiron(III) forms and the fully saturated form.[16] Alternatively, partially saturated transferrin was digested with trypsin. The protein not protected by binding of iron was digested, thus allowing the identification of metal binding at the N-terminal or C-terminal half of the polypeptide. This has shown that site A is associated with the C-terminal region and site B with the N-terminal region.[17] Addition of Fe(III) nitrilotriacetate to human transferrin, followed by trypsin hydrolysis gives a carbohydrate-containing iron-binding fragment of molecular weight 43,000. In contrast, addition of Fe(III) as the citrate, chloride, or ascorbate followed by trypsin digestion gave a 36,000 fragment with no carbohydrate; this is the N-terminal binding site or site B.

It appears[18] that the A site is more strongly binding than the B site, but is not necessarily more accessible to all complexes of iron, as noted above. The A site is the site that retains iron at low pH.

Transferrin Receptor Sites of the reticulocyte

The reticulocyte membrane contains specific receptors for transferrin, which are now being isolated from different species as transferrin-carrying components of solubilized ^{125}I-labelled transferrin-saturated membranes. It is suggested[12] that the transferrin receptor of rabbit reticulocyte is a protein with complex subunit structure and a molecular weight near 350,000. It is uncertain whether iron is released into the cell after the binding of transferrin to the reticulocyte receptor site or whether transferrin is taken into the cell and the iron released internally. One suggestion is that the iron–transferrin complex releases its iron at sites on the mitochondrial membranes.

Release of iron from transferrin

The considerable stability of the iron(III)–transferrin complex makes transferrin an excellent carrier molecule, but presents problems in terms of releasing the iron. Excellent chelators for iron have little effect as *in vitro* mediators of iron release,[19] pyrophosphate being the most effective. ATP and ADP are also effective at pH 6. In view of the essential role of the anion in binding iron to transferrin, it seems reasonable that removal of the anion should underlie any route for the release of iron, but there seems to be no correlation between the ability of anions to replace carbonate in the transferrin complex and their effectiveness in mediating iron release. In microbial transport systems release of iron from the carrier is brought about by reduction of the iron to iron(II), but it appears that iron is released from transferrin as iron(III). Other factors under discussion at present include the possibility that iron is released preferentially from particular sites.

The suggestion that transferrin releases iron at mitochondrial membranes has led to the examination of the effects of a range of ligands on the uptake of iron from transferrin by isolated rat liver mitochondria.[20]

FERRITIN

Iron is stored in a number of mammalian organs in two forms, ferritin and hemosiderin. Hemosiderin is not well understood and is probably a product of the breakdown of ferritin. Ferritin, however, is now quite well characterized. It is a water-soluble protein made up of 24 equivalent subunits. These are arranged in a hollow spherical shell of external diameter about 125 Å, and giving an inner cavity of around 70 Å in diameter. This inner cavity is filled with an inorganic, micellar core made up of an oxohydroxophosphatoiron(III) complex of composition[21] $(FeO.OH)_8(FeO.OPO_3H_2)$. On this basis the precentage of iron in the micelle is 57%. The micelle may contain up to 4500 iron atoms, although the ferritin may not necessarily be fully saturated with iron. The protein shell[22] has six channels, of approximately 10 Å diameter, passing through it. These channels are assumed to be utilized in the uptake and release of iron.

Uptake of iron is associated with the catalysis of oxidation of Fe(II) to Fe(III) by apoferritin, while mobilization of iron involves reduction to Fe(II) by reduced flavins. Synthesis of the apoferritin is rapidly stimulated in most cells by the presence of iron; in rat liver the subunits are synthesized within 2–3 minutes. The discussion that follows is based on horse spleen ferritin, but ferritins from plant and animal sources have similar amino acid compositions although their molecular weights and iron storage capacities may very somewhat. However certain abnormal ferritins (isoferritins) are associated with tumour cells and other diseased states.

Other structural aspects

The structure of horse spleen apoferritin at 2.8 Å resolution[23] shows that the subunits are arranged in cubic symmetry, as represented in Figure 8.1. The subunits approximate to cylinders of diameter about 27 Å and length 54 Å. The protein shell binds several cations on or near its outside and inside surfaces. These sites may represent sites for binding of iron during iron accumulation.

The structure of the core cannot be determined by X-ray diffraction techniques, as the iron core does not have a definite orientation with respect to

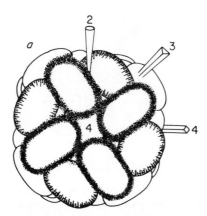

Figure 8.1 Schematic arrangement of the subunits in horse spleen apoferritin viewed down a four-fold axis. Some of the symmetry axes are shown. (Reproduced by permission of Nature from P. M. Harrison et al., Nature, **271**, 282 (1978)).

248

the outer shell. However EXAFS measurements have been carried out on the core in ferritin.[24] By comparison with an Fe glycine model compound, it was found that the iron centres are surrounded by 6.4 ± 0.6 oxygens at 1.95 ± 0.02 Å distance, probably in a distorted octahedral arrangement. This rules out an earlier suggestion that Fe^{3+} in ferritin is surrounded tetrahedrally by O^{2-} and OH^- ligands, a suggestion already queried on spectroscopic evidence.[25] The EXAFS data indicates that each iron has 7 ± 1 iron neighbours at an average of 3.29 ± 0.05 Å. On combining these results with stoichiometry and density data, it has been concluded[24] that the iron core involves a layer arrangement, with iron in the interstices between two nearly close-packed layers of oxygens, and that these compact O—Fe—O layers are only weakly bound to adjacent layers. The sheet is terminated by the binding of the phosphate groups, two groups being utilized for each eighteen iron(III) centres to maintain the stoichiometry. Such a

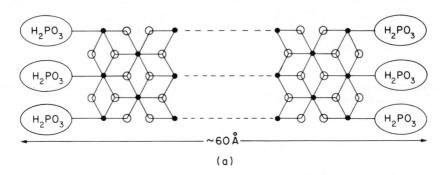

(a)

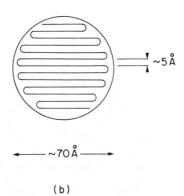

(b)

Figure 8.2 The structure of the iron-containing core in ferritin (Reproduced by permission from *J. Amer. Chem. Soc.*, **101**, 67 (1979), Copyright (1979) American Chemical Society. (a) One way of using the phosphorus atoms to terminate the two-dimensional sheets into strips. The solid circles are iron and the larger open circles are 0 or OH. From the stoichiometry there are 9 Fe per P, giving a width of ~ 60 Å. The length of the strip depends on the amount of iron in the micelle. (b) Schematic drawing of the folding of the strip into a 70-Å diameter micelle.

strip will be approximately 60 Å wide, consistent with the space available in the cavity of apoferritin. This strip is then folded upon itself to give the core material, as represented schematically in Figure 8.2. The length of the strip determines the extent of occupancy of the core.

Models for the ferritin core have been prepared by the hydrolysis of salts of Fe(III), such as the nitrate and citrate.[26] The so-called 'Spiro-Saltman' ball, prepared as above, is about 70 Å in diameter and has a molecular weight of about 150,000. In this structure the Fe(III) is octahedrally coordinated.[25] EXAFS studies[24] on the model, which has a stoichiometry of $[Fe_4O_3(OH)_4(NO_3)_3]1.5H_2O$, show that it is not a good model for ferritin, but a second form of the model may be similar to the ferritin core.

Incorporation of iron into ferritin

In the presence of air, apoferritin catalyses the oxidation of Fe(II) prior to its incorporation into the ferritin. One view of this process is that it involves pairs of Fe(II) ions which bind dioxygen with transfer of two electrons to give a Fe(III)—O_2^{2-}—Fe(III) complex, which then undergoes hydrolysis and transport to the interior of the molecule. An alternative view is that binding of Fe(II) on the inner surface of the protein accelerates its oxidation, the resulting bound Fe(III) forming nucleation centres to which further iron atoms are added. Support for the double-site theory appears to come from the X-ray studies which show the presence of pairs of Tb(III) sites, but these lie some 30 Å away from the entry channels.

Some discussion has centred on the stoichiometry of the Fe(II)/[O_2] reaction, and on the possible presence of peroxide and superoxide products. These species have not been identified.[27-29] It is claimed that four Fe(II) are oxidized per dioxygen,[27] although it appears[28] that this ratio depends upon [Fe(II)], possibly reflecting the probability that a dioxygen radical encounters Fe(II) or some other electron donor. These results appear to argue against the bridging peroxo theory, but are not conclusive.

ABSORPTION AND EXCHANGE OF IRON[30]

Iron is unique amongst the trace elements in that there is no secretory system. Thus its concentration levels are regulated primarily by absorption and not by excretion. The effects of various factors on iron uptake have been assessed, but little is known about the way in which uptake is controlled. Work in this area has been concerned largely with the search for carrier molecules in both membrane and cytosol fractions of mucosal homogenates. In summary it appears that Fe(III) binds to specific receptor proteins in the brush border of the mucosal cell and is then transported into the mucosal cell, where it will be in equilibrium with iron-poor ferritin. On reaching the serosal surface of the cell it becomes attached to transferrin for transport in the plasma. A number of smaller molecules have also been implicated in iron transport.

Interrelationships between transferrin, ferritin, and synthetic sites are of interest. Transferrin is found in plasma and ferritin is almost completely localized within cells, so it is interesting to speculate if the two proteins ever interact with each other *in vivo*. *In vitro* studies[31] have shown that transfer of iron occurs between the two proteins in the presence of citrate, ATP, or ascorbate in both directions but the net flow is always from ferritin to transferrin. Ferritin which is 30–50% saturated with iron appears to be most reactive. It has been noted that transferrin may carry iron to the mitochondrial membrane, implying it can enter cells, while it appears that transferrin enters developing red blood cells to deliver iron for hemoglobin synthesis, thus there is a possibility that transferrin and ferritin interact directly. Iron only crosses[32] the mitochondrial inner membrane as Fe(II), and it has been suggested as a result of studies on copper-deficient systems, that cytochrome oxidase may play a role in reducing Fe at the inner mitochondrial membrane.

IRON TRANSPORT IN MICROBES

Most aerobic and facultative microorganisms have efficient transport processes for taking up iron in controlled amounts. The iron present outside the organism is present as insoluble iron(III) compounds at physiological pH values and so is not readily available. In any case iron could not be allowed to enter freely into the cell as this would result in toxic overloads. Accordingly, in response to shortages of iron inside the cell, the organism synthesizes and releases ligands which are highly specific for iron(III). These are the siderophores or iron-carrying ligands. It appears that the intact iron(III) complex is transported back into the cell *via* specific receptor sites on the cell surface, although there may be other pathways of different types. The iron is released from the complex by being enzymatically reduced to Fe(II), which binds weakly to the ionophore, although in some cases this reduction may be preceded by partial hydrolysis of the siderophore. The iron in the cell is then delivered to an appropriate point for biosynthetic purposes. As well as these highly specific transport processes organisms usually have additional, less specific uptake pathways of lower affinity for iron.

Siderophores

A range of these chelating ligands has been described in detail.[33] With one exception they provide a set of oxygen donors for Fe(III) and all complexes appear to involve six-coordinate Fe(III). The complexes are all kinetically labile, a fact which has complicated studies on their transport. This problem has been overcome in some cases by the use of inert Cr(III) complexes.[33]

It is convenient to classify siderophores into phenolates (strictly, catecholates) and hydroxamates. A few examples are known in which siderophores contain both phenolate and hydroxamate groups. Thus mycobactin has two hydroxamate groups, each of which is bidentate, a simple phenolate group, and uniquely a nitrogen donor atom from a substituted oxazoline group. In general the phenolates are found in bacteria and the hydroxamates in higher organisms such as fungi and yeasts.

$$R-\overset{O}{\underset{}{C}}-\overset{}{\underset{}{N}}-R'$$

Hydroxamic acid

Catechols

These ligands bind strongly to iron(III). Acetohydroxamic acid ($R = CH_3$, $R' = H$), a simple example, binds iron(III) with $\log\beta_2 \sim 21$ and $\log\beta_3 \sim 28$. In contrast it binds iron(II) with $\log\beta_2 \sim 8$. The naturally occurring siderophores, the ferrichromes, which provide three hydroxamate groups have $\log K$ values in the range 30–32.

Hydroxamate siderophores

Most naturally occurring hydroxamic acids have three hydroxamic acid groups per molecule and so form neutral complexes with iron(III). The **ferrichromes**[34, 35] involve a cyclic peptide with three hydroxamic side chains, while in the **ferrioxamines** these three groupings are part of linear or cyclic molecules. The ferrichromes are produced by fungi such as *Ustilago sphaerogena* while the ferrioxamines are produced by *Nocardia* and *Streptomyces* species. Some structures[34, 35] are given in Figure 8.3. A range of ligands is available depending upon the nature of the substituents R^n; for ferrichrome $R' = R'' = R''' = H$, $R = CH_3$.

Other classes of hydroxamate siderophores are those related to rhodotorulic acid, fusarinine, and aerobactin, all of which are depicted in Figure 8.3. Aerobactin contains a carboxylate group that could possibly bind to Fe(III).

Phenolate siderophores

Examples are shown in Figure 8.4. Enterobactin (sometimes termed enterochelin) is a tricatechol siderophore that is widespread in bacteria. The iron(III) complex will be negatively charged. Its structure[36] is also shown in Figure 8.4, and this is strongly supported[37] in solution by resonance Raman data which suggest that iron is bound exclusively to the phenolate oxygen atoms. Several siderophoric compounds of lower molecular weight have been isolated from low-iron media in which organisms have been grown. Thus compounds **I**, **II**, and **III** in Figure 8.4 have been synthesized by *Azotobacter vinelandii*, *E. coli*, and *Bacillus subtilis* respectively. In some cases the interaction of iron(III) with these ligands has not been fully characterized. It is interesting that itoic acid (**III**) was the first phenolate siderophore to be isolated.[32]

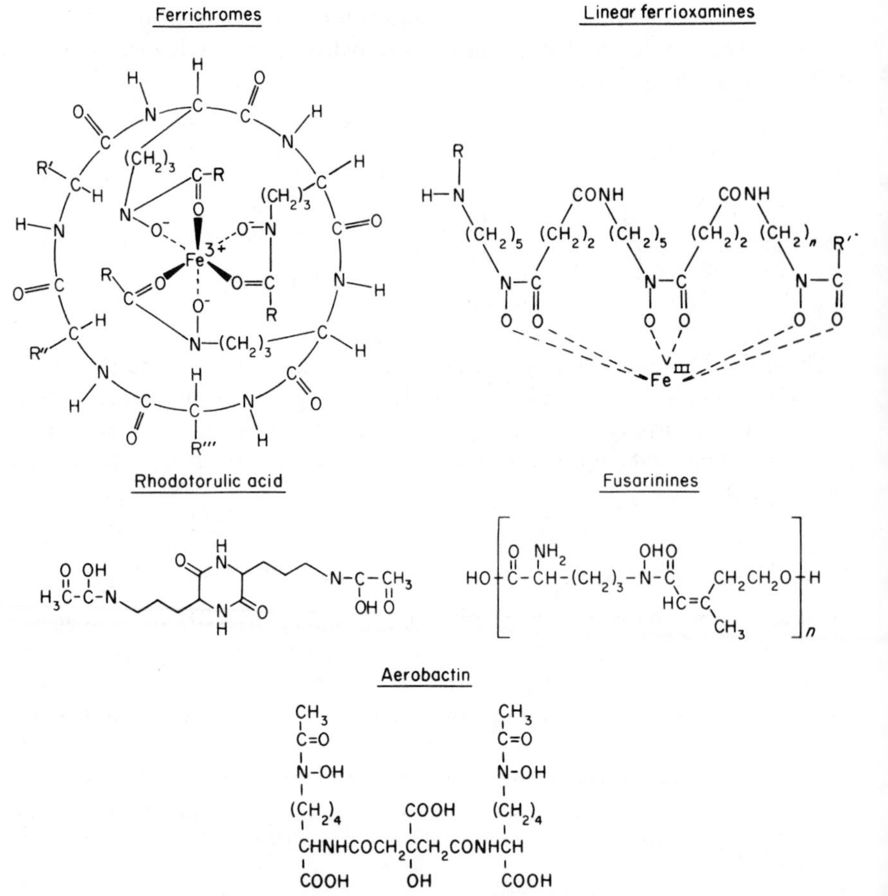

Ferrichromes

Linear ferrioxamines

Rhodotorulic acid

Fusarinines

Aerobactin

Figure 8.3 Hydroxamate siderophores.

Iron release from siderophores

It appears that different mechanisms hold for the release of iron from enterobactin and ferrichrome-type siderophores. Release of iron from the latter ligand (of hydroxamate structure) results from the reduction of iron(III) to iron(II). The hydroxamate chelates are kept intact and will be used again by the cell. In contrast, in the case of enterobactin, the cyclic triester linkages are cleaved by a specific esterase prior to the release of iron(III) *via* the reduction reaction. The reason for this difference is illustrated by recent electrochemical studies.[38] Cyclic voltammograms of ferrichrome A and ferrioxamine B show reversible one-electron waves with potentials –446 and –454 mV respectively. Thus the reduction of hydroxamates by a range of physiological reducing agents is quite feasible. Ferric enterochelin gives a potential of -750 mV at pH 7. Thus reduction of iron(III) enterochelin by physiological agents is not feasible—hence the need for preliminary hydrolysis of the ligand.

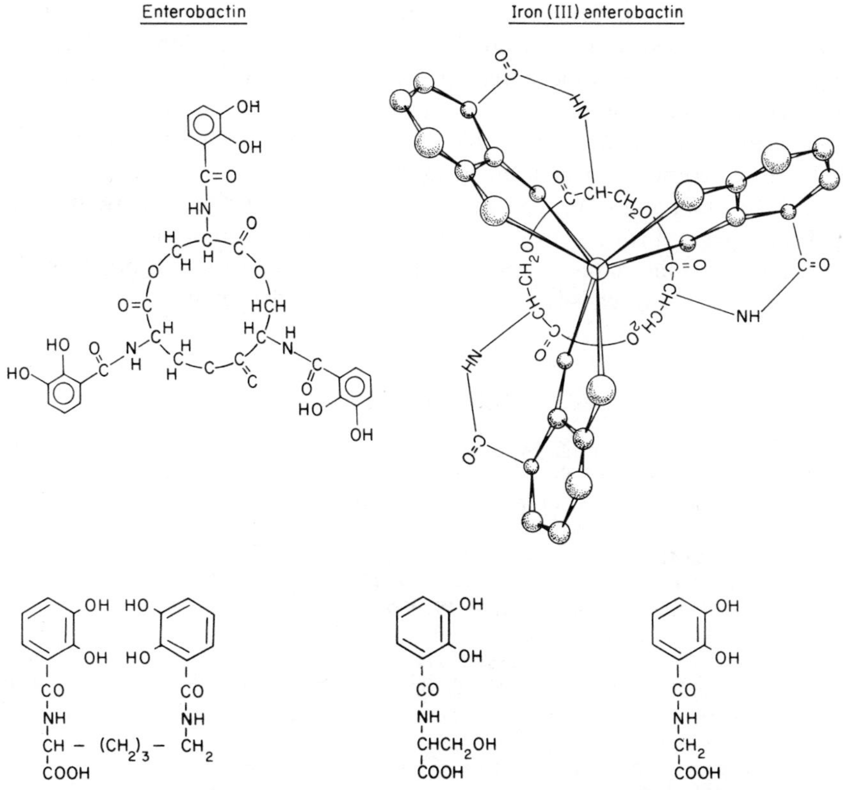

Figure 8.4 Phenolate siderophores (Structure of iron (III) compound reproduced by permission from K. N. Raymond *et al. J. Amer. Chem. Soc.*, **98**, 1763 (1976). Copyright (1976) American Chemical Society).

In vivo microbial transport of iron

There is still some uncertainty as to whether the intact iron(III)–siderophore complex is transported into the cell. Experiments with labelled iron and labelled siderophore certainly showed that both groups entered the cell, but this does not rule out the possibility that the components of the complex were transported separately. In the case of ferrichrome transport in *E. coli* the use of the kinetically inert Cr(III) complex of ferrichrome has allowed the positive demonstration that the intact complex does enter the cell. In other cases the situation is not resolved and there is the possibility that parallel pathways exist.

Enteric bacteria such as *E. coli* utilize enterobactin for iron transport. But even though they do not synthesize ferrichrome they are able to utilize it in iron transport, as in the example described above. Thus there appears to be a specific receptor site and transport system for a siderophore that is not synthesized in the cell.

Attention has been focussed in recent years on the membrane transport processes associated with iron uptake[39] and with the biosynthesis of siderophores. Interest in the former topic was stimulated by the finding of a common locus for the binding of siderophores, phages, and the antibiotic colicins on the outer membrane of *E. coli*. Thus these toxic agents have exploited the siderophore receptors to gain entry to the cell. It appears therefore that in addition to supplying iron, the siderophore complexes protect enteric bacteria against phages and colicins by competing for the outer-membrane receptor site. Other protective mechanisms appear to involve cellular reactions promoted by iron, implying a further role in protection for the siderophores.

REFERENCES

1. E. B. Brown, P. Aisen, J. Fielding, and R. R. Crichton (Eds.), *Proteins of Iron Metabolism*, Grune and Stratton, New York, 1977.
1(a). R. R. Crichton (Ed.), *Proteins of Iron Storage and Transport*, North Holland, 1975.
1(b). N. D. Chasteen, *Coord, Chem. Revs*, **22**, 1 (1977).
2. K. N. Raymond (Ed.), *Adv. Chem. Ser.*, **162**, 1, 33, 104 (1977)
3. R. J. P. Williams, *Chem. in Britain*, **1979**, 555.
4. R. T. A. MacGillivray and K. Brew, *Science*, **190**, 1306 (1975).
5. C. K. Luk, *Biochemistry*, **10**, 2838 (1971).
6. B. P. Gaber, V. Miskowski, and T. G. Spiro, *J. Amer. Chem. Soc.*, **96**, 6868 (1974).
7. Y. Tominatsu, S. Kint, and J. R. Scherer, *Biochem. Biophys, Res. Comm.*, **54**, 1067 (1973).
8. T. B. Rogers, R. A. Gold, and R. E. Feeney, Ref. 1, p. 161.
9. R. C. Woodworth, R. J. P. Williams, and B. M. Alsaadi, Ref. 1, p. 211.
10. B. P. Gaber, W. E. Schillinger, S. H. Koenig, and P. Aisen, *J. Biol. Chem.*, **245**, 4251 (1970).
11. M R. Schlabach and G. W. Bates, *J. Biol. Chem.*, **250**, 2182 (1975).
12. P. Aisen and A. Leibmann, Ref. 2. p. 104.
13. J. L. Zweier and P. Aisen, *J. Biol. Chem.*, **252**, 6090 (1977).
14. D. C. Harris, *Biochemistry*, **16**, 560 (1977).
15. E. J. Zapolski and J. V. Princiotto, Ref. 1, p. 250.
16. D. G. Makey and U. S. Seal, *Biochim. Biophys. Acta*, **453**, 250 (1976).
17. R. W. Evans and J. Williams, *Biochem. J.*, **173**, 543 (1978).
18. P. Aisen, A. Leibman, and J. Zweier, *J. Biol. Chem.*, **253**, 1930 (1978).
19. S. Pollack, G. Vanderhoff, and F. Lasky, *Biochim. Biophys. Acta*, **497**, 481 (1977).
20. K. Konopka, *FEBS Letters*, **92**, 308 (1978).
21. J. L. Farrant, *Biochim. Biophys. Acta*, **13**, 569 (1954).
22. P. M. Harrison, *Semin. Haematol.*, **14**. 55 (1977); P. M. Harrison and A. Treffry, *Inorganic Biochemistry*, Specialist Per. Report. Chem. Soc. (Ed. H.A.O. Hill), **1**, 120 (1979).
23. S. H. Banyard, D. K. Stammers, and P. M. Harrison, *Nature*, **271**, 282 (1978).
24. S. M. Heald, E. A. Stern, B. Bunker, E. M. Holt, and S. L. Holt, *J. Amer. Chem. Soc.*, **101**, 67 (1979).
25. H. B. Gray, *Adv. Chem. Ser.*, **100**, 365 (1971).
26. B. A. Sommer, D. W. Margerum, J. Renner, P. Saltman, and T. G. Spiro, *Bioinorg. Chem.*, **2**, 295 (1973); G. Bates, J. Hegenauer, J. Renner, P. Saltman, and T. G. Spiro, *Bioinorg. Chem.*, **2**, 311 (1973); T. G. Spiro, L. Pape, and P. Saltman, *J. Amer. Chem.*

Soc., **89**, 5555 (1967); E. M. Holt, S. L. Holt, W. F. Tucker, R. O. Asplund, and K. J. Watson, *J. Amer. Chem. Soc.*, **96**, 2621 (1974).

27. G. Melino, S. Stefanini, E. Chiancine, E. Antonini, and A. Finazzi Agro, *FEBS Letters*, **86**, 136 (1978).
28. A. Treffry, J. M. Sowerby, and P. M. Harrison, *FEBS Letters*, **95**, 221 (1978).
29. M. Wauters, A. M. Michelson, and R. R. Crichton, *FEBS Letters*, **91**, 176 (1978).
30. *"Iron Absorption and Nutrition"*, *Fed. Proc.*, **36**, 2015 (1977).
31. D. C. Harris, *Biochemistry*, **17**, 3071 (1978).
32. T. Flatmark and I. Romslo, *Adv. Chem. Ser.*, **162**, 78 (1977).
33. J. B. Neilands, *Inorganic Biochemistry* (Ed. G. Eichhorn), Elsevier, Amsterdam 1973; p. 167; K. N. Raymond, *Adv. Chem. Ser.*, **162**, 33 (1977).
34. A. Zalkin, J. D. Forrester, and D. H. Templeton, *J. Amer. Chem. Soc.*, **88**, 1810 (1966).
35. M. Llinas, *Structure and Bonding*, **17**, 135 (1973).
36. S. S. Isied, G. Kus, and K. N. Raymond, *J. Amer. Chem. Soc.*, **98**, 1763 (1976).
37. S. Salama, J. D. Strong, J. B. Neilands, and T. G. Spiro, *Biochemistry*, **17**, 3781 (1978).
38. S. R. Cooper, J. V. McArdle, and K. N. Raymond, *Proc. Natl. Acad. Sci. U.S.A.*, **75**, 3551 (1978).
39. J. B. Neilands, *Adv. Chem. Ser.*, **162**, 3 (1977).

Note added in proof

A ferritin-like molecule has now been isolated from several bacterial species, demonstrating for the first time the possibility of iron storage in bacteria. The cytochrome $b_{557.5}$ from *A. vinelandii* is associated with large amounts of iron, now recognised to be a ferritin. (E. I. Steifel and G. D. Watt, *Nature*, **279**, 81 (1979)). This bacterio ferritin-cytochrome has an iron content of 13–20%, and an electron-dense core of 55 Å. It is suggested to serve as an iron store for nitrogenase. It is now recognised that other bacteria contain iron storage compounds of a type distinct from ferritin, while iron stores will also be found in magnetotactic organisms, (L. Frankel et al, *Science*, **203**, 1355 (1979)).

CHAPTER NINE

THE ALKALI METAL AND ALKALINE EARTH METAL CATIONS IN BIOLOGY

A considerable amount of information is available[1] on the role of the Group IA and Group IIA metal ions in biological processes. Some of this material can now be brought together[2] in a fairly cohesive manner, while current developments in experimental approaches to studying the roles of these ions suggest that considerable advances will be made in the coming years.

Some references have been made in Chapter 1 to the role of these elements, while the function of Mg^{2+} and Ca^{2+} as activators of hydrolytic enzymes has been discussed briefly in Chapter 4. To set the scene for this chapter it is necessary to bring forward a number of facts concerning the occurrence and role of the four important metal ions, Na^+, K^+, Mg^{2+}, and Ca^{2+}. One other important problem that must be considered is the way in which these metal ions may be studied. Their electronic and neclear properties do not lend themselves readily to this, although ^{23}Na and ^{43}Ca NMR have been used.

THE DISTRIBUTION OF INORGANIC IONS

In general the concentration of inorganic ions varies in a systematic way inside and outside the cell. The cells are in contact with a fluid which supplies all essential materials. These substances must pass through the membrane that separates the inside of the cell from its external environment. In the other direction must pass the various waste products of the cell. It is clear, though, that the cell membrane does not in general allow molecules or ions to pass through by simple diffusion. It is extremely selective and only certain species may pass through. In some cases ions may only be allowed to travel in one direction, out but not in, for example. That this selectivity of the membrane must operate is shown by the fact that the ionic composition inside the cell is different from that outside it. Thus the interior of the cell contains large amounts of K^+ but relatively little Na^+. ($[K^+]/[Na^+]$ lies in the range $\frac{3}{1} \rightarrow \frac{20}{1}$.) Similarly there is more Mg^{2+} and phosphate inside than outside. The reverse situation holds for Na^+, Ca^{2+}, and Cl . Here the concentration is greater outside the cell than inside. These differences could not exist if the alkali metal cations freely diffused in and out of the cells, and represent positive activity on the part of the cell membrane with due expenditure of energy. The way in which the cell membrane does this is an important area of research. It is particularly important in animal cells, more so than in bacterial cells where the membrane is much less permeable.

The distribution of the metal cations in this way inside and outside the cell allows a number of generalizations. Thus Mg^{2+} and K^+ are associated with intracellular activity and Ca^{2+} and Na^+ are involved in processes occurring outside the cell, and the two sets of ions may not replace each other. Mg^{2+} and K^+ are important activators of enzymes occurring inside the cell, while Ca^{2+} activates extracellular enzymes. The exclusion of Ca^{2+} from the cell under normal conditions explains why a sudden influx of Ca^{2+} acts as a trigger. Na^+ in general will only activate a few enzymes. These may not be activated by K^+, and Na^+ will not activate the K^+-dependent enzymes, though NH_4^+ will. Cell Mg^{2+} and K^+ are very important in the stabilization of RNA and the RNA/DNA synthetic systems. Thus in resting bacterial cells (i.e. the effect produced by limiting the supply of certain requirements) the in-cell K^+, Mg^{2+}, and phosphate is lower than normal. Once the limitation is removed, the cells grow but the first step prior to this is to increase the concentrations of Mg^{2+}, K^+, phosphate. This is an example of ion pumping. Once the metals are inside the cell, they will be bound by appropriate ligands in accordance with their known chemical reactivity. In general, we expect Na^+ and K^+ to be bound to a single multidentate ligand, while Ca^{2+} and Mg^{2+} will act as a bridge between two ligands. Ca^{2+} is particularly well suited for this.

The fact that calcium is not utilized by the cell is associated with its extensive use as a structural factor, in bones and teeth for example. The concentration of calcium is controlled by two hormones, calcitonin and parathyroid hormone. The skeleton is the major reservoir of calcium, calcitonin inhibiting the release of calcium ions and parathyroid hormone mobilizing the element from the bones. Together they are probably responsible for the remodelling and maintenance of bone structure.

The existence of the sodium–potassium concentration gradient across the cell membrane has important implications in physiology. The difference in concentration results in the generation of a potential difference between the two sides of the membrane. Thus the inside of the cell is some $80\,mV$ negative with respect to the external surface of the membrane. The existence of this potential allows the nerve fibres to conduct impulses and hence muscles to contract.

The control of the permeability of the membrane is also dependent on the presence of divalent metal ions; probably for electrostatic interaction with two carboxylate groups, which is necessary to maintain the integrity of the membrane.

Some details of the occurrence of metal ions are given in Table 9.1. Before considering detailed examples of the role of these metal ions in biology, we shall consider aspects of the chemistry of the Group IA and IIA cations, paying particular attention to complex formation and stability constants. It seems reasonable that the first stage in the transport of a cation throught the cell membrane is the binding of that cation by components of the cell wall. It is worthwhile looking at the factors which determine the binding of these pairs of cations by model ligands. In particular we will discuss certain macrocyclic ligands, which will, for example, reverse the usual binding order for Na^+ and K^+ exclusively, and which are of considerable significance in model studies on transport processes.

TABLE 9.1 The distribution of IA and IIA cations in some biological systems

	Na^+	K^+	Mg^{2+}	Ca^{2+}
Bacteria* (dried)				
E. coli		310	247	2.5
Aerobacter aerogenes		410	124	5.0
B. subtilis		1260	124	2.5
B. subtilis spores		231	206	400
B. cereus		1180	450	7.5
B. cereus spores		51	124	475
E. coli (wet)	80	250	20	5
Yeast cells*	10	110	13	1.0
Skeletal muscle*	27	92	22	3
Human red cell	11	92	2.5	0.1
Whole blood	85	44	1.57	
Squid nerve (outside)	~440	22	55	10
Squid nerve (inside)	10	300	7	5×10^{-4}
Crab leg nerve (outside)	510	12		
Crab leg nerve (inside)	52	410		
Frog sartorius muscle (outside)	120	2.6		
Frog sartorius muscle (inside)	15	130		
Giant algal cells				
Nitella translucens				
cytoplasm	55	150		
vacuole	65	75		
outside solution	1.0	0.1		
Tolypella intricata				
chloroplasts	36	340		
vacuole	7	100		
outside	1.0	0.4		

Concentrations: m mol kg^{-1} (*) otherwise m mol dm^{-3}

THE CHEMISTRY OF THE s BLOCK ELEMENTS

These elements are listed in Table 9.2, together with some physical data. Their chemistry is dominated by the tendency to attain the noble gas configuration, with the formation of ions M^+ and M^{2+} in Groups I and II respectively. Their chemistry is therefore essentially ionic chemistry, with the exception of lithium and particularly beryllium, as indicated by the ionization potential data. The small size of the 'cations' of these two elements makes them good polarizers. Hence the chemistry of beryllium is covalent, while that of lithium shows some covalent properties. We are not, of course, particularly concerned with these, but magnesium is similar to lithium in many ways. This is an example of the diagonal relationship, and may be relevant to the use of lithium salts in the treatment of certain mental disorders, as discussed in Chapter 10.

The importance of the ionic radius of the cation cannot be overemphasized. It is reflected in solution and structural aspects of the chemistry of the Group IA and IIA cations, and is intimately connected with the whole problem of cation

TABLE 9.2 Some properties of the IA and Group IIA cations

		Ionic radii (Å)	First ionization potential (eV)	Common coordination nos.
Lithium	Li^+	0.60	5.39	4, 6
Sodium	Na^+	0.95	5.14	6
Potassium	K^+	1.33	4.34	6, 8
Rubidium	Rb^+	1.48	4.18	8
Caesium	Cs^+	1.69	3.81	8
Ammonium	NH_4^+	1.45		
Beryllium	Be^{2+}	—	9.32	2, 4
Magnesium	Mg^{2+}	0.65	7.64	6
Calcium	Ca^{2+}	0.99	6.11	6, 8
Strontium	Sr^{2+}	1.10	5.69	6, 8
Barium	Ba^{2+}	1.29	5.21	6, 8

selectivity in biological systems, where the size effect is manifested in several aspects of this phenomenon. Values of ionic radii are listed in Table 9.2, and are based upon the hard-sphere model for ionic structures with $(r^+ + r^-)$ equal to the equilibrium internuclear distance. The ionic radius is a limiting factor in determining the coordination number of a cation, as expressed in the radius-ratio rule, while it also determines the extent of hydration (or solvation in general) and hence (with the lattice energy) the solubility of salts. For cations of similar charge, the hydration energy will increase with decrease in cation radius, while the Group IIA cations will be solvated more strongly than the Group IA cations, in view of their smaller size and increased charge, which makes them better polarizers.

We are particularly interested in factors that result in ligands binding selectively Na^+ or K^+, and Mg^{2+} or Ca^{2+}, and also in distinguishing between cations of similar size but different charge, for example, Na^+ and Ca^{2+}. Some complex biological ligands such as certain antibiotics, together with the synthetic macrocyclic and macrobicyclic ligands, the crown ethers and cryptands respectively, distinguish between cations directly and simply on a size basis by providing a 'hole' or cavity which is better suited for one cation than another. It is clearly a more complicated matter to distinguish between the IA and IIA cations of similar radii; here selectivity arises from their different charges, reflecting polarizing power and solvation phenomenon. These factors are also important for ligands other than macrocycles in distinguishing between cations of different size.

For simple, small ligands, particularly the anions of weak acids, binding to cations is determined by the polarizing power of the cation, and decreases with increase in the radius of the cation. This order can be changed by increasing the complexity of the ligand, as is shown in Table 9.3 by comparing data for acetate and nitrilotriacetate. For complex formation with acetate, formation constants

TABLE 9.3 Formation constant data (log K) for IA and IIA cations at 25°C.

Ligand	Solvent*	Li^+	Na^+	K^+	Rb^+	Cs^+	Tl^+	Mg^{2+}	Ca^{2+}	Sr^{2+}	Ba^{2+}
EDTA	W	2.8	1.7	1.0			5.84	8.9	10.7	8.8	7.9
SO_4^{2-}	W	0.64	0.71	0.96				2.23	2.28		
NO_3^-	W	−1.5	−0.5	−0.1		0.1	−0.3		0.28	0.82	0.92
$P_2O_7^{4-}$	W	2.4	2.3	1.5		2.3	1.69	5.79	5.0	4.66	
Glycine	W							3.4	1.4	0.9	0.8
Acetate	W							0.82	0.77	0.44	0.41
Nitrilotriacetate	W			1.0			4.4	5.3	6.4	5.0	4.8
Nonactin	acetone		4.84	4.85		4.0					
Nonactin	wet acetone		2.32	4.30		2.60					
Nonactin	E		3.38	3.58	3.53	2.95					
Valinomycin	M	1.28	0.67	4.90	5.26	4.41	2.72	1.20	2.95	2.65	3.34
Cryptand 2.1.1	W	5.30	2.80	<2.0	<2.0	<2.0					
2.2.1	W	2.5	5.40	3.95	2.55	<2.0					
2.2.2	W	<2.0	3.9	5.4	4.35	<2.0					
3.2.2	W	<2.0	1.65	2.2	2.05	2.2					
3.3.2	W	<2.0	<2.0	<2.0	≦0.7	<2.0					
3.3.3	W	<2.0	<2.0	<2.0	≦0.5	<2.0					
18-Crown-6	W		0.3	2.06							
15-Crown-5	W		0.70	0.74			1.23				
Dicyclohexyl-18	W		1.21	2.02	1.52	0.96	2.44			3.24	3.57
-crown-6 (A)	M		4.08	6.01							

*Solvents W = water, M = methanol, E = ethanol.

Sources: J. M. Lehn, *Structure and Bonding*, **16**, 1 (1973).

R. M. Izatt, D. J. Eatough, and J. J. Christensen, *Structure and Bonding*, **16**, 161 (1973).

Spec. Pub. Chem. Soc., Nos. 17, 25.

follow the normal sequence of $Mg^{2+} > Ca^{2+} > Sr^{2+} > Ba^{2+}$, but for nitrilotria-cetate Ca^{2+} is bound more strongly than Mg^{2+}. In general, the normal sequence is also reversed for anions of strong acids, i.e. $K^+ > Na^+$ and $Ca^{2+} > Mg^{2+}$. Such deviations reflect the size of the ligand, which result in several water molecules being displaced from the cation, a phenomenon which favours complex formation with large cations rather than small ones.

Selectivity towards cations of similar radii but different charge, can be rationalized in terms of the hydration of the cation, the ligand, and the complex, in addition to any size effect. Low dielectric constant media will destabilize the complex with higher positive charge more than those with low charge, and so the use of ligands with hydrophobic exteriors should result in selectivity towards monovalent cations. On the other hand, the use of negatively charged ligands, particularly those with hydrophilic properties, should favour M^{2+} over M^+.

Complexes of the IA and IIA cations

Interest in this topic has largely come about as a direct result of attempts to account for the selectivity of biological transport processes. Group IA cations form complexes weakly, usually as solvates or with oxygen donors, although species such as $[Na(NH_3)_4]^+ I^-$ were known. In general, conventional coordination complexes[3] either involve neutral hard ligands with a simple anion, or chelating anionic ligands, such as the β-diketones, together with neutral oxygen donors, such as the phosphine oxides. A good number of these complexes have been studied by X-ray diffraction techniques, which show that high coordination numbers are a feature of the heavier members of Group IA. Thus the complex with isonitrosoacetophenone involves seven-coordinate K^+, while that formed between potassium nitrophenolate and isonitrosoacetophenone has an irregular structure in which K^+ is surrounded by seven oxygen atoms and one nitrogen atom.

The alkaline earth cations form a much wider range of complexes, reflecting their higher charge and smaller radius. Mg^{2+} tends to be bound by nitrogen donors, but the other Group IIA cations prefer oxygen donors, although ammine and amine complexes of Ca^{2+} are known. However, in biological systems, Ca^{2+} is coordinated by oxygen atoms, usually to several negative carboxylate groups, the presence of a nitrogen donor ligand leading to preferential binding of Mg^{2+}. There are interesting differences in the structures of Mg^{2+} and Ca^{2+} compounds that are probably relevant to the differing biological roles of the two cations. Mg^{2+} forms six-coordinate compounds of regular structure, but Ca^{2+} often has higher coordination numbers (7 or 8) and shows irregular geometry in its complexes. The flexibility thus implied for Ca^{2+} in its binding requirements means that it will bind more strongly to the irregular sites offered by biological ligands. In particular it will be more effective in acting as a bridge to cross-link ligands in a reversible manner. In contrast Mg^{2+} although it may act as a cross-linking agent in polymers in intracellular situations, will be much less effective. One example of a stronger binding of Mg^{2+} to an anionic polymer is with teichoic acids in cell walls of Gram positive bacteria,[4] which may be acting as a specific pathway for the assimilation of Mg^{2+}.

Reference has been made to situations in which simple ligands are able to reverse the normal stability sequence and preferentially complex with K^+ rather than Na^+, and Ca^{2+} rather than Mg^{2+}. However, this property of selective binding of cations is shown to a much more marked degree by some groups of macrocyclic ligands that have attracted considerable attention in recent years in a number of contexts. These are (a) the cyclic eithers, (b) the cryptands, and (c) some naturally occurring ionophores, usually antibiotics of serveral structural types, together with synthetic analogues. Antibiotics such as valinomycin have long been known to induce the transport of K^+ into mitochondria, hence the term 'ionophore'. Valinomycin is a cyclic peptide that is probably the best ligand known for the selective complexation of K^+ compared to Na^+. Indeed, valinomycin and nonactin, a second example, have been used in the development of metal-selective membrane electrodes which can be used to estimate K^+ in the presence of Na^+. The antibiotic actinomycin, on the other hand, binds Na^+ rather than K^+. The use of macrocyclic ligands to bind differentially the divalent metal ions has long been postulated.[5] Each of the three groups will now be considered. It will be seen that they have certain features in common, including a cavity for a cation, a hydrophobic exterior, and a flexible structure that allows the stepwise replacement of water molecules on the metal.

The Cyclic Ethers

These were originally synthesized by Pedersen.[6] Some examples are schematically represented in Figure 9.1. For obvious reasons they have been termed 'crown ethers'! Individual examples are named by a prefix giving the number and kind of hydrocarbon rings, followed by a number giving the total number of atoms in the polyether ring, the name 'crown', and then the total number of oxygen atoms in the polyether ring. The nomenclature is easily understood by considering the examples in Figure 9.1.

A wide range of crown ethers with the number of oxygen atoms varying from three to twenty have been synthesized. Many of them with between five and ten oxygen atoms form stable complexes[3] with a range of cations, including main group and transition metals. In most cases a 1 : 1 stoichiometry holds, but other stoichiometries are found when, for example, the cation is too large to fit into the ether cavity. These include 2 : 1 and 3 : 2 ether : metal complexes (designated 'sandwich' and 'club sandwich' respectively), although the latter stoichiometry has not been found in the solid state. Some of the larger crown ethers will bind two cations. All these cyclic ether complexes have a hydrophilic interior and a hydrophobic exterior, which allow them to dissolve in non-aqueous solvents.

The formation constants for interaction of a particular crown ether with cations vary considerably from cation to cation, depending upon the relative sizes of cation and cavity. For the alkali metal cations there is a useful correlation between ionic radius and the number of oxygen atoms in the crown ether that

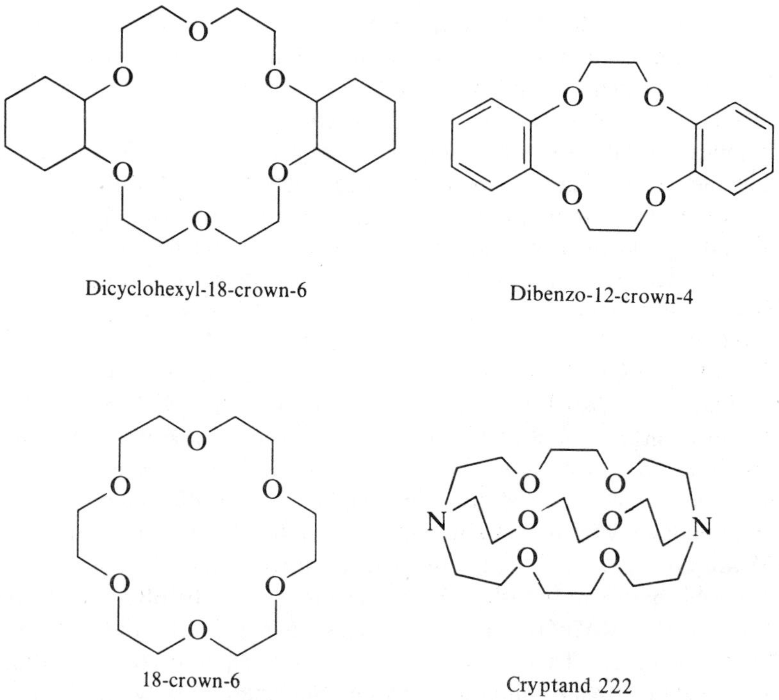

Dicyclohexyl-18-crown-6

Dibenzo-12-crown-4

18-crown-6

Cryptand 222

Figure 9.1 Some oxygen donor ligands. The macrocyclic ethers and cryptands.

provides the best fit, namely Li^+ (4), Na^+ (5), K^+ (6), and Cs^+ (8). Large ring ethers also form stable complexes by wrapping around the cation, for example[7] dibenzo-30-crown-10 with K^+.

Formation constant data[8] thus demonstrate the importance of cavity size in controlling the specificity of the ether for a cation. They also show that complex formation is dependent on the nature of the solvent and the presence of hydrophilic or hydrophobic substituents on the crown ether. The presence of hydrophobic groupings will favour binding of M^+ over M^{2+}, as the latter cation, being a better polarizing agent, will lose water molecules with greater difficulty. This indicates how cations of similar size but different charge (such as Na^+ and Ca^{2+}) may be distinguished between.

Equilibrium constants[8,9] for the formation of $1:1$ complexes in methanol between IA cations and the various isomers of dicyclohexyl-18-crown-6 are considerably less than those for complex formation with 18-crown-6. This must reflect stereochemical differences between 18-crown-6 and the dicyclohexyl ligand and the fact that the various non-covalent interactions are sensitive to small conformational differences.[8a]

It is important to distinguish between strength of binding and the selectivity of a crown ether for a cation. Thus while other crown-5 ethers interact more strongly with Na^+, the most specific one for Na^+ in terms of competition with other cations is dicyclohexyl-16-crown-5. However, comparison of the interaction of crown-5 ligands with Na^+ and K^+ is complicated by the fact that K^+ gives 2:1 complexes and so formation constants appear to be similar.

Complex formation involves stepwise substitution of the cation's solvation shell by the crown ether oxygen atoms. Such reactions, which will be sensitive to the nature of the solvent, are probably preceded by a conformational change in the ligand so that the donor atoms point in towards the cation.

Considerable effort is being expended[9] at present on the synthesis of a wide range of cyclic ethers with varying heteroatoms and substituents on the ether. Thus the normal selectivity of 18-crown-6 for K^+ over Na^+ can be controlled[10] by electron-withdrawing groups on the ligand, which result in a marked decrease in its ability to extract K^+ salts from water into methylene dichloride but has little effect on the extraction of Na^+. Eventually therefore, the normal selectivity of 18-crown-6 ethers to K^+ is reversed. This probably results from asymmetric complexation of Na^+ by the ether (Na^+ is too small) so that Na^+ does not interact with the oxygen atoms most influenced by the electronic effect.

Crown ethers are of value in several other contexts. Thus the outer hydrophobic nature of the ether allows its complexes to dissolve in hydrophobic solvents, so that $KMnO_4$ can be solubilized in benzene by the addition of a suitable crown ether. This approach has allowed the quantitative oxidation of a whole range of organic compounds by permanganate in benzene, for example olefins to acids, and active methylene groups to carbonyl functions.[11] The anions in such solvents are unsolvated, so that they are often very effective nucleophiles, thus facilitating reactions under mild conditions. Examples of the use of such 'naked anions' include superoxide,[12] carboxylate,[13] halide,[14] and hydroxide.[15] Modified polyethers have also been used[16] as 'host' compounds, in which structural features are designed to complement those of specific 'guest' compounds. Such host–guest relationships have been used as models for interactions such as those between substrate and enzyme.

The Cryptands

An example of these macrobicyclic ligands is shown in Figure 9.1. They have been synthesized and studied in particular by Lehn.[17] The cryptands are named by giving the number of oxygen atoms in each of the three chains linking the bridgehead nitrogen atoms. The cryptands provide three-dimensional selectivity towards cations, compared to the two-dimensional selectivity provided by the crown ethers, in terms of the size of the molecular cavity. They thus bind cations more specifically to give complexes (cryptates) of 1:1 composition. The structures of a number of cryptates with groups IA and IIA cations have been determined,[17] and show that the dehydrated cation is enclosed by the ligand. The heteroatoms are all directed towards the cation inside the cage. It should be noted however that the free ligand can take up a number of conformation.

Formation constants for interaction with a range of cations have been measured: some are given in Table 9.3. The size of the cavity can be controlled through the number of oxygen atoms present in the cryptand. The cryptand in Figure 9.1 (222) has six oxygen donors (two in each chain) and two nitrogen donors. Maximum formation constants for complex formation with Na^+ and K^+ are obtained for cryptands (221) and (222), for which the ratio of the ionic to cavity radii is the same. Interestingly, K^+ combines[18] about one hundred times faster with (222) than does Na^+ with (221). This may result from the lower flexibility of the smaller cryptand, which results in a lower concentration of the kinetically active form of the cryptand, in which the lone pairs of the bridgehead nitrogen point towards the cavity. For a particular cryptand, the rate of the formation reaction is inversely proportional to the radius of the cation, showing that loss of the coordinated water is important in or before the rate-determining step of complex formation.

Cryptates are of considerable stability. Thus $K^+/(222)$ is 10^4 times more stable than the K^+/valinomycin complex and 10^5 times more stable than the K^+ complex with an analogous crown ether. The cryptates usually give a higher selectivity, except that valinomycin is more selective for K^+ against Na^+. They also shown selectivity towards Group IA and IIA cations (e.g. K^+ and B^{2+}) which are of similar size. They normally prefer to bind to the M^{2+} cation, but this can be modified by substituents on the cryptands, or by replacing oxygen donor atoms by CH_2 groups. Thus for the cryptand (222), and the cryptand with two oxygen atoms in one chain replaced by CH_2 groups, the ratio of Ba^{2+}/K^+ binding is $\sim 10^4$ and $\sim 10^{-2}$ respectively, i.e. a change of factor of 10^6. The higher charge of the alkaline earth cation makes it more sensitive to a reduction in donor sites. Benzene derivatives of (222) have an enhanced 'thickness' and a greater hydrophobic shield which decreases the stability of the M^{2+} complex over the M^+ complex.[17]

Crystal structures of cryptates formed between cryptand (221) and NaSCN and KSCN have been determined.[19] This ligand prefers Na^+ over K^+ and there is a close correspondence between the sizes of Na^+ and the cavity. In both structures the cation is held inside the cavity by interaction with all the heteroatoms of the ligand, but in the potassium cryptate the metal ion also interacts with the thiocyanate ion. The Na^+ occupies a central position, unlike the potassium ion which lies in the cavity of the eighteen-membered ring of (221). Detailed consideration of the structural data shows that the fit between K^+ and this cavity is better than the fit of Na^+. This factor plus the presence of a K^+–thiocyanate bond actually leads of ΔH of complexation of (221) with K^+ being higher than that for complexation with Na^+. Thus the preference of ligand (221) for Na^+ over K^+ is not due to cavity size effects but to a favourable entropy of complexation. On the other hand, for ligand (222) the misfitting of Na^+ to the cavity of the ligand is such that cavity size effects do account for the preferential binding of K^+ over Na^+. This serves to demonstrate the complexity of metal ion binding to these ligands.

The cryptands have found applications in similar areas to the crown ethers. Cryptand (222) solubilizes $KMnO_4$ in benzene, and dissolves barium sulphate in water (up to $50 \, \text{g dm}^{-3}$). Current developments include the syntheses of macrotricyclic compounds. Three cavities are provided by these ligands, although only two are usually occupied. Ligands have been synthesized that bind selectively a different metal ion in each of two cavities, a situation that has several interesting applications.

The Ionophore Antibiotics

These are naturally occurring compounds[20] that have a marked effect on the permeability of membranes to cations. There are two classes of ionophores. One class is the channel-forming ionophores, which are proteins that span the membrane providing a hydrophilic channel through which cations pass. The second class, the carrier ionophores, which is the subject of this discussion, bind cations prior to transport of the cation–ionophore complex through the membrane. One group of carrier ionophores involves monobasic carboxylic acid derivatives of polyethers, while a larger class involves neutral macrocyclic molecules that are usually selective to K^+. The structure of valinomycin, a cyclic dodecadepsipeptide, is shown schematically in Figure 9.2. The conformation in solution is quite different and is solvent dependent. X-ray structural determinations show that the K^+–valinomycin complex contains the dehydrated cation, with the ester carbonyl groups directed inwardly towards it, giving a distorted octahedral geometry around the metal. Infrared studies show in contrast that the Na^+ complex has non-equivalent carbonyl groups, reflecting the misfitting of Na^+ due to its smaller size.

Another neutral K^+–selective ionophore, enniatin B, is a cyclohexadepsipeptide, which coordinates to K^+ via its six carbonyl groups. The complex is best represented by charged discs, which are able to stack above each other, a phenomenon of possible significance to transport processes. The remaining characterized group of neutral ionophores is the macrotetrolide actins (Figure 9.2), a number of closely related species being known. Nonactin provides a cavity for K^+, the ligand wrapping around the cation like the seam of a tennis ball, and providing four carbonyl and four furan oxygen atoms as donors. Once again Na^+ misfits due to its size.

The interaction of Na^+, K^+, Cs^+ with nonactin has been studied by 220 MHz proton magnetic resonance spectroscopy[21] in hexadeuteroacetone and hexadeuteroacetone–water mixtures. Formation constants have been calculated and are listed in Table 9.3. The main point of interest is that although all three ions bind to nonactin to similar extents in dry acetone, the binding of Na^+ and Cs^+ is dramatically reduced when water is present. Thus the formation constants in wet acetone for Na^+, K^+, and Cs^+ are 210, 2×10^4, and 400 respectively. The binding of K^+ to nonactin is obviously highly favoured in the more aqueous medium.

The results in dry acetone indicate that the free (i.e. non-hydrated) ions are all bound to the same extent by nonactin. In wet acetone the alkali metal must be stripped of its hydration shell before it may be bound in the nonactin cavity. Only

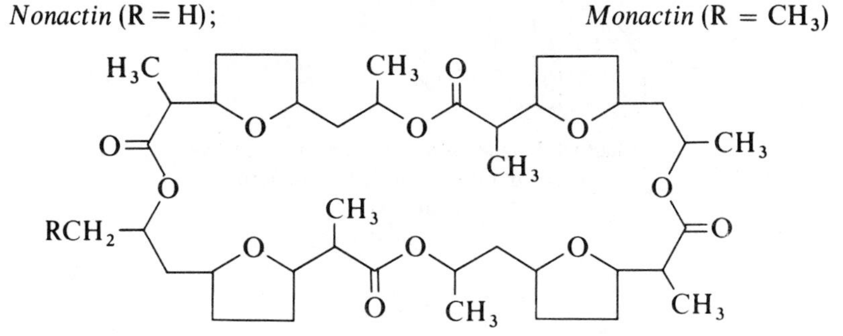

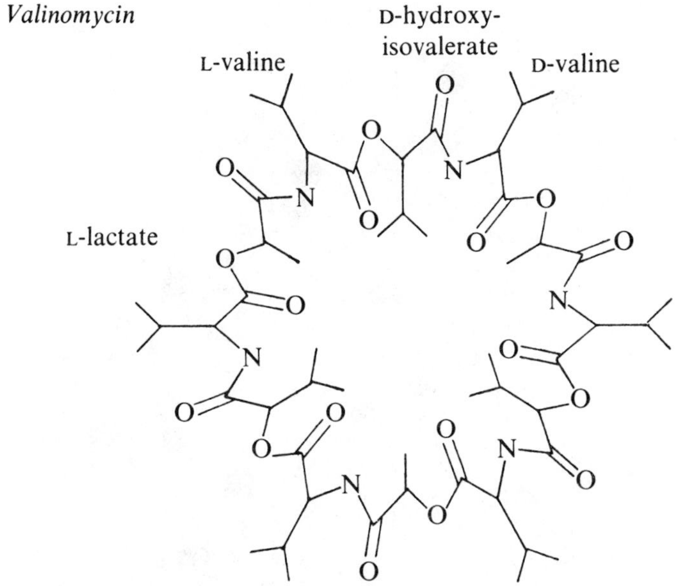

Nonactin (R = H);   *Monactin (R = CH₃)*

Valinomycin

Figure 9.2 Some macrocyclic antibiotics.

in the case of K⁺ is this favoured. This serves to emphasize the importance of the consideration of hydration energy in this context and in turn the question of ionic size and coordination number.

Eigen and his coworkers[22] have shown that the formation of K⁺–monactin complex in methanol occurs very rapidly. They have suggested that the fast replacement of the solvent shell of the alkali metal cation is a stepwise process, in which the twisted ring system opens up and then closes to the normal compact form. These conformational changes are represented in Figure 9.3. Their occurrence has been confirmed for nonactin by NMR studies.[21]

The structures of several polyether monocarboxylate ionophores have been determined, usually as the thallium or silver salt. In general they take up conformations, *via* hydrogen bonding involving the carboxylate group, which provide a cavity for the cation. An example[23] is shown in Figure 9.4. Monensin

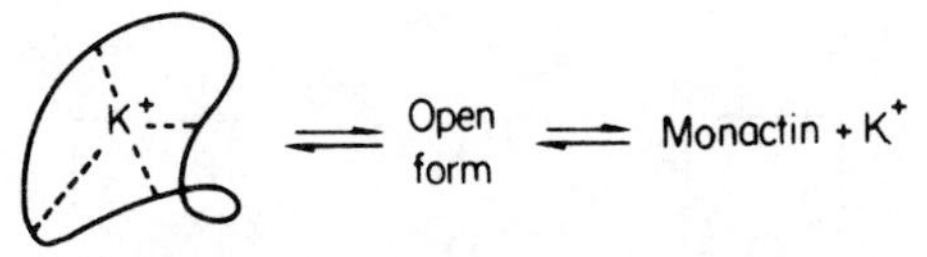

268

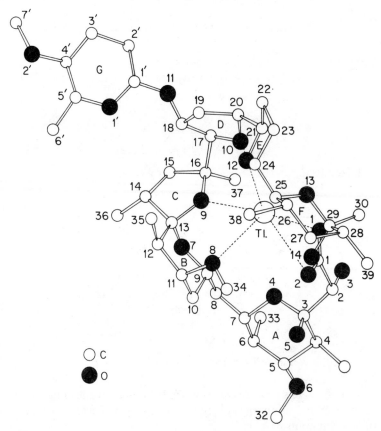

$K^+ \rightleftharpoons$ Open form $\rightleftharpoons$ Monactin + K^+

Figure 9.3 Schematic representation of the conformation of the K^+-monactin complex.

○ C
● O

Figure 9.4 The structure of the thallium salt of antibiotic-6016 (Reproduced by permission from N. Otake *et al*, *J.C.S.Chem. Comm.*, **1978**, 875).

binds Na^+ rather than K^+; nigericin is selective for K^+ giving a five-coordinate cation, while the greater flexibility of dianemycin results in it binding IA cations without discrimination. Ionophores X-573A (Lasalocid A) and A23187 are selective for Ca^{2+} over Mg^{2+}. X-573A has a linear structure in polar solvents and a cyclic conformation in non-polar solvents, in which the outer surface of the ionophore is hydrophobic. X-573A also forms complexes with monovalent cations. If formed using non-polar solvents, the complexes are dimeric with two linked X-573A/M$^+$ groups, a situation confirmed in solution by NMR measurements. However the structure of the Na$^+$ salt of X-573A from methanol is monomeric,[24] in which the Na$^+$ cation is complexed by five oxygen donor atoms (but no carboxylate coordination) and the sixth position is filled by a solvent molecule that caps the cation. The same five oxygen donor atoms are utilized in the dimer for each Na$^+$, but the 'capping' of the cation is completed by dimer formation. These results may indicate that in the polar exterior of the lipid bilayer monomeric complexes are formed, but that these are transported across the lipid bilayer as dimers.

MEMBRANES, IONOPHORES AND CATION TRANSPORT

As noted in Chapter 1, an essential feature of membranes is their lipid bilayer, which presents a hydrophobic barrier to the straightforward transport of hydrated IA and IIA cations. The presence of specific mechanisms for transport of cations through this barrier thus allows entry of cations to be controlled in a specific fashion. Such translocation of cations occurs by two basic processes.

(a) By active transport against a concentration gradient with due expenditure of energy *via* hydrolysis of ATP. This involves an enzyme-linked ion pump, examples of which will be discussed later. The total energy involved in maintaining these concentration gradients by active transport is a sizeable fraction of the energy requirements of the resting cell.

(b) By facilitated diffusion following a concentration gradient; this will continue until concentrations either side of the membrane are equal. This process depends on the presence of ion carriers or ionophores. As noted earlier, these ionophores may be carriers or channel formers. Carrier ionophores will bind selectively to certain cations and carry them as liposoluble complexes across the membrane. Channel-forming ionophores are proteins that span the bilayer and offer a water-filled channel for transport of hydrophilic species. These channels, particularly the naturally occurring pores in membranes, may have highly specialized gating mechanisms to control entry and which will distinguish between cations. Examples of channel-forming ionophores are gramicidin and F.30 alamethicin, which are linear peptides that transport cations M$^+$ and M^{2+} respectively. It is usually possible to classify an ionophore. Mobile carriers are much more dependent on membrane fluidity than channel formers. Thus a mobile carrier cannot function below the transition temperature of the phospholipids in the membrane as they will be frozen. On the other hand, a channel-forming ionophore can only function if it spans the membrane. So if an ionophore ceases to function in thicker membranes it is probably a channel former. The kinetics of ion transport are also relevant. If an ionophore operates at a rate of greater than 10^4 ions per second, then it cannot be a mobile carrier, as the rate of diffusion of the carrier complex across the membrane will be too slow to provide this level of ion flux.

This division of cation transport into these two classes is probably an oversimplification, as many transport systems are interlinked in various ways. Thus the sodium concentration gradient resulting from the active transport of Na^+ out of the cell is utilized in the transport of amino acids, which are carried as Na^+ complexes across the membrane into the cell. The complex relationships between concentration gradients in cell explains why the ionophoric antibiotics are toxic. Thus valinomycin transports K^+ and may cause the collapse of a negative membrane potential. Nigericin exchanges H^+ for K^+ and so in the presence of external K^+ can cause the collapse of a proton gradient. Thus proton ionophores (uncouplers of oxidative phosphorylation) and ionophores for M^+ and M^{2+} are important tools in cell physiology.

So far we have only considered the crown ethers, cryptands, and ionophore antibiotics as models for selective binding. Clearly they are also models for transport: thus they all have a cavity which will bind a cation with varying selectivity, while the resulting complex has a hydrophobic exterior and so can cross the membrane in response to a concentration gradient. The study of model transport systems is usually carried out using black bilayer lipid membranes, which are bilayers of about 70 Å thickness formed from various lipid material such as lecithin from eggs. The transport of the cation is monitored conductometrically. Other methods for monitoring ionophoric activity include: *equilibrium extraction*, in which the effect of an ionophore on the distribution of a cation between water and an organic solvent is tested; the use of *reconstituted vesicles*, although problems arise from the range of vesicle sizes produced, and the *swelling of mitochondria* through the accumulation of cations brought about by the ionophore. Ionophore activity can be demonstrated successfully by a simple experiment involving a U-tube containing two aqueous layers separated by an immiscible solvent such as chloroform that corresponds to the membrane. The ionophore material is in the organic phase and a coloured alkali salt placed in the aqueous phase in one arm of the U-tube. The second aqueous phase is periodically tested for the appearance of the cation. A positive result confirms (in a non-selective way) the ionophoric property of the compound under test.

Several illustrative examples of recent work in this area will now be given. Synthetic peptides, (Leu-Ser-Leu-Gly)$_n$, where $n = 6, 9, 12$ have been used[25] as models for pore-forming ionophores. They have been tested conductometrically for the induction of permeability to cations in lipid bilayer membranes. Positive results were found only for the peptide with $n = 12$, the only example long enough to span the membrane. This formed a helix four turns in length, with a hydrophobic exterior resulting from the leucine side chains, and a hydrophilic interior due to the serine hydroxyl groups, a situation similar to that postulated for naturally occurring pore channels.

Particular attention has been paid to gramicidin A, a linear pentadecapeptide antibiotic, that dimerizes to form transmembrane channels, of about 5 Å in diameter, in biological and synthetic membranes. These channels are permeable to alkali metal cations and are blocked by Ca^{2+}. The length of the channel is about 32 Å, but the binding of Cs^+ or K^+ (and probably other IA cations) induces conformational changes with a decrease in the channel length to 26 Å and an increase in its diameter to 6.8 Å. Recent X-ray studies have shown that each dimeric channel contains two equivalent cation-binding sites. This is in accord with the dependence of the channel conductance on cation concentration.[26] Other work on gramicidin has involved the synthesis of analogues with chemical modifications at strategic points, for example a charged group at the channel mouth or chemically dimerized gramicidin molecules. The use of compounds such as gramicidin as models for natural channels does however fail to reproduce the high ionic selectivity of these channels to the physiologically important cations. It is therefore of the utmost significance that natural pores are now being isolated and studied. Thus the matrix protein (porin) of *Escherichia coli* has now been isolated.[27] This forms channels is lipid bilayers, which interact co-operatively within aggregates. The channel diameter has been estimated as 9 Å.

The following examples illustrate recent studies on carrier ionophores. The ionophoric activity of some crown ethers to K^+ has been tested.[28] 18-Crown-6 derivatives, although known to bind K^+ selectively are not effective as ionophores because the two sides of the complex are exposed to the hydrophobic membrane. The ill-fitting of K^+ to 15-crown-5 derivatives that results in the formation of complexes of $1:2$ stoichiometry means that the cation is shielded more effectively as $CKC,^+$ and so these crowns are much more successful as ionophores. However the most effective ionophore overall for several M^+ cations is di-t-butyldibenzo-30-crown-10, because this completely encapsulates the cation, presenting a completely hydrophobic entity to the lipid bilayer.

The dimeric synthetic peptide shown in Figure 9.5 binds K^+ selectively.[29] On reduction a monomer is formed which will not form a complex. This allows the setting up of a model system where transport of K^+ is linked to the redox state of

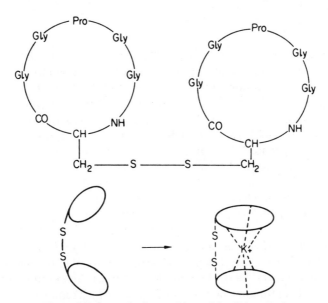

Figure 9.5 A synthetic peptide and its K^+ complex.

the carrier. It is suggested that the conformational change depicted in Figure 9.5 occurs when K^+ is bound by the dimer. Further examples have been reviewed.[9]

The study of the alkali metal and alkaline earth metal cations

Certain of these ions have been followed in physiological and other studies by using their radioactive isotopes, although ^{42}K has a low half-life of 12.4 hours. Their concentrations may be monitored by standard techniques such as atomic absorption analysis or ion-selective electrodes, but certain specialized techniques have been developed, particularly for Ca^{2+}. The value of these techniques depends not only on detection limits but also on the response time. Concentrations of Ca^{2+} have been measured particularly by use of the indicator arsenazo III[30] and by photoproteins such as aequorin,[31] which emit light on interaction with Ca^{2+}, detection limits being $10^{-7}\,mol\,dm^{-3}$ and $6 \times 10^{-8}\,mol\,dm^{-3}$ respectively. Ca^{2+}-sensitive electrodes have detection limits of $10^{-8}\,mol\,dm^{-3}$ but response time of $\sim 2\,s$.

The study of the environment of the bound cation is much more difficult, as apart from ^{23}Na, ^{43}Ca (and possibly ^{36}K) NMR, no instrumental techniques can be utilized. The topic of ^{23}Na NMR in biological systems has been reviewed.[32] Accordingly it is important that metal-for-metal substitution be undertaken, hopefully using transition metal ions. As noted in Chapter 3, any replacement must be isomorphous replacement, with the probe ion occupying the same site, i.e. size and chemistry must be similar. Some probes are given in Table 3.6, together with details of ionic radii. There is no probe for Na^+; probes for the other cations are discussed below.

Potassium

K^+ may be replaced by a series of singly charged cations (NH_4^+, Rb^+, Cs^+) and useful mechanistic conclusions may be obtained from a comparison of these cations. $^{86}Rb^+$ is a more convenient radioactive isotope than $^{42}K^+$ in view of its longer half-life of 18.7 days. Tl^+ has attracted attention as a probe for K^+. It binds more strongly to ligands than K^+ and as a result is more effective in activating the enzymes pyruvate kinase,[33] diol dehydratase,[34] and yeast aldehyde dehydrogenase.[35] However it is a softer cation, and there is always the possibility that it may bind to additional protein sites causing inhibition, as observed for aldehyde dehydrogenase.[35] In view of the fact that satisfactory Tl^+-for-K^+ substitution has been achieved in a number of biological systems, it is probably a satisfactory probe at low concentration, subject to the above qualification. Its main relevant probe property is that of NMR (^{205}Tl, spin $= \frac{1}{2}$), but the shift in the UV band and the quenching of fluorescence on complex formation offer alternative techniques, while the electron microscope may also be used to locate bound Tl^+ under favourable conditions. An example of the use of Tl NMR is work on the enzyme pyruvate kinase[33] which requires a divalent and a monovalent cation for activation. A double-probe experiment was carried out using Mn^{2+} and Tl^+, and indicated that the binding sites for these two cations are very close together.

Magnesium

Ionic radii considerations suggest that Ni^{2+} would be a suitable probe for Mg^{2+}, but the different chemistry of the two cations results in such substitutions, except in one example, being unsuccessful in terms of biological activity. Mn^{2+} and Mg^{2+} are very similar chemically speaking, and so despite their different sizes Mn^{2+} has been used very effectively by Cohn[36] to study Mg^{2+}-dependent enzymes.

Calcium

Mn^{2+} and Eu^{2+} are bivalent ions of appropriate radii for the replacement of Ca^{2+}. Mn^{2+} is not really suitable however in view of its chemical differences from Ca^{2+}. Eu^{2+} has useful probe properties (paramagnetism and Mössbauer spectroscopy) and has been used successfully, for example in concanavalin, but problems may arise from redox reactions. If the requirement for a bivalent metal ion is relaxed, then the lanthanide elements in the III oxidation state may also be used as probes.[37] These will bind more strongly than Ca^{2+} and complications may result from this. Eu(III), Gd(III), and Nd(III) have all been considered, although reservations have been expressed over the use of EPR spectra to discuss the chemical nature of Gd(III)-binding groups.[38] The comparison of the spectra of neodymium model compounds with those of neodymium bound to bovine serum albumin has led to the conclusion that the metal site involves carboxylate ligands, a view that is consistent with the known behaviour of calcium. Examples of the use of lanthanides to probe Ca^{2+} sites on trypsin and other enzymes have been given in Chapter 4.

Other studies have been concerned particularly with the transport of Ca^{2+} and its behaviour in membranes. Thus it is thought that Ca^{2+} affects the conductance of cations through nerve membranes by binding to the membrane, replacing Mg^{2+} and hence activating certain enzymes. These changes are blocked in lobster giant axons by the action[39] of La^{3+}, La^{3+} acting 'as if it were equivalent to an extraordinarily high Ca^{2+} concentration'. This is an illustration of the greater binding properties of the triply charged La^{3+}. Again[40] the efflux of ^{45}Ca from the squid giant axon is inhibited by added $LaCl_3$, the lanthanum ions competing with external calcium for binding sites on the cell membrane. Other examples are known in which La^{3+} competes successfully with Ca^{2+} for calcium-binding sites.

The particular advantage of the lanthanide ions is that they offer a range of ions having slightly varying properties such as size. Replacement of Ca^{2+} by a series of these ions should then allow much insight into the correlation of biological activity, physical properties of the probe, and the environment of the metal ion and so in turn allow greater insight into the role of Ca^{2+}.

ACTIVE TRANSPORT OF IONS THROUGH MEMBRANES

Reference has been made to ion pumps that are associated with the active transport of cations across membranes against a concentration gradient that may be extremely large. This translocation process requires expenditure of energy, which is derived from the hydrolysis of ATP at the inner surface of the membrane. The sodium pump in human red cells involves the entry of two moles of K^+ and the efflux of three moles of Na^+ for each mole of ATP consumed. If cells are cooled to $0°C$ active transport of ions ceases and the membrane behaves as a normal Donnan membrane in which there is a flow of ions towards an equalization of ion concentrations inside and outside the cell. The same effect results from the addition of inhibitors for enzymatic phosphorylation such as the cardiac glycoside ouabain.

This fundamental topic of ion transport has been much studied in recent years and while definitive answers cannot yet be given to many questions, considerable progress has still been made. Transport systems will be found in the outer membranes of cells and also in the membranes of internal components. For these organelles (such as the mitochondria) have an important role in the control of intracellular cation concentrations, for example in the rapid release and subsequent reaccumulation of Ca^{2+} in trigger action. It should be noted that several transport systems may operate for a particular cation and that pumps also exist for anions such as chloride. Pumps may also function in different modes from the normal mode by suitable adjustment of conditions. An extensive review of various transport ATPases is available.[41]

Entry of a cation into a cell is usually balanced in some way in order to maintain electroneutrality. Thus entry of Ca^{2+} may be associated with influx of phosphate, which is then described as a calcium porter; this overall process is termed 'symport'. Alternatively influx of one cation may be balanced by the efflux of another cation; this is termed 'antiport'.

In this section we will consider one transport system (the sodium pump) in some detail, while transport systems for Ca^{2+} will be referred to in the next section but without detailed consideration of mechanisms. It is also important that some comments on proton-transport pathways be presented, in view of current interest in chemiosmosis.

Chemiosmosis

The presence of ion gradients, which arise from membrane activity, is of supreme importance in cell physiology. Proton concentration gradients across membranes provide the simplest, but particularly important example of such a cation gradient. This is related to the fundamental problem of how synthesis of ATP from ADP and inorganic phosphate may be linked to the electron-transfer reactions of oxidative and photosynthetic multienzyme chains. The chemiosmotic theory[42] postulates that these two processes are coupled through protons. A proton-translocating redox enzyme sets up a difference in proton concentration (i.e. proton motive force) across the membrane. This proton motive force then drives H^+ through the membrane *via* an ATPase, thereby driving synthesis of ATP. Thus the redox reaction is coupled to the ATPase reaction by a flow of protons across the membrane, although the actual mechanisms of proton translocation (by both components) have not yet been resolved.[43] The membrane will contain various specific carrier systems for other molecules and ions that will respond to the electrochemical proton difference set up across the membrane by these two processes.

It appears that cytochrome *c* oxidase in mitochondrial membranes functions as a redox-linked proton pump, in which protons are translocated across the inner mitochondrial membrane from the inside to the outside. The translocation of charge has been shown to exceed that of electron transfer by an amount that can be accounted for by proton efflux.[44] Current approaches[45] to establishing the mechanism of proton movement, involve the study of proton transfer through hydrogen-bonded chains. These allow rapid transfer of protons from a region of high electrochemical potential across a membrane with little loss in energy. Before being released into a solution of low electrochemical potential, such high-energy protons can clearly be made to drive a chemical reaction (such as ATP synthesis) in an exactly analogous fashion to electrochemical reactions. Such a proton-transport chain could result in membranes if the protein of the redox enzyme were able to take up a configuration so that it spanned the membrane and had a continuous chain of hydrogen-bonded side chains (hydroxyl groups of serine, threonine, and tyrosine and carboxyl group of aspartic and glutamic acids).

Several methods are available for measuring the proton concentration gradients across membranes. Most of these depend on the measurement of the distribution of protonated amines across the membrane in response to the gradient. Provided the pK_a of the amine is high enough to ensure that the amine is fully protonated then $[RNH_3^+]_{in}/[RNH_3^+]_{out} = [H^+]_{in}/[H^+]_{out}$. The use of a fluorescent amine offers a convenient method for determining concentrations, as the fluorescence is quenched inside the cell. Such techniques have been used to follow the kinetics of development of pH gradients in chloroplasts.[46]

The sodium pump (Na$^+$,K$^+$-activated ATPase)[47]

This involves a membrane enzyme, Na$^+$,K$^+$-ATPase, that catalyses the hydrolysis of ATP and which requires Na$^+$ and K$^+$ as cofactors. The first example of such a transport ATPase was discovered by Skou[48] in 1957. The transport ATPase binds and releases Na$^+$ and K$^+$ at certain stages in the enzymatic reaction as the affinities of their sites changes as the reaction proceeds. Conformational changes that take place in the enzyme during the reaction result in the cations being released at the opposite surface of the membrane from which thay were taken up, so that the overall result is hydrolysis of ATP and translocation of the cations. Mg^{2+} is also required for maximum activity. It should be noted that the enzyme catalyses other M$^+$-dependent reactions which do not result in translocation of cations.

Purification of the enzyme

The enzyme has been purified from a number of membrane preparations from tissues rich in Na^+, K^+-ATPase activity. It is usual practice to solubilize the purified enzyme with sodium dodecyl sulphate and to determine the molecular weights of the component peptides. Two peptides are usually found, a large one of about 100,000 mol. wt containing the phosphorylation site and the ouabain-binding site, and a small one of mol. wt about 50,000 which is a glycopeptide. Controlled hydrolysis with trypsin of the catalytic protein and correlation with loss of ATPase activity gives different results in the presence of K^+ and Na^+, showing that there are two different conformational forms of the enzyme in the presence of these cations. The ouabain site is lost when the protein is cleaved to give a 85,000 fragment. An interesting technique has been used[49] to locate the ouabain site, namely by the use of 2-nitro-5-azidobenzoyl derivatives of ouabain and their photolysis when bound to the enzyme. This gives highly reactive nitrenes which insert into nearby C—H bonds, so providing a covalent tag for the site.

There is evidence to show that the glycopeptide is a component of the pump, and it is now thought that the ATPase is a dimer made up of two pairs of the large and small peptides. The dimeric structure is supported by the identification under certain conditions of two phosphorylation sites, two ATP-binding sites and two ouabain-binding sites. Under other conditions only one ATP site and one ouabain site may be found, suggesting that the two halves of the dimer interact so that binding of ATP or ouabain to one half lowers the affinity of the other half for these molecules. This clearly has important mechanistic implications.

Mechanism of translocation of cations

This could involve a carrier ionophore or a channel with a suitable gate to confer selectively, but most models for cation transport postulate some form of internal transfer mechanism, involving a transition between two conformations of the protein. For many years support was given to a one-site circulating carrier mechanism in which a phosphorylated carrier responsible for moving Na^+ outwards was converted by loss of phosphate at the outer face of the membrane into a carrier that moved K^+ inwards. Binding of Na^+ at an appropriate number of sites at the inner surface was suggested to initiate a rotation of the protein to expose these sites at the outer surface. Dephosphorylation changed the affinity of the site so that Na^+ was liberated, K^+ was bound, and the reverse rotation initiated. This theory has been abandoned in view of the identification of internal and external sites for both Na^+ and K^+ before phosphorylation. An alternative simple view is to assume that the large protein spans the membrane and has binding sites at each end. Such a postulate is consistent with the size of this protein. The phosphorylation site must be in contact with the inside surface of the membrane at some stage in the cycle. When the outer sites are filled with K^+ and the inner with Na^+ the molecule rotates giving rise to a change in affinity so that the new outward-facing sites now prefer K^+ with resulting loss of Na^+. This theory accounts for the coexistence of both Na^+ and K^+ sites but appears unlikely on structural grounds.

A more realistic suggestion exploits the dimeric nature of the enzyme and suggests two carriers coupled together in a 'flip-flop' scheme.[50] Related to this is the proposal that only half of the sites are active, that one set of the dimer binds phosphate and the other ATP. The phosphorylation of one subunit by ATP is accompanied by dephosphorylation of the other, although at any particular time only a minor part of the enzyme is phosphorylated. Such 'half-of the-sites-reactivity' is known for alkaline phosphatase from *E. coli*.

Other aspects of the mechanism

It is now accepted that Na^+-dependent phosphorylation of the Na^+, K^+ ATPase occurs at an aspartyl group, and that this is followed by a K^+-dependent phosphatase reaction in which the phospho group is transferred to water. The overall reaction is therefore one of hydrolysis of ATP *via* the phosphoenzyme. In the presence of Na^+ the phosphoenzyme transfers its phosphate group back to ADP regenerating the ATP. The following model, due to Albers and Post,[51] accounts for many features of the sodium pump. E_1 is the Na^+ form of the enzyme and E_2 is the K^+ form of the enzyme, both representing stable forms of the unphosphorylated enzyme. E_1 binds ATP strongly, and addition of Mg^{2+} brings about phosphorylation to $E_1P.ADP$. E_1P loses ADP and undergoes a slow conformational change to E_2P, which is hydrolysed readily in the presence of K^+ to give the dephosphoenzyme E_2K. E_2 is converted back to E_1 in a slow conformational change, catalysed by ATP binding at a low-affinity site. This model has been linked to cation transport.[52]

$$ATP + E_1 \rightleftharpoons E_1ATP \xrightarrow{Na^+, Mg^{2+}} E_1P.ADP \rightleftharpoons E_1P + ADP$$

$$ATP \updownarrow \qquad \qquad K^+ \qquad \qquad \updownarrow$$

$$P_i + E_2K \rightleftharpoons E_2P$$

In addition to its role in the hydrolysis of the phosphoenzyme, K^+ is also involved in its formation, which is inhibited by K^+. It appears that this observation, which appears to be inconsistent with the above mechanism involving the phosphoenzyme, reflects the occlusion of K^+ in the unphosphorylated enzyme. Thus the rephosphorylation of Na^+, K^+-ATPase that has been dephosphorylated in the presence of one of K^+, Rb^+, or Cs^+ actually varies depending upon which cation was used for the dephosphorylation. The suggestion[51] that this reflects the occlusion of these cations in the un-phosphorylated enzyme has now been confirmed experimentally.[53] These results are important in substantiating the role of the phosphoenzyme as an intermediate.

Approaches to the elucidation of the mechanism of the sodium pump have involved studies on the role of the membrane lipids on transport (as these can affect the fluidity of the membrane and hence any rotational activity), the binding of nucleotide analogues,[52] the use of lithium magnetic resonance, the effect of a wide range of inhibitors, and standard kinetic and binding studies under a range of conditions. One general problem in working with the fragmented ATPase is that Na^+ and K^+ have access to both sides of the membrane and so compete at the internal Na^+ and external K^+ sites. Much more work is now being done on the ATPase incorporated in vesicles. Full transport and ATPase activity have been observed for such vesicles, confirming the effective reconstitution of the enzyme. Inside-out membrane vesicles may also be prepared. The use of lithium magnetic resonance techniques to probe cation sites seems doubtful as it is difficult to prove that Li^+ will bind specifically to any site.

Anomalous kinetic behaviour for Na^+, K^+-ATPase, observed over a number of years, is now recognized to result from an impurity in one commercial source of ATP, which dramatically increases the sensitivity of the enzyme to inhibition by K^+. This impurity is vanadate which competes with ATP for its binding sites. Vanadate binds at both high- and low-affinity ATP sites, which have low and high affinity for vanadate respectively.[54] Thus the use of vanadate offers a means of distinguishing between these sites, and furthermore there is the intriguing possibility that vanadate may be a physiological regulator for ATPase.

CALCIUM IN CONTROL PROCESSES

The concentrations of Ca^{2+} inside and outside the cell are approximately 10^{-6} and 10^{-3} mol dm^{-3} respectively. A range of biochemical and/ or physiological processes may thus be triggered off by entry of Ca^{2+} into the cell or by release of Ca^{2+} from internal organelles. Such transport systems must then have the capacity to reaccumulate the high concentration of Ca^{2+} as rapidly as possible. Each control system requires a storage capacity for Ca^{2+} and a means of translocating Ca^{2+} rapidly, either by a suitable ion pump (for accumulation against a concentration gradient) or by passive release for efflux of Ca^{2+}.

Binding, transport and accumulation of Ca^{2+}

For various types of muscle, the sarcoplasmic reticulum is well established to be the Ca^{2+} regulator. Ca^{2+} is taken up by the Ca^{2+}, Mg^{2+}-ATPase, a transport ATPase with a number of similarities to Na^+, K^+-APTase, which is present in the sarcoplasmic reticulum (SR) membrane. The Ca^{2+} is stored in the SR on calcium-binding proteins, particularly calsequestrin, which has a high affinity for Ca^{2+} and binds some 43 moles of Ca^{2+} per mole of protein (44,000). Muscle contraction is associated with the release of Ca^{2+} from the SR and its binding to sites on the muscle fibres (to be discussed later). The concentration of Ca^{2+} in the sarcoplasm rises a 100-fold in milliseconds! Efflux from the SR is necessarily very fast, and is probably too fast to be accommodated by a reversal of the calcium pump. It is suggested therefore that this rapid efflux is a passive release of Ca^{2+}

down the concentration gradient, through hydrophilic channels in the SR membrane formed by aggregation of the ATPase protein. The Ca^{2+}, Mg^{2+}-ATPase is known to form such aggregates in vesicles.[55] Thus the ATPase enzyme acts as a pump in one direction. and under certain conditions can act as a rapid efflux channel in the opposite direction. Immediately after releasing Ca^{2+} the SR starts pumping it back again *via* the Ca^{2+}, Mg^{2+}-ATPase.

The endoplasmic reticulum of other mammalian cells is able to take up Ca^{2+} at a rate appropriate for a regulator, but its storage capacity is limited, so it appears unlikely to be an important regulator of Ca^{2+} concentration. On the other hand, mitochondria seem well qualified for this role, having a considerable capacity for Ca^{2+} uptake and an effective Ca^{2+} pump for the accumulation of Ca^{2+}.

Concentrations of Ca^{2+} inside possible regulatory components have been determined[9] in a number of systems by the use of aequorin and obelin, the Ca^{2+}-activated photoproteins, and by various dyes. Thus mitochondria in squid axons were studied[56] *in situ* with the dye arsenazo III which showed that mitochondria in an intact axon can buffer about 99.5% of an imposed Ca^{2+} load in the range $0.050-2.50$ mmol dm^{-3}, so that mitochondria could maintain the axoplasmic $[Ca^{2+}]$ at $0.003-0.005$ mmol dm^{-3} despite the substantial amounts of Ca^{2+} entering. Mitochondria also appear[57] able to control intracellular Ca^{2+} in the heart, and so may be relevant to the regulation of relaxation and contraction. In the mammalian heart some 60 nmol of Ca^{2+}/g of tissue have to be removed during relaxation. But this is only a small fraction of the total capacity of heart mitochondria for Ca^{2+}. At present, however, kinetic data are needed to show that the uptake of Ca^{2+} by mitochondria is fast enough for it to have a role in the regulation of heart beat.

The uptake and release of Ca^{2+} in these regulatory systems has been much studied by the use of lanthanide probes and ruthenium reds. The latter compounds are polymeric species that inhibit Ca^{2+} transport, probably by binding preferentially to Ca^{2+} sites. In many cases regulatory systems have specific calcium-binding proteins of high affinity for Ca^{2+}, which are also found in certain enzyme systems (e.g. cyclic nucleotide phosphodiesterase) and in vitamin D linked Ca^{2+} uptake. These calcium-binding proteins show many similarities, in molecular weights and amino acid sequence for example, and to some extent are interchangeable. They have been reviewed on a number of occasions,[58] and feature prominently in the following discussion.

Calcium and muscle contraction

Some of the best understood Ca^{2+}-binding regulatory proteins are found in muscle. Initiation of contraction results from the arrival of a nerve impulse at a motor nerve ending in a fibre, which causes Ca^{2+} to be released from its stores. This Ca^{2+} then interacts in a specific manner with a regulatory protein, although this varies with type of muscle. In vertebrate skeletal muscle, Ca^{2+} interacts with troponin, but in fish and amphibian muscle the sarcoplasm contains large amounts of parvalbumins, which are water-soluble low molecular weight proteins which bind Ca^{2+} and are probably equivalent in overall function to troponin. Other smooth muscles of certain molluscs do not contain either troponin or parvalbumins. Ca^{2+} control of these muscles is through a Ca^{2+}-dependent phosphorylation of a low molecular weight protein chain (or light chain) of the myosin present in the tissue. In some cases the troponin/parvalbumin-type mechanism exists alongside the myosin-linked regulatory system, as in the nematode worm, *Caenorhabditis elegans*, on which interesting studies are being carried out at present with mutants deficient in one or the other system.[59]

In all cases contractile activity is controlled by the regulation in some way of the Mg^{2+}-ATPase of actomyosin. The presence of Ca^{2+} results ultimately in the relief of the inhibition of this Mg^{2+}-ATPase, so that ATP is hydrolysed and contraction takes place.

Before discussing the various regulatory systems it is necessary to outline the process by which muscle contracts and relaxes.

The sliding filament model of muscle contraction

The sarcoplasmic reticulum, referred to above, is a structure in muscle which surrounds the myofibrils, which are the components in muscle fibre that contract. The muscle fibre is a long cell enclosed by the plasmalemma outer membrane. The major constituents of the fibres are the proteins myosin and actin, plus the regulatory proteins troponin and tropomyosin, and other species.

Each muscle fibre is made up of some 1000 fibrils. Myosin and actin are organized in the fibrils in the form of two filaments, one thick and one thin. Myosin is a long, thin molecule of overall length 150 nm and includes a thicker portion or head 20 nm long. The thick filaments are made up of either 180 or 360 myosin molecules. The thin filaments are made up of double-stranded chains of actin molecules. Muscle contraction and relaxation results from the sliding of one filament over the other; the actin and myosin components interact *vis* cross-linking, resulting in the locking of the muscle. Associated with the myosin heads is a site for ATP hydrolyses, thus providing the energy for contraction. The addition of Mg^{2+} and ATP to filament systems does not lead to the splitting of ATP. Although Mg^{2+} is the cofactor for the ATPase it is not able to function until some inhibition is removed by the addition of Ca^{2+}. Therefore in the absence of Ca^{2+} the MgATP complex binds to the site, preventing the interaction of myosin and actin, the filaments sliding past each other without interaction. But in the presence of Ca^{2+} actomyosin formation, ATP hydrolysis and muscle contraction proceeds.

Ca^{2+} regulation via troponin

Troponin is located at intervals of 38.5 nm along the thin filaments, and involves three subunits troponin-C (TN-C), troponin-I (TN-I), and troponin-T (TN-T).

TN-I inhibits the Mg^{2+} actomyosin ATPase. Binding of Ca^{2+} to troponin-C causes structural changes which are transmitted through TN-I and TN-T subunits so that tropomyosin shifts from an inhibitory position to one that allows actomyosin formation.

The binding of Ca^{2+} to TN-C has been much studied; two high-affinity and two low-affinity sites have been characterized. High-resolution proton magnetic resonance measurements show that TN-C exists in three conformational states, corresponding to the binding of 0, 2 and 3 moles of Ca^{2+}.

Studies on fragments of TN-C show that one of these, the CB-9 peptide containing residues 85-134, includes at least one of the high-affinity sites for Ca^{2+}. The fragment is able to carry out some of the functions of TN-C, such as the formation of a Ca^{2+}-dependent complex with TN-I which overcomes the inhibition by TN-I of the Mg^{2+}-stimulated ATPase, while conformational changes in CB-9 on binding to calcium have been observed by magnetic resonance techniques.[60]

Binding of Ca^{2+} to the low-affinity sites on TN-C labelled with the fluorescent probe dansylaziridine produces an enhancement of fluorescence that has allowed the direct monitoring of the removal of Ca^{2+} from TN-C by the SR, so providing the first simulation of this regulatory process in muscle.[61]

The Parvalbumins

These low molecular weight acidic proteins ($\sim 12,000$) contain two Ca^{2+}-binding sites. The binding of Ca^{2+} by fragments of parvalbumin containing residues 1–75 and 76–108 has been confirmed, suggesting that the calcium-binding domains are present in these fragments. Calcium magnetic resonance techniques have been used in this context.[62] Parvalbumins from carp can activate rat brain phosphodiesterase in a Ca^{2+}-dependent manner, and so evidently may replace the calcium-binding protein modulator of the enzyme.[63]

Myosin-linked regulation

Control of actin-myosin in smooth muscle involves the phosphorylation of myosin light chains by a Ca^{2+}-dependent kinase. Ca^{2+} does not bind to the light chain, but a Ca^{2+}-binding regulator protein is present, although its role is unclear at present. This phosphorylation step initiates activation by actin of the Mg^{2+}-ATPase activity. When Ca^{2+} is removed the phosphate groups are removed from the myosin light chains by a phosphatase. The Ca^{2+}-binding protein has been found in a number of systems; thus the fourth subunit of rabbit skeletal muscle kinase is identical with this regulator protein. The protein is distinct from TN-C. Cyclic nucleotides may have a regulatory effect on the contraction of smooth muscle as the light-chain kinase may itself be phosphorylated by c-AMP-dependent protein kinase with resulting decrease in the activity of the myosin light-chain kinase.[64]

Calcium and secretion

In many secretory cells, materials such as hormones, neurotransmitters and other control and defence molecules are stored in vesicles or granules. Granules arise from the concentration of secretory material in vacuoles, and often dominate the ultrastructural appearance of endocrine or exocrine organs. After an appropriate stimulus these granules migrate to the periphery of the cell and are extruded in the process of exocytosis. Ca^{2+} is implicated at some stage in the stimulus–secretion coupling. Examples of secretion by exocytosis include histamine (from mast cells), insulin (from pancreatic islets), medullary hormone (from adrenal medullary chromaffin cells), enzyme and other defence chemicals such as the prostaglandins (from polymorphonuclear leukocytes), and various growth factors and meiosis-triggering hormone. All these processes involve the expenditure of energy and the presence of Ca^{2+}. An attempt has been made to link the presence of Ca^{2+} with phosphorylation of several proteins in the mast cell.[65] Thus use of the secretagogue 40/80, which mobilizes Ca^{2+} from intracellular stores, or the ionophore A23187, to promote entry of Ca^{2+} into the cell, stimulates phosphorylation of proteins at a time when the secretory response is developing, showing that the two phonomena could be linked.

Alternatively Ca^{2+} could function by binding to the outer membranes of granules (sites have been characterized) and facilitating membrane fusion by decreasing repulsion between the membranes of cell and granule by neutralizing negative charge. More specifically Ca^{2+} could bridge the two membranes. However this view is probably oversimplified. Thus in chromaffin granules a protein, synexin, has been isolated that causes a Ca^{2+}-dependent aggregation of granules.[66] It will be interesting to see if this or similar proteins are found in other secretory tissues.

Several mechanisms hold for the way in which Ca^{2+} is supplied to trigger off secretion. An interesting example is the release of histamine from mast cells, where antigen stimulation of the membrane leads to the formation of a Ca^{2+}-selective channel.

Fusion of vesicles in cells allows their contents to be discharged inside or outside the cell as necessary. Thus the transmission of nerve impulses across synaptic gaps requires the release of the neurotransmitter acetylcholine.[67] This is contained in presynaptic vesicles, and on release diffuses across to the end plate membrane of the next axon and initiates a fresh impulse before being destroyed by an acetylcholine esterase. The membrane of the acetylcholine vesicle is stabilized by Mg^{2+}, but the depolarization of the nerve membrane, associated with the transmission of the nerve impulse, allows the entry of Ca^{2+}. These ions displace Mg^{2+} and cause the membrane of the vesicle and the outer membrane of the cell to fuse together with ejection of the neurotransmitter. Release of acetylcholine has also been attributed to a Ca^{2+}-dependent phosphorylation.

The role of Ca^{2+} in catalysing vesicle fusion has been explored[68] using artificial phospholipid vesicles and some novel techniques. Thus addition of $5 \, \text{mmol} \, \text{dm}^{-3}$ of Ca^{2+} to liposomes causes their average diameter to increase from 63 to 120 nm as a result of vesicle fusion (a reaction different from the example given above). Several questions could be asked about this reaction. Do the vesicles open, for example, before fusion? Fusion may be followed by having one set of vesicles containing firefly ATPase and another Mg^{2+} and ATP. As the vesicles fuse so a luminescence is seen. By carrying out these experiments in a medium containing reagents that quench the luminescence, and observing that this had no effect on that phenomenon, it could be shown that the two vesicles underwent fusion without contamination of their contents by the medium. It is probable that fusion is initiated by reversible vesicle interaction, followed by Ca^{2+} bridging of phosphatidic acid residues in the vesicles.

Calcium in blood clotting mechanisms

Coagulation of blood occurs to prevent excessive bleeding when tissues are damaged. The overall mechanism is complex and involves a cascade process, many steps of which are dependent on Ca^{2+}. This cascade mechanism involves a number of proteins which are normally present in blood as inactive or precursor forms. When clot formation is necessary there is a sequential activation of these zymogens. In the scheme given below,[69] the coagulation pathway is initiated by the interaction of a number of species with factor **XI** in the presence of a surface, and leads ultimately to the formation of thrombin which converts fibrinogen to fibrin, the cross-linking of which is catalysed by factor **XIIIa**.

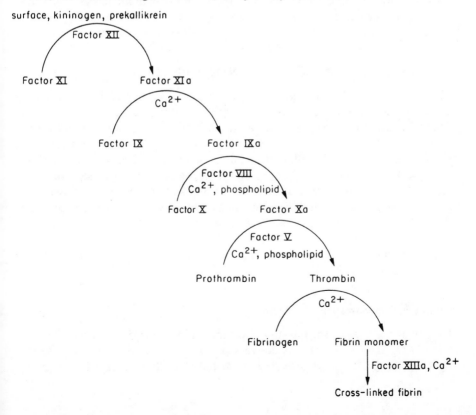

Many of these clotting factors have been characterized. They are glycoproteins and usually contain hexose, hexosamine, and neuraminic acid. Several of the active factors are serine proteinases, formed by cleavage of the precursor, but others have a more complex role, particularly when phospholipid is required for activity.

Factors **IX**, **X**, **VII**, and prothrombin require vitamin K for their biosynthesis. This is associated with the presence in these proteins of a number of γ-carboxyglutamate residues, as in the absence of vitamin K these are replaced by glutamate residues and the proteins are inactive. These γ-carboxyglutamate residues appear to be implicated in the binding of Ca^{2+}, and indeed are involved as such in other Ca^{2+}-dependent proteins. For this reason interaction of Ca^{2+} with peptides containing γ-carboxyglutamate residues has been studied.[70]

Prothrombin contains some ten binding sites for Ca^{2+} per mole, and interestingly contains ten γ-carboxyglutamate residues. On activation of prothrombin, two fragments (1 and 2) are cleaved off. Fragment 1 contains all ten γ-carboxyglutamate groups,[71] although the binding of Ca^{2+} to these fragments appears at present to be not fully characterized. Binding of Ca^{2+} to fragment 1 induces two conformational transitions that lead to binding of phospholipid and dimerization.

Factor **XIII** is a a_2b_2 tetramer and shows no activity until cleaved by thrombin to give factor **XIIIa**. This is a 36- or 37-residue protein that requires Ca^{2+} for activity, and functions as a glutamine-lysine endo-γ-glutamyltransferase that catalyses the cross-linking of monomeric fibrin. Ca^{2+} binds strongly to the a subunit and this site is probably associated with enzyme activity.

CALCIFICATION AND MOBILIZATION

The deposition of calcium salts is an essential feature of the development of extracellular structures such as shell, bones, and teeth. On the other hand, deposition in the incorrect location can result in stone formation, osteoarthritis, cataracts, and arterial disorders. A complex system is available for the control of the mobilization and deposition of calcium, and includes parathyroid hormone, vitamin D, calcitonin, calsequestrin, and osteocalcin. Some of these regulatory factors have been reviewed recently.[72] Calcium is often stored in the tissues as granules, which may be mobilized for shell formation and other processes. The calcium salts are present as small crystals so that deposition and reabsorption rates will be rapid and equilibria will be maintained. The protein and polysaccharide environment or matrix of these crystals ensures that they do not grow beyond a certain size.

Uptake and mobilization of calcium

Uptake of Ca^{2+} from the intestine is promoted by the 1,25-dihydroxy from of vitamin D_3. This metabolically active form of vitamin D is obtained from 25-hydroxyvitamin D_3 by hydroxylation *via* the renal 25-hydroxyvitamin D_3-1-

hydroxylase in a reaction which is controlled by Ca^{2+}, parathyroid hormone and phosphate. The production of this active form of the vitamin is probably the limiting factor in absorption of Ca^{2+} in the intestine, but absorption also depends upon a calcium-binding protein, which participates in transport of Ca^{2+} across the intestinal epithelium. This protein has been isolated from a number of avian and mammalian species, and is probably necessary to maintain transport of Ca^{2+} rather than to initiate it. Vitamin D also stimulates the activities of alkaline phosphatase and Ca^{2+}-ATPase, and there may be a role for the latter enzyme in supplying inorganic phosphate.

$1,25\text{-}(OH)_2D_3$ is also associated in bone mobilization, although little is known about this phenomenon. High-affinity binding proteins for vitamin D_3 metabolites have been found in bone cytosols.[73] Depletion of Ca^{2+} from the skeleton is caused by cadmium. Itai-itai disease in Japan resulted from the consumption of rice grown in soil contaminated by cadmium. This condition usually affects middle-aged women, probably because their body stores of calcium have already been depleted by child birth. This condition was, not unexpectedly, made worse by deficiencies of calcium and vitamin D in the diet. Another toxic element, lead, is also stored in the skeleton and may be released from it later in time in response to a number of stimuli. Mobilization of calcium may be required for medical reasons as noted elsewhere. This may be accomplished by chelating agents such as EDTA, nitrilotriacetate, and various phosphonates.

Deposition of calcium

Ca^{2+} salts are usually readily precipitated from body fluids. Now human salivary secretions are supersaturated with basic calcium phosphate to allow recalcification and protection of dental enamel, yet it is extremely difficult to precipitate the phosphate. This necessary physiological situation results from the presence in saliva of precipitation-inhibiting proteins. One of these is statherin (5380), which is a tyrosine-rich acidic protein which also contains many proline and glutamic acid residues.[74]

γ-Carboxyglutamic acid residues have been identified in the matrix of bone and dentin, and in certain pathological conditions. Acidic polypeptides which inhibit the precipitation of calcium oxalate in urine also contain γ-carboxyglutamate residues. These residues are also present to a considerable extent in osteocalcin, a small calcium-binding protein found in bone whose synthesis is dependent on vitamin K.[75]

The formation of calcium phosphate crystals in the build-up of bone structure can in theory[2] be controlled by specific proteins, provided the repeat distances between certain groups on the protein match exactly the repeat distances in the crystal. Matrix vesicles do in fact facilitate the precipitation of calcium phosphate, the solubility product $[Ca^{2+}][P_i]$ being lowered in the presence of these vesicles. The calcification of pre-osseous cartilage to produce bone involves such initiation sites. Nevertheless it must be stressed that it is difficult to establish with certainty mechanisms for such nucleation of precipitation, and also that the solubility product can be a misleading concept when dealing with the small crystals precipitated by protein nucleation.

IA and IIA CATIONS IN MICROORGANISMS

Many aspects of the physiology of microorganisms are dependent[76] upon the cation concentration gradients set up by various transport processes. Thus the proton motive force (PMF) set up by the translocation of H^+ through the redox enzymes of the respiratory chain can be linked to oxidative phosphorylation, pyridine nucleotide transhydrogenase, flagellar activity and so on. These physiological processes may thus be explored by the use of uncouplers (which carry protons through membranes and therefore 'uncouple' oxidative phosphorylation) and ionophores, which can be used to control concentration gradients of other cations. Under conditions where such electrochemical gradients across the plasma membrane are 'short circuited' by the use of suitable ionophores, the organism *Streptococcus faecalis* was still able to grow,[77] showing that these concentration gradients were not essential for the synthesis of macromolecules and various structural elements, DNA replication or cell division.

One recent development in the study of microorganisms is the application of high-resolution NMR (for several nuclei) to investigate living cells. The binding of alkali metal ions has been examined with particular reference to the genus *Halobacterium*, which grows in highly saline media (e.g. ~ 3.5 mol dm^{-3} NaCl), and has high intracellular ion contents of about 4 mol dm^{-3}[K$^+$] and 1.5 mol dm^{-3}[Na$^+$]. Bound and free intracellular cations have been identified by ^{23}Na, ^{87}Rb, and ^{133}Cs NMR techniques.[78] There should be considerable developments in this area.

Cations as growth-limiting factors in chemostat cultures

The need for particular cations for growth of microorganisms is often studied in cation-limited chemostat culture, where supply of a cation can be made a growth-limiting factor. For bacteria, the growth rate has been demonstrated[79] to be dependent upon [Mg^{2+}] and [K$^+$]. Thus in the former case when Mg^{2+} was made the limiting factor, the growth of bacteria was directly proportional to [Mg^{2+}]. The plot of cell numbers against [Mg^{2+}] may be extrapolated back through the origin. The RNA content varied in a similar fashion, due to changes in the cellular ribosome content. It appears that a large proportion of the intracellular Mg^{2+} and K$^+$ may be associated with the ribosomes. Both cations are known to affect the stability of this structure *in vivo* and *in vitro*. *In vitro* studies suggest that the [K$^+$]/[Mg^{2+}] ratio is also important. A low ratio is necessary for tighter cross-links in the RNA and ribosomal aggregation. A high ratio tends to result in disassociation of the ribosomes into subunits and the unfolding of the RNA chain.

An absolute requirement for potassium in growth has also been demonstrated for the yeast *Candida utilis*.[80] Here however the results were different from those obtained with prokaryotic organisms in that the potassium content of *C. utilis* varied markedly with the nature of the growth limitation, and that there was little correlation with RNA content. It is suggested that in the yeast the functioning of the cell can be modified to accommodate K^+ deficiency more readily than is the case with the prokaryotic organisms, or alternatively that some of the functions of K^+ could be met by other cations. Nevertheless, it is clear that there are fundamental differences in the physiological factors relating to the requirement of K^+ by eukaryotic and prokaryotic cells.

Transport of IA and IIA cations

Both Ca^{2+} and Na^+ are normally expelled from the cytoplasm of bacteria. Up until recently this was thought to occur by antiport with protons, and such a Na^+/H^+ antiport has been identified for *Azotobacter vinelandii*.[81] Na^+ efflux has also been coupled to Ca^{2+} uptake in *Halobacterium halobium* vesicles,[82] while transport of Na^+ has been linked to the accumulation of amino acids and sugars, in a symport mechanism. However it now appears[83] that ATP-linked transport of Ca^{2+} is present in *Streptococcus faecalis*, while transport ATPase have been found in yeasts and other more complex organisms.

The uptake of potassium has been particularly well studied. No ATPase akin to the Na^+, K^+-ATPase of mammalian systems has yet been positively identified. K^+ accumulation in *Escherichia coli* may occur by four K^+-transport systems, which have been demonstrated by genetic and kinetic investigations on different strains. A combination of suitable growth conditions and mutant species allows them to be studied individually.[84] The Kdp system has been most studied. This is a high-affinity transport system ($K_m = 2 \times 10^{-6}\,mol\,dm^{-3}$) driven by ATP, and dependent on a periplasmic protein component. The TrkF system is driven by proton motive force (PMF) and has no protein dependence, while the TrkA depends on both ATP and PMF. The fourth system TrkD, like TrkA, is a low-affinity system.

The three structural proteins of the Kdp transport system have been located in the inner membrane of *E. coli*, while a K^+-stimulated membrane ATPase has been identified as part of the Kdp system. It seems probable therefore that further studies will confirm the existance of an ATPase transport system similar to those in mammalian cells.[85]

Specific examples of functions of IA and IIA cations

K^+ is an essential growth factor for all microorganisms. Its high concentration in most bacteria ($> 0.1\,mol\,dm^{-3}$) may be attributed largely to its role in ribosome function and osmotic regulation. Nevertheless there is good evidence to show its involvement in protein synthesis, carbohydrate metabolism, and the uptake of phosphates, while it has specific roles in the activation of many enzymes, as described later.

The role of M^{2+} cations in bacterial cell wall structure has been explored by the use of metal chelating agents. Thus Ca^{2+}, Mg^{2+}, and Zn^{2+} are known to be components of the cell wall of *Pseudomonas aeruginosa*, and are assumed to bridge carboxylate groups, a role for which Ca^{2+} is particularly well suited, as discussed earlier in this chapter. Incubation of *P. aeruginosa* with EDTA produces osmotically fragile species, due, it is suggested, to removal of the metal by EDTA and loss of the structural integrity of the cell wall. On addition of multivalent cations osmotically stable forms were restored. Monovalent cations were ineffective, as these cations could not bridge effectively.

When lysozyme and EDTA were used together, osmoplasts were formed which could not be restored with Ca^{2+}. This suggests that the impaired cell walls were more susceptible to lysis. It has been suggested that a lipopolysaccharide component is liberated on attack by EDTA. This is presumed to be a subunit cross-linked to other subunits and to other compounds *via* cations. On removal of the cations the subunit is liberated.

The nature of the Mg^{2+}-binding anionic polymers in cell walls varies with growth conditions. Cells of *Bacillus subtilis* produce walls containing teichoic acid or teichuronic acid under different conditions. These anionic polymers have different chemical composition but nevertheless very similar surface charge distribution. The binding of Mg^{2+} to both polymers involves a $1:2$ stoichiometry of $Mg^{2+}:CO_2^-$ group, but their binding constants differ, $K_{Assoc} = 0.61 \times 10^3$ and 0.3×10^3 dm^3 mol^{-1} for teichoic acid and teichuronic acid respectively. In conditions of limited $[Mg^{2+}]$, the cells produce walls with teichoic acid which are better able to extract Mg^{2+} from a limited supply. In conditions of abundant $[Mg^{2+}]$ the walls contain teichuronic acid, as a lower-affinity metal-binding site is then adequate.

The role of Ca^{2+} in microorganisms is particularly interesting. This is often associated with the release of Ca^{2+} into the cytoplasm to trigger physiological reactions. *Paramecium* is able[76] to reverse its direction of travel on encountering an obstacle and then to resume its forward motion in a slightly different direction. This results from a temporary reversal of the direction of ciliary beating over the cell surface, as a result of an influx of Ca^{2+} through the membrane which raises the cytoplasmic $[Ca^{2+}]$ to greater than 10^{-6} mol dm^{-3}. This influx is a passive one, running down the concentration gradient, and is probably initiated by the opening of a gate that controls entry to the channel. The presence of Ca^{2+} reverses the direction of cilia beat, but only briefly ($\sim 50ms$) as the calcium pump reduces the calcium level back to the original level. Forward movement is then resumed. Mutant species are available in which Ca^{2+} influx cannot occur (as shown by electrophysiological measurements). Such species are unable to reverse. Organisms whose outer membranes have been disrupted can still swim when given ATP. At $[Ca^{2+}]$ in the range $10^{-7}–10^{-8}$ mol dm^{-3} they swim forward and at higher levels they reverse.

The chemotactic behaviour of bacteria is controlled in a similar fashion by Ca^{2+}. *Bacillus subtilis* either swims smoothly or tumbles, the latter motion resulting in a reorientation for the next swimming activity. When heading towards repellants, bacteria tumble more frequently to increase the probability of swimming away, in which case tumbling will then cease. Tumbling is caused by the clockwise rotation of flagella and swimming by anticlockwise rotation. Studies with the ionophore A23187 show that the direction of flagellar movement is controlled by the levels of Ca^{2+} in the cytoplasm. High concentrations result in tumbling. Repellants which cause tumbling increase the flow of Ca^{2+} across the plasma membrane.[86]

Although Ca^{2+} is normally excluded from bacterial cells, sporulating *Bacilli* accumulate Ca^{2+} in certain stages, which are also associated with the biosynthesis of dipicolinic acid. Spores of most bacterial species contain large amounts of Ca^{2+} and dipicolinic acid on a 1 : 1 basis (up to 40% dry weight) suggesting that calcium dipicolinate is present in the spore. The presence of this compound is suggested to confer high heat resistance and the maintenance of dormancy to the spore. The inflow of Ca^{2+} to the cell has been attributed to the morphological changes taking place in sporulation which result in the outer surface of the outer membrane of the fore-spore being formed from the inner surface of the cell plasma membrane, so that overall the polarity of the membrane is reversed.[87]

Ca^{2+} has also been implicated in the control of morphogenesis in the eggs of algae[76] and is well known for this role in fungi.[88]

SOME FURTHER ROLES FOR IA CATIONS

So far emphasis has been placed on the Na^+, K^+-ATPase, but many other enzymes are activated by K^+. Indeed the extent of this phenomenon is probably not fully appreciated as many reaction media contain K^+ and its effect has not been fully assessed. In addition the IA cations stabilize a range of structures, while they act as charge carriers in nerve impulses and are important in osmotic balance and in the provision of physiologically active concentration gradients. Many of these topics have already been illustrated, but some additional examples are discussed briefly.

Transmission of nerve impulses

The nerve impulse may be regarded as an electrical impulse conducted along the membrane of the nerve fibre or axon as a result of the breakdown of the normal ion selectivity of the membrane which allows an influx of sodium ions. Thus the excitability of isolated nerve fibres is lost when Na^+ is absent from the external medium. Nerve transmission may be studied directly in the giant nerve fibres of the squid as they have a diameter large enough (up to 1 mm) to allow electrodes to be inserted. The change in intracellular cation distribution has been observed directly by the use of microelectrodes selective for Na^+ and K^+.

Normally there will be a large excess of K^+ inside the cell and an excess of Na^+ and Cl^- outside. The resting cell membrane may be treated as a Cl^- or K^+ electrode. Thus for K^+, the potential will be given by $E = (RT/F) \log_e [K^+]_{out}/[K^+]_{in}$. The data given in Table 9.1 for the squid axon shows that the resting potential should be -74 mV. On depolarization the value

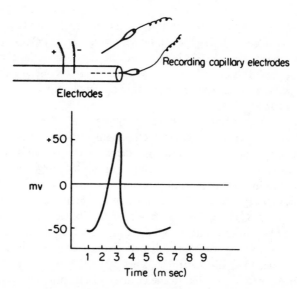

Figure 9.6 The recording of nerve impulses.

will be + 55 mV. In Figure 9.6 the method of recording these potentials is shown and a typical result given. If a nerve is stimulated some distance away from the recording electrodes, a delay occurs while the impulse is conducted to the electrodes. There is then a rapid change in the potential difference which dies away again as the impulse moves on past the electrodes. This is the action potential, while the shape depicted in Figure 9.6 is termed a spike. It may be seen that the total change in potential is close to the calculated value $(74 + 55 \, \text{mV})$. The phenomena may be understood in terms of the change in potential difference, produced when an electrical impulse is used to depolarize the membrane, causing a locally depolarized region that affects the regions on either side of it. While the influx of Na^+ into the cell is rapid, the amount transferred is very small. The Na^+ will then be pumped out by the sodium pump, causing the fall of the spike.

For the spike to be propagated there has to be a mechanism for the amplification of a partial depolarization of the membrane potential. However, this may be understood. As the nerve membrane is depolarized, so sodium ions will enter the axon and by virtue of their positive charge will reinforce the lowering of the initial resting potential. This will cause an increase in the sodium permeability and a further reinforcement of the lowering of the potential.

Transmission of the impulse across synaptic gaps between nerve fibres (15–20 nm) and between nerve and muscle (50–100 nm) is carried out by the neurotransmitter acetylcholine as described earlier.

Structure stabilization and control in nucleic acids

Alkali metal ions have large and specific effects on the structures of nucleic acids which may well be of considerable importance in their biological function. Some recent examples that demonstrate the ordering effect of these cations on nucleotides will be given. NMR studies[89] show that the nucleotide 5'-guanosine monophosphate has an ordered structure in D_2O, which results from base stacking and hydrogen bonding. It appears however that this self-assembly is dependent upon the alkali metal cation present. The dilithium and dicesium salts do not form such a structured complex, as in these cases the size of the cation is inappropriate. However the spectra of K_2(5'-GMP), Na_2(5'-GMP), and Rb_2(5'-GMP) are all different from each other, showing the presence of different structures. That of the K^+ salt survives at higher temperatures than that of the Na^+ salt, indicating it is more stable. This view is confirmed by ^{39}K and ^{23}Na NMR results.[90]

The formation[91] of the polynucleotide structure poly(I) is strikingly dependent on the size of the alkali metal ion which occupies the central cavity. It appears that two binding sites are available; Li^+ and Na^+ occupy a site which is 2.2 Å away from the centre of four carbonyl oxygen atoms, while K^+ and Rb^+ occupy a site that is 2.8 Å away from eight carbonyl oxygens. In these cases Li^+ and Cs^+ are probably too small and too large respectively to bind effectively.

It appears that alkali metal ions do not differ markedly in their affinities for nucleic acid helices. The normal double helical structure does not provide specific sites. But it is clear that certain polymer structures are able to do this—possibly through a conformational change. Specific metal binding could control an equilibrium between two structural conformations and thus have an effect on stabilities and interactions.

Enzyme activators (other than Na^+, K^+-ATPase)

A large number of enzymes depend upon K^+ as an activator,[92] and many more will probably be found. In these cases it seems unlikely that the substrate will bind to K^+, and while ideas on mechanisms are rather vague it seems probable that K^+ functions by controlling the conformation of the enzyme or preventing subunit dissociation. Such conclusions have usually been reached on the basis of UV, CD, and fluorescence spectroscopy. The requirement for K^+ is not an absolute one in that other monovalent cations can usually act but with lowered activity. Rb^+ usually replaces K^+ most effectively, while Tl^+ has been much used as a probe, but with limited success.

Enzymes requiring K^+ usually fall into one of two general classifications—those involving phosphoryl transfer and those involving hydrolytic or elimination reactions. One of the best studied examples from the first group is pyruvate kinase. Rabbit muscle pyruvate kinase is made up of four subunits and also requires a divalent cation, which is thought to act as a bridge, $E-M-S$. The reaction catalysed is the transfer of phosphate from phosphoenolpyruvic acid to ADP. The presence of conformational changes on adding K^+ has been shown spectroscopically. The use of Tl^+ has been of value in this case.[93] Its stronger binding to the enzyme allowed the use of equilibrium dialysis techniques, which confirmed the binding of 4 moles of Tl^+ per mole of enzyme, thus giving a $1:1:1$ relationship in $Tl^+:Mn^{2+}:PEP$.

Tl$^+$ NMR studies[94] in the presence of Mn^{2+} showed a broadening of the Tl resonance to an extent dependent on [Mn^{2+}], suggesting that Mn^{2+} and Tl$^+$ are bound close enough for them to interact. The Tl—Mn distance was estimated to be 8.2 Å. One possible complication in this work is the presence of dioxygen, as this paramagnetic species could also cause line broadening.

Other examples of phosphoryl-transfer enzymes with a requirement for K$^+$ are the aspartokinases,[95] phosphofructokinase and fructokinase,[96] acetic thiokinase, pyruvate, and other carboxylases,[92] and formyltetrahydrofolate synthetase.[97]

Examples of enzymes catalysing elimination and hydrolytic reactions include phosphatases,[98] pyridoxal phosphate enzymes catalysing an α–β elimination or addition reaction,[99] and certain B$_{12}$ coenzyme-dependent enzymes that catalyse C—O and C—N bond cleavage.[100]

A small number of membrane-bound enzymes (in addition to Na$^+$, K$^+$-ATPase) require Na$^+$ for activity, e.g. oxaloacetate decarboxylase.

Several K$^+$-dependent enzymes have been studied using Tl$^+$ as an activator. This technique must be used carefully, and spectroscopic measurements examined in the light of binding and kinetic studies. Thus Tl$^+$ activates yeast aldehyde dehydrogenase more effectively than does K$^+$, and gives lower K_m values for substrate binding, but at higher concentrations it inhibits as a result of binding to additional sites on the protein. This has important implications in the interpretation of Tl NMR data. At present some 50 mmol dm^{-3} [Tl$^+$] is necessary for measurement of longitudinal and transverse relaxation times. At this concentration Tl$^+$ is already binding substantially to inhibitory sites on the aldehyde dehydrogenase, with resulting complications in the NMR spectrum. Alternatively if a low enough [Tl$^+$] is used so that binding at inhibitory sites is negligible, then the enzyme protein is not stable.[35]

This enzyme, aldehyde dehydrogenase, is involved in the metabolism of alcohols. The first stage, the oxidation of alcohol to aldehyde is catalysed by the zinc metalloenzyme alcohol dehydrogenase, which has been much studied. In contrast the aldehyde dehydrogenases have not been well characterized. However the enzyme from *Saccharomyces cerevisiae* is now being studied.[101] This differs from other NAD(P)$^+$-dependent oxidoreductases in that NAD(P)$^+$ is the first substrate to bind to the enzyme. The role of K$^+$ in this enzyme is that of conformation control.

REFERENCES

1. H. H. Ussing, *The Alkali Metal Ions in Isolated Systems and Tissues, Handbuch der Experimentellen Pharmakologie,* Springer-Verlag, Berlin, 1960, p. 1; P. Kruhoffer, J. H. Thaysen and N. A. Thorn, *The Alkali Metal Ions in the Organism, Handbuch der Experimentellen Pharmakologie,* Springer-Verlag, Berlin, 1960, p. 196; P. R. Kernan, *Cell—Potassium,* Butterworths, London, 1965; C. P. Bianchi, *Cell—Calcium,* Butterworths, London, 1968.
2. R. J. P. Williams, *Quart. Rev. Chem. Soc.,* **24**, 331 (1970). C. J. Duncan(Ed.), *Calcium in Biological Systems,* Soc. Exp. Biol. Symp, **XXX**, 1976.

3. N. S. Poonia and A. V. Bajaj, *Chem. Rev.*, **79**, 389 (1979).
4. J. E. Heckels, P. A. Lambert, and J. Baddiley, *Biochem. J.*, **162**, 359 (1977).
5. R. J. P. Williams, *Analyst*, **78**, 586 (1953).
6. C. J. Pedersen, *J. Amer. Chem. Soc.*, **89**, 7017 (1967); **92**, 386, 391 (1970).
7. M. A. Bush and M. R. Truter, *J. Chem. Soc. Perkin II*, **1972**, 345.
8. R. M. Izatt, D. J. Eatough, and J. J. Christensen, *Structure and Bonding*, **16**, 161 (1973).
8a. A. C. Coxon, D. A. Laidler, R. B. Pettman, and J. F. Stoddart, *J. Amer. Chem. Soc.*, **100**, 8260 (1978).
9. M. N. Hughes, in *Specialist Periodical Report on Inorganic Biochemistry*, (Ed. H. A. O. Hill), **1**, 88 (1979).
10. K. H. Pannell, W. Yee, G. S. Lewandos, and D. C. Hanbrick, *J. Amer. Chem. Soc.*, **99**, 1457 (1977).
11. F. D. Cima, *J. Chem. Soc. Perkin II*, **1973**, 55.
12. E. J. Corey, K. C. Nicolaou, M. Shibasaki, Y. Machida, and C. S. Shiner, *Tetrahedron Letters*, **37**, 3183 (1975).
13. D. Dehm and A. Padwa, *J. Org. Che.*, **40**, 3139 (1975).
14. D. J. Sam and H. E. Simmons, *J. Amer. Chem. Soc.*, **96**, 2252 (1974).
15. D. J. Sam and H. E. Simmon, *J. Amer. Chem. Soc.*, **94**, 4024 (1972).
16. D. J. Cram and J. M. Cram, *Acc. Chem. Res.*, **11**, 8 (1978).
17. J. M. Lehn, *Structure and Bonding*, **16**, 1 (1973); *Acc. Chem. Res.*, **11**, 49 (1978); *Pure Appl. Chem.*, **50**, 871 (1978).
18. K. Henco, B. Timmler, and G. Maass, *Angew. Chem. Int. Ed.*, **16**, 538 (1977).
19. F. Mathieu, B. Metz, D. Moras, and R. Weiss, *J. Amer. Chem. Soc.*, **100**, 4412 (1978).
20. W. Simon, W. E. Morf, and P. C. Meier, *Structure and Bonding*, **16**, 113 (1973). D. E. Fenton, *Chem. Soc. Rev.*, **6**, 325 (1977).
21. J. H. Prestegard and S. I. Chan, *Biochemistry*, **8**, 3921 (1969); *J. Amer. Chem. Soc.*, **92**, 4440 (1970).
22. H. Diebler, M. Eigen, G. Igenfutz, G. Maass, and R. Winkler, *Pure Appl. Chem.*, **20**, 93 (1969).
23. N. Otake, T. Ogita, H. Nakayama, H. Miyamae, S. Sato, and Y. Saito, *J. C. S. Chem. Comm.*, **1978**, 875.
24. C. C. Chiang and I. C. Paul, *Science*, **196**, 1441 (1977).
25. S. J. Kennedy, R. W. Roeske, A. R. Freeman, A. M. Watanabe, and H. R. Bensch, *Science*, **196**, 1342 (1977).
26. R. E. Koeppe, J. M. Berg, K. O. Hodgson, and L. Stryer, *Nature*, **279**, 723 (1979); B. W. Urban, S. B. Hladky, and D. A. Haydon, *Fed. Proc.*, **37**, 2628 (1978).
27. R. Benz, K. Janko, W. Boos, and P. Lauger, *Biochim. Biophys. Acta*, **511**, 305 (1978); H. Schindler and J. P. Rosenbusch, *Proc. Natl. Acad. Sci. U.S.A.*, **75**, 3751 (1978).
28. E. J. Harris, B. Zaba, M. R. Truter, D. G. Parsons, and J. N. Wingfield, *Arch. Biochem. Biophys.*, **182**, 311 (1977).
29. R. Schwyzer, A. Tun-Kyi, M. Caviezel, and P. Moser, *Helv. Chim. Acta*, **53**, 15 (1970).
30. R. Miledi, I. Parker, and G. Schalow, *Proc. Roy. Soc.* **B198**, 201 (1977).
31. O. Shimomura and E. H. Johnson, in *Calcium in Biological Systems* (Ed. C. J. Duncan), SEB Symp., **XXX**, 40 (1976).
32. P. Laszlo, *Angew. Chem. Int. Ed.*, **17**, 254 (1978).
33. F. J. Keyne and J. Reuben, *J. Amer. Chem. Soc.*, **92**, 220 (1970).
34. J. P. Manners, K. G. Morallee, and R. J. P. Williams, *Chem. Comm.*, **1970**, 965.
35. K. Bostian, G. F. Betts, W. K. Man, and M. N. Hughes, *FEBS Letters*, **59**, 88 (1975).
36. M. Cohn, *Quart. Rev. Biophys.*, **3**, 61 (1970).
37. E. Nieboer, *Structure and Bonding*, **22**, 1 (1975).
38. E. C. N. F. Geraldes and R. J. P. Williams, *J. Chem. Soc. Dalton*, **1977**, 1721.

39. M. Takata, W. F. Pickard, J. Y. Lettvin, and J. W. Moore, *J. Gen. Physiol*, **50**, 461 (1966).
40. C. Van Breeman and P. De Weer, *Nature*, **226**, 760 (1970).
41. Y. Kagawa, *Biochim. Biophys. Acta*, **505**, 45 (1978).
42. P. Mitchell, *FEBS Letters*, **78**, 1 (1977); *Nature*, **191**, 144 (1961); R. J. P. Williams, *FEBS Letters*, **85**, 1 (1978); *Biochem. Biophys. Acta*, **505**, 1 (1978); *J. Th. Biol.*, **1**, 1 (1961).
43. P. D. Boyer, B. Chance, L. Ernster, P. Mitchell, E. Racker, and E. C. Slater, *Ann. Rev. Biochem.*, **46**, 955 (1977).
44. M. K. F. Wikström, *Nature*, **266**, 271 (1977); J. Moyle and P. Mitchell, *FEBS Letters*, **88**, 268 (1978); M. Wikström and K. Krab, *FEBS Letters*, **91**, 8 (1978); E. Sigel and E. Carafoli, *Eur. J. Biochem.*, **89**, 119 (1978).
45. J. F. Nagle and H. J. Morowitz, *Proc. Natl. Acad. Sci. U.S.A.*, **75**, 298 (1978).
46. M. Avron, *FEBS Letters*, **96**, 225 (1978).
47. I. M. Glynn and S. J. D. Karlish, *Ann. Rev. Physiol.*, **37**, 13 (1975)
48. J. C. Skou, *Biochim Biophys. Acta*, **23**, 394 (1957); *Physiol. Rev.*, **45**, 596 (1965).
49. B. Forbush, J. Kaplan and J. F. Hoffman, *Biochemistry*, **17**, 3667 (1978).
50. K. R. H. Repke and R. Schon, *Acta Biol. Med. Ger.*, **31**, K19 (1973).
51. R. L. Post, C. Hegyvary, and S. Kume, *J. Biol. Chem.*, **247**, 6530 (1972). R. W. Albers, in *The Enzymes of Biological Membrances* (Ed. A. Martonosi), Vol. 3, Wiley, New York, 1976, p. 283.
52. S. J. D. Karlish, D. W. Yates, and I. M. Glynn, *Biochim. Biophys. Acta*, **525**, 230, 252 (1978).
53. L. A. Beauge and I. M. Glynn, *Nature*, **280**, 510 (1979).
54. L. C. Cantley, G. Cantley, and L. Josephson, *J. Biol. Chem.*, **253**, 7361 (1978).
55. M. le Maire, K. E. Jørgensen, H. Røigaard-Petersen, and J. V. Møller, *Biochemistry*, **15**, 5805 (1976); T. Chyn and A. Martonosi, *Biochim. Biophys. Acta*, **468**, 114 (1977).
56. J. F. Brinley, J. F. Tiffert, L. J. Mullins, and A. Scarpa, *FEBS Letters*, **82**, 197 (1977).
57. L. Mela, *Current Topics in Membranes and Transport* (Eds. F. Bonner and A. Kleinzeller), **9**, 321 (1977).
58. E. Carafoli *et al.*, (Eds.), *Calcium Transport in Contraction and Secretion*, North Holland, 1975; C. J. Duncan (Ed.), *Calcium in Biological Systems*, S.E.B. Symp. **XXX**, Cambridge University Press, 1976; R. H. Wasserman *et al.*, (Eds.), *Calcium Binding Proteins and Calcium Function*, North Holland 1977.
59. H. E. Harris, M. Y. W. Tso, and H. F. Epstein, *Biochemistry*, **16**, 859 (1977).
60. E. R. Birnbaum and B. D. Sykes, *Biochemistry*, **17**, 4965 (1978).
61. J. D. Johnson, J. H. Collins, and J. D. Potter, *J. Biol. Chem.*, **253**, 6451 (1978); J. D. Johnson and A. Schwartz, *J. Biol. Chem.* **253**, 5243 (1978).
62. J. Derancourt, J. Haiech, and J-F. Pechere, *Biochim. Biophys. Acta*, **532**, 373 (1978); J. Parello, H. Lilja, A. Cave, and B. Lindman, *FEBS Letters*, **87**, 19 (1978).
63. J. D. Potter, J. R. Dedman, and A. R. Means, *J. Biol. Chem.*, **252**, 5609 (1977).
64. R. S. Adelstein, M. A. Conti, D. R. Hathaway, and C. B. Klee, *J. Biol. Chem.*, **253**, 8347 (1978).
65. W. Sieghart, T. C. Theoharides, S. L. Alper, W. W. Douglas, and P. Greengard, *Nature*, **275**, 329 (1978).
66. C. E. Creutz, C. J. Pazoles, and H. B. Pollard, *J. Biol. Chem.*, **253**, 2858 (1978).
67. P. F. Baker, *Sci. Progr. Oxf.*, **64**, 94 (1977).
68. T. D. Ingolia and D. E. Koshland, *J. Biol. Chem.*, **253**, 3821 (1978).
69. *Biochem. Soc. Trans.*, **5**, 1241 (1977).
70. P. Robertson, R. G. Hiskey, and K. A. Kochler, *J. Biol. Chem.*, **253**, 5880 (1978).
71. J. W. Bloom and K. G. Man, *Biochemistry*, **17**, 4430 (1978).
72. Parathyroid hormone, Calcitonin and Vitamin D, *Fed. Proc.*, **37**, 2557 (1978).
73. B. E. Kream, M. Jose, S. Yamada, and H. F. DeLuca, *Science*, **197**, 1089 (1977).
74. D. H. Schlesinger and D. I. Hay, *J. Biol. Chem.*, **252**, 1689 (1977).

75. P. V. Hauschka and M. L. Reid, *J. Biol. Chem.*, **253**, 9063 (1978).
76. F. M. Harold, *Ann. Rev. Microbiol.*, **31**, 181 (1977).
77. F. M. Harold and J. Van Brunt, *Science*, **197**, 372 (1977).
78. H. T. Edzes, M. Ginzberg, B. Z. Ginzberg, and H. J. C. Berendsen, *Experientia*, **33**, 732 (1977); M. Goldberg and H. Gilboa, *Biochim. Biophys. Acta*, **538**, 268 (1978).
79. D. W. Tempest, *Symp. Soc. Gen. Microbiol.* **19**, 87 (1969).
80. H. Aiking and D. W. Tempest, *Arch. Microbial.*, **108**, 13 (1976).
81. P. Bhattacharyya and E. M. Barnes, *J. Biol. Chem.*, **253**, 3848 (1978).
82. J. W. Belliveau and J. K. Lanyi, *Arch. Biochem. Biophys.*, **186**, 98 (1978).
83. H. Kobayashi, J. Van Brunt, and F. M. Harold, *J. Biol. Chem.*, **253**, 2085 (1978).
84. D. B. Rhoads and W. Epstein, *J. Biol. Chem.*, **252**, 1394 (1977).
85. W. Epstein, V. Whitelaw, and J. Hesse, *J. Biol. Chem.*, **253**, 6666 (1978). L. A. Laimins, D. B. Rhoads, K. Altendorf, and W. Epstein, *Proc. Natl. Acad. Sci. U.S.A.*, **75**, 3216 (1978).
86. G. W. Ordal, *Nature*, **270**, 66 (1977).
87. C. Hogarth and D. J. Ellar, *Biochem. J.*, **176**, 197 (1978).
88. S. E. Cameron and H. B. Le John, *J. Biol. Chem.*, **247**, 4729 (1972).
89. T. J. Pinnavaia, C. L. Marshall, C. M. Mettler, C. L. Fisk, H. T. Miles, and E. D. Becker, *J. Amer. Chem. Soc.*, **100**, 3627 (1978).
90. P. Laszlo and A. Paris, *C. R. Hebd. Seances Acad. Sci. Ser D*, **286**, 717, 781 (1978).
91. H. T. Miles and J. Frazier, *J. Amer. Chem. Soc.*, **100**, 8037 (1978).
92. C. H. Suelter, *Science*, **168**, 789 (1970); H. Meyer and F. Meyer, *Biochemistry*, **17**, 825 (1978).
93. F. J. Kayne, *Arch. Biochem. Biophys.*, **143**, 232 (1971).
94. J. Reuben and F. J. Kayne, *J. Biol. Chem.*, **246**, 6227 (1971).
95. D. E. Wampler and E. W. Westhead, *Biochemistry*, **7**, 1661 (1968).
96. R. D. Mavis and E. Stellwagon, *J. Biol. Chem.*, **245**, 674 (1970).
97. R. E. MacKenzie and J. C. Rabinowitz, *J. Biol. Chem.*, **246**, 3731 (1971).
98. H. W. Behrisch, *Biochem, J.*, **121**, 399 (1971).
99. W. E. Newton and E. E. Snell, *Proc. Natl. Acad. Sci. U.S.A.*, **51**, 382 (1964); C. H. Suelter and E. E. Snell, *J. Biol. Chem.*, **252**, 1852 (1977).
100. J. J. Bakes and T. C. Stadtman, *Fed. Proc.*, **31**, 494 (1972).
101. K. A. Bostian and G. Betts, *Biochem. J.*, **173**, 773, 787 (1978).

Note added in proof

It is becoming clear that the calcium-binding regulatory proteins found in many Ca^{2+}-dependent processes are identical. This protein has been called 'Calmodulin', in view of its role as the intracellular calcium receptor. For a review see A. R. Means and J. R. Dedman, *Nature*, **285**, 72 (1980).

For a comprehensive review of Na^+, K^+-ATPase see: *Na^+, K^+-ATPase. Structure and Kinetics* (Eds. J. C. Skoy and J. G. Nørby), Academic Press (1979).

CHAPTER TEN

METAL IONS AND CHELATING AGENTS IN MEDICINE

The concentrations of the naturally occurring metal ions in living systems are carefully controlled within fine limits. This control in general is exercised by certain proteins and hormones. Clearly, disorders must arise if this balance is upset. Conversely, the interrelationships between metal ions and binding substances in the body are so complex that disorders or diseases involving the metal-binding substances may result in the presence of high or low concentrations of metal ion, compared with that normally present. Accordingly, analysis of body fluids and tissues for trace metal ions could become an important diagnostic procedure. It does appear that a number of major diseases are associated with changes in concentration of the trace metal ions in certain tissues and body fluids, while it has long been appreciated that the balance of alkali metals in and out of the cell is affected in many diseased conditions.

A trace metal ion that is essential for the activity of enzyme systems may well become toxic if the concentration is raised above certain limits. The level beyond which a metal ion becomes toxic is dependent upon its location. Thus a doubling of the extracellular concentration of K^+ leads to heart disorder and possibly death, but this concentration is still well below the normal concentration of intracellular K^+. A deficiency of any metal ion, for example zinc or the transition elements, will result, in the appropriate cases, in reduced enzyme activity with a breakdown of the normal metabolic processes.

A range of other metal ions, particularly the heavy metal ions, have toxic effects. These often result from environmental and industrial problems. In such cases it is necessary to design the most effective chelating agent to remove the metal ion. The toxic effect of certain metal complexes has been exploited in the preparation of anti-cancer drugs, while several other diseased conditions are treated by metal chemotherapy. Furthermore, certain drugs are good ligands and it is difficult to escape from the possibility that they might function by binding to native metal ions.

In this chapter the role of metal ions in medicine will be considered under the headings described above, namely (a) trace metal ions, (b) metals in chemotherapy, (c) metal ion toxicity, and (d) drugs that may function through interaction with metal ions.

TRACE METAL IONS

Trace metals are involved in many aspects of metabolism, in enzymes, in transport, in synthesis, in control, and in bone formation, so the interrelationship between ligand and metals and health is a complex one.[1] Imbalances of trace metals may result from nutritional problems, while there are also problems in analytical techniques and in determining what is a 'normal' concentration level. Metal ions are absorbed from the small intestine but little is known about mechanisms in this context,[2] although certain proteins seem to be associated with specific uptake of certain metals. Endeavours are now being made on the one hand, to quantify the form in which metal ions are distributed in body fluids,[3] and on the other, to relate deficiencies is particular elements to specific disorders and to understand the detailed mechanisms for this. Metal ion deficiency usually arises from dietary deficiencies and it is sometimes difficult to relate a diseased state to the deficiency of one particular element. Some illustrative examples will be given of clinical problems arising from deficiencies of metal ions and also of the opposite problem, where there is an excess imbalance of a metal due to some malfunction of the body's control system.

Deficiency in trace metals

The best known example is probably that of anaemia, in which there is a deficiency of iron, the red cells of the blood containing less hemoglobin than is normal. Pernicious anaemia results from a deficiency in vitamin B_{12}, while sickle cell anaemia is associated with a genetically controlled defect of hemoglobin synthesis. Patients with sickle cell anaemica have abnormal hemoglobin, differing in a single amino acid residue from normal hemoglobin. Sickle cell anaemia confers resistance to malaria. Anaemia is treated with ferrous sulphate, sometimes with ascorbic acid to aid absorption. Chelate complexes have been used in the hope that iron would be released slowly, avoiding the toxic effects of a build-up of concentration.

In growing animals, zinc deficiency results in loss of appetite, skeletal and hair abnormalities, skin lesions, and inhibition of sexual maturation. Several studies have suggested that dwarfism in pre-adolescent humans can be attributed also to zinc deficiency, but doubt has now been cast on these results, as it seems that other limiting nutritional factors may also have contributed to this. In a few cases in humans, zinc deficiency has caused gross abnormalities in sensory perception, particularly with perverted taste (dysguesia), and perverted smell (dysomia). Such people develop anorexia and psychological difficulties, but the symptoms may be relieved by zinc supplementation. The role of zinc compounds in wound healing has attracted attention. Zinc deficiency inhibits this due to decreased synthesis of protein and collagen. It appears however that zinc supplementation to improve wound healing is of doubtful value in zinc-repleted individuals but is of definite value in zinc-deficient patients.[4]

Manganese deficiency in humans has only recently been reported. It induces transient dermatitis, changes in hair colour and growth, and hypocholesterolemia. Little is known about the biochemistry of these problems, although the effect of Mn^{2+} on skeletal abnormalities and reproduction is better characterized. A remarkable observation is the fact that the addition of manganese supplements to the diets fed to pregnant mutant mice, who develop congenital ataxia, resulted in ataxia-free offspring. The mutation was not abolished because offspring of the ataxia-free mice developed ataxia. Nevertheless it is of great interest that an essential nutrient, managanese, can prevent the development of a genetically predetermined phenotype.

Much more information is available[1] on copper uptake and transport in the body; most of the plasma copper is bound in ceruloplasmin, while in the cell copper is bound to a metallothionein, polynucleotides, and certain other proteins. The widespread distribution of copper means that copper deficiency in humans is unlikely. A progressive brain disease in infants (Menkes' kinky hair syndrome) seems to be associated with copper deficiency, as brain tissue in this disease was largely lacking in the copper-containing enzyme cytochrome-c oxidase, while serum copper and ceruloplasmin were subnormal. Some improvement in patients was produced by copper.

Copper deficiency produces defects in the structural stabilization of the fibrous protein of connective tissues in a number of animals. A number of abnormalities are associated with this. Of particular note is the fact that animals of a number of species fed on copper-deficient diets died as a result of structural lesions of major arteries. This structural weakness arises from the absence of cross-linking between polypeptide chains. These cross-linkages result from condensation reactions following the oxidative deamination of amino groups of lysyl side chains. It is thought that this latter reaction is catalysed by a copper-containing enzyme, monoamine oxidase, and so is inhibited by copper deficiency.

Deposition and local excess of metal ions

Body copper content is normally between 100 and 150 mg, with highest concentrations found in the locus of the brain stem. In Wilson's disease, copper concentrations up to one hundred times greater than normal have been found. The copper is accumulated in a number of tissues but in particular is found in the liver, brain, and kidneys. It may be seen as brown or green rings in the cornea. It is now established that the excess copper is accumulated first in the liver and then in the central nervous system, disorders in the liver and nervous system being manifested in this sequence. The symptoms of Wilson's disease (hepatolenticular degeneration) include hepatic cirrhosis, lack of coordination, severe tremors, and progressive mental deterioration. The extent of the symptoms is dependent upon the amount of copper present, and the symptoms may be reversed by the use of chelating agents to remove excess body stores of copper. The fact that the symptoms disappear after such treatment implies that the brain damage is a biochemical rather than a structural one.

The reason for the deposition of copper in the tissues is not certain, although it is known that the disease is genetically induced. Copper is stored in certain copper proteins in the liver, and it has been suggested that Wilson's disease is associated with a failure to synthesize apoceruloplasmin, so that transfer of copper to this protein cannot then take place with the result that copper is available for deposition elsewhere. It is clear however that this cannot be the only factor, as some patients have only slightly reduced ceruloplasmin levels. Furthermore copper is found in large amounts in the liver of newborn babies, where it is stored in a protein having 2% content of copper. This drops to the normal level at three months when the liver has developed the ability to synthesize the protein of ceruloplasmin. Another view is that in Wilson's disease, the liver storage protein metallothionein has an abnormal structure that leads to enhanced binding of copper ions, which leads in turn to a disturbance of the body's storage and transport system for copper. The enhanced binding of copper by the metallothionein from patients with Wilson's disease has been demonstrated.

Treatment of Wilson's disease involves a copper-restricted diet and the use of chelating agents, particularly penicillamine (Table 10.3) and $Na_2CaEDTA$. The calcium salt of EDTA is used to prevent the labilization of Ca^{2+} from the skeleton. Pencillamine, as the D isomer, is the favoured treatment at present. A dose of *ca.* 1 g daily will lead to the excretion of 8–9 mg of copper in new patients, although this will fall as the treatment progresses and copper is excreted. Factors that determine the choice of chelating agent and problems with their use will be discussed later under the topic of chelation therapy.

A further example involves the treatment of calcium deposition. The blood is rich in calcium salts, as they are rejected by the cell, but these salts are not very soluble and so are precipitated, resulting in the formation of stones, and hardening of arteries, and the accumulation of calcium in cataracts in the eye. Cholesterol is also involved in the formation of gall stones and the development of atherosclerosis. These conditions have all been treated with EDTA, with some measure of success, in order to remove the calcium ions by chelation.

Metal ion concentration as a diagnostic test

Of the metal ions that have been examined in detail in this context, copper appears to be the most valuable pointer to diseased conditions. Thus for infectious hepatitis, serum copper levels up to three times the normal value of $350 \mu g/100$ ml have been observed. This is due to an accumulation of ceruloplasmin. Other diseases associated with high copper concentrations in the blood are leukemia, lymphomas, rheumatoid arthritis, psoriasis, cirrhosis, nephritis, and Hodgkin's disease. High copper levels are associated with other phenomena, but clearly the presence of a high serum copper concentration is of diagnostic value when used in conjunction with other tests. Copper analysis can also be used to help monitor the effectiveness of treatment for certain of these conditions, as the copper level is directly proportional to the severity of the disease. This situation holds for hepatitis and for malignancies. In the latter case it is also noteworthy that the increase in serum copper level appears before any clinical changes are apparent.

METAL IONS AND CHELATES IN CHEMOTHERAPY

The role of metals and their compounds in the treatment of disease goes back over four millenia. Gold was used for this purpose in China in 2500 BC, while in the middle ages certain of its compounds were thought to be effective against leprosy. Plant products used at that time to alleviate mental disorders are known now to contain lithium salts which are employed today in the treatment of affective disorders.

In this century gold compounds have long been used in the treatment of rheumatoid arthritis, while Albert[5] and Dwyer[6] have shown that some metal complexes were very effective against certain microbial infections. An impetus to much current work on metal chemotherapy has come from the discovery of Rosenberg[7] that some simple platinum complexes were effective against specific tumours.

Synthetic metal chelates as antimicrobial agents

Certain metal chelates are active at low concentration against a range of bacteria, fungi, and viruses. Thus the organism *Staphylococcus pyrogenes*, found in wound infections, is very resistant to many standard antibiotics and is a great problem in hospitals. However it succumbs dramatically to the complexes $[Fe(Me_4phen)_3]^{2+}$ and $[Ru(Me_4phen)_2(acac)]^+$. These complexes are coordinatively saturated and so operate by physical interactions. The size, charge distribution, shape, and redox potentials of metal chelates can be varied regularly. They are, therefore, useful reagents with which to investigate pharmacological problems. It is clear that metal chelates can act in a number of ways. Thus they may inactivate a virus by occupying sites on its surface which would normally be utilized in the initiation of the infection of the host cell. The first step in the infection would be an adsorption reaction involving electrostatic interactions. Alternatively, the complex cations may penetrate the cell wall and prevent virus reproduction.

The importance of the receptor site principle is demonstrated by the fact that the toxicity of certain of these cationic complexes is matched by onium salts of the ligand. It is also demonstrated in the toxic action of certain chelates towards mice and rats. Thus tris-(2,2'-bipyridyl)iron(II) sulphate and tris-(1,10-phenanthroline)ruthenium(II) iodide cause paralysis with general symptoms similar to those of the curareform drugs.

The active component of curare is D-tubocurare which deactivates the membrane of the motor end plate of a nerve to the transmitter substance acetylcholine, so causing paralysis.

Acetycholine has a charged head: $Me_3NCH_2CH_2OCOMe$. The effective ionic radius of the head is about 3.2 Å. The receptor site for acetylcholine is assumed to include a depression to receive the cationic head, with maximum interaction, together with suitable groups to interact with the carbonyl oxygen and to hydrogen bond with the ether oxygen. D-Tubocurare is a large molecule, having two positively charged centres. Its toxic action is probably due to a blocking of the acetylcholine receptor site.

The cationic complexes again probably act by competing with acetylcholine for the receptor site. Neutral complexes have no effect. The effectiveness of any cationic complex is associated with the overall peripheral charge of the complex. Thus $[Co(NH_3)_6]^{3+}$ is twice as potent as $[Co(NH_3)_5NO_2]^{2+}$

Gold and its compounds

The major role of gold compounds in chemotherapy involves the treatment of rheumatoid arthritis with gold(I) compounds such as the thiomalate ('Myocrisin'), thiosulphate ('Sanocrysin'), and the thioglucose derivative ('Solganal'), (Table 10.1). Gold(I) readily disproportionates into gold(o) and gold(III) in aqueous solution, and the use of these thiolato and other sulphur ligands reflects the stabilization of the very soft gold(I) oxidation state by the soft sulphur ligand. Other gold(I) compounds have been used against bacterial infections, including TB, while the gold(III) compound (d^8) [Au(5-diazouracil)$_2$Cl$_2$]Cl has anti-tumour activity in mice. There are several other aspects of the biological chemistry of gold (such as the uptake of gold by microorganisms and the use of colloidal gold) but these fall outside the scope of this chapter. They have been summarized by Sadler,[8] who also reviewed the relevant chemistry of gold.[49]

Gold compounds were first used in 1927 for the treatment of rheumatoid arthritis and are still used at the present time. Gold thiomalate is thought to compare favourably with other drugs in the treatment of difficult cases, while some 50% of patients are reported either to be cured or improved markedly as a result of its use. On the other hand, a substantial percentage (around 40%) show toxic side effects, the toxicity being accumulative. Traces of gold compounds reach many cells and remain in the body for many years. The dose of Myocrisin is 50 mg in 0.5 ml H_2O and is injected intramuscularly each week until about 1 g has been administered. It is estimated that about 30% of each dose is excreted within one week, and that about 85% of the total gold in the body is bound to serum albumin.

The actual role of Au(I) in controlling rheumatoid arthritis is uncertain. It is known that in treatment the gold is concentrated within macrophages and further localized in the lysosome. Such lysosomes have been termed 'aurosomes'.

TABLE 10.1 Gold compounds and ligands used in chemotheraphy

The lysosome contains hydrolytic enzymes which could be responsible for joint damage in rheumatoid arthritis when released within the joint. It has been suggested therefore, in view of this characteristic accumulation of the gold drug in the lysosome, that the gold compound functions by inhibiting these hydrolytic enzymes. Other suggestions for the activity of gold include its action on the infectious agent that causes rheumatoid arthritis and an effect on the joint structure through stabilization of the collagen by cross-links.

Studies[9] on the structures of Au(I) thiols show that they are polymers in aqueous solution, thus allowing the Au(I) to form a linear two-coordinate system. Bridging by sulphur is probable in aurothiomalate and aurothioglucose as it seems unlikely that the carboxylate groups will bind to Au(I). In a parallel situation infinite Ag—S frameworks are known in silver(I) cyclohexanethiolate, with three-coordinate silver. NMR studies on gold(I) thiomalate at high ionic strength show three types of coordinated thiomalate, which are sensitive to acid. The Au(I) ion can bind further thiolate groups and this is suggested to give three-coordinate gold. It is important to note the ease of thiol exchange in these compounds as this may be relevant to the ready distribution of Myocrisin in the body and for some of its toxic side effects. Polymerization of these gold thiolates may be necessary before they are taken up by macrophages, as this process is probably related to the size of the particle.

Gold compounds are usually tested in animals against adjuvant arthritis, initiated by injection of *Mycobacterium butyricum*. A new drug, the phosphine complex Et_3PAuCl, is effective when taken orally in suppressing the inflammatory lesions of adjuvant arthritis in rats. Myocrisin and Solganal are both ineffective when taken orally. These phosphine complexes also have less harmful side effects on the kidneys and so may be promising advances. Recently sodium bis(*N*-methylhydantoinato)gold(I) tetrahydrate has been characterized by X-ray crystallography.[10] This interesting compound is the first Au(I) compound in which Au(I) is coordinated to nitrogen ligands only, and was prepared by treating $NaAuCl_4$ with *N*-methylhydantoin. In the older literature tetrasuccinimidogold(III) was claimed to be very effective against microorganism-induced arthritis, but the compound could not be reprepared. It appears possible therefore that this 'gold(III)' complex may have been the $1 : 2$ gold(I) complex of similar structure to the *N*-methylhydantoinato complex. Results are not yet available for anti-inflammatory tests with sodium bis(*N*-methylhydantoinato) gold but these are likely to be promising if the suggestion of Sadler[10] relating to the succinimido complex is correct.

Lithium and mental health

Lithium has been used for some thirty years in the treatment of the manic phase of manic depressive patients. One in two thousand people in the U.K. receive such lithium treatment. Manic depressive psychoses involve alternating phases of depression and mania or over-excitement. Sometimes mood swings take place over a cycle of a few weeks, while in other cases psychotic episodes occur at intervals of a year or so. The subject of lithium and effective disorders is well documented[11] but very little is known about the mechanism of lithium action. It is nevertheless much used in the treatment of such disorders and its effect is noted if drinking water contains lithium salts. Thus there is an inverse correlation between the level of lithium salts in drinking water in various Texas cities and admissions to mental hospitals.

Lithium is usually administered as the carbonate and is supplemented with thyroxin to treat possible side effects of Li^+ on thyroid function.[12] Care has to be taken to ensure that blood concentrations of Li^+ do not become too high. At one time lithium chloride was used as a salt substitute ('Westral') and a number of fatalities resulted from this. Toxic effects of Li^+ are associated with thyroid and kidney (where particularly high concentrations of lithium are likely to be found), while Li^+ is also retained in the bone so that treatment with lithium salts leads initially to a loss of Ca.[2]

Although a precise mechanism for the action of lithium is not known, a number of hypotheses are currently being considered. Lithium is able to pass through all membranes readily and so has access to many important control sites. This is one reason why so many mechanisms have to be considered. From a general point of view it is possible that Li^+ could be interfering with aspects of Na^+, K^+, Mg^{2+}, or Ca^{2+} metabolism. The existence of the diagonal relationship suggests that competition between Li^+ and Mg^{2+} for Mg^{2+} sites must be considered as a priority.

Formation constant data will indicate the likelihood of Li^+ competing with the four cations cited above. Approximate concentrations of all five cations in the cytoplasm of cells during lithium therapy are as follows (mol dm^{-3}); Li^+ (10^{-3}), Na^+ (10^{-2}), Mg^{2+} (10^{-3}), Ca^{2+} (10^{-7}), while ATP, an important ligand for these species, is also present at 10^{-3} mol dm^{-3}. A simple comparison[13] of formation constant data suggests that Li^+ would be unable to compete with any of the other cations, but this conclusion is misleading[13] as the other cations may be firmly bound by ligands such as ATP so that in vivo Li^+ could compete for other biologically important ligands, as the effective concentration of the other cations is now reduced. A comparison of such 'conditional' formation constants suggests that biological ligands with a set of O^-, N, O^-, O^-, donor atoms would bind Li^+ rather than Na^+, K^+, or Ca^{2+}, and that while such ligands would form more stable complexes with Mg^{2+} than with Li^+, Li^+ would still be able to compete for some 20–30% of the binding sites. This would be sufficient for a whole range of biological processes to be affected. It must be stressed that this line of argument does not prove that Li^+ disturbs Mg^{2+} metabolism—but it does demonstrate that this attractive hypothesis is chemically feasible.

One detailed theory of lithium action involves the suggestion that it acts as an inhibitor of hormone-stimulated adenylate cyclase. This enzyme catalyses the formation of cyclic AMP from ATP. This reaction is central to many processes as the cyclic AMP is an intracellular mediator (the so-called 'second messenger'). Li^+ inhibits a range of such hormone-stimulated cyclases including the adrenaline-stimulated adenylate cyclase in man.[14] Clearly inhibition of this process could control the manic phase, although so far clinical tests have not confirmed this hypothesis.

Metal complexes as anti-tumour agents

In 1964 Rosenberg, when studying the effect of an electric current on *E. coli*, found[7] that cell division was inhibited although growth was not. The bacterium thus grew as long filaments. The effect was finally attributed to the formation of platinum complexes in trace amounts (about 10 p.p.m.), the platinum originating from the electrode. The most active complexes were found to be the *cis* isomers of dichlorodiammineplatinum(II), $Pt(NH_3)_2Cl_2$ and tetrachlorodiammineplatinum(IV), $Pt(NH_3)_2Cl_4$. *Trans* complexes were found to be inactive. This observation led to the testing of these compounds against a number of tumours, and to the widespread current use of *cis*-$Pt(NH_3)_2Cl_2$ (abbreviated to *cis*-DDP) in the treatment of ovarian, testicular, and other forms of cancer. *Trans* complexes are toxic but do not have anti-tumour properties.

A great deal of research is currently directed[15-19] towards the elucidation of the mechanism of action of these compounds. Although the main requirements for activity seem now to be recognized, and while there is also some agreement on the mode of binding of DDP to DNA, the actual mechanism of cytotoxicity is still unclear.

A wide range of complexes of platinum and other metals has been tested in which the ligands have been systematically varied. The results of such screening procedures are described in terms of lethal doses (LD_{50} is the dose to kill 50% of the test animals), inhibitory doses (ID_{90} is the minimum dose that causes tumour regression), and therapeutic index ($TI = LD_{50}/ID_{90}$).

Structural features associated with anti-tumour properties

Active complexes (Table 10.2) always involve a *cis* geometry, with a pair of good

TABLE 10.2 Some anti-tumour complexes.[16, 17]

Complex	$LD_{50}(mg\ kg^{-1})$	$ID_{90}(mg\ kg^{-1})$	TI
cis-$Pt(NH_3)_2Cl_2$	13.0	1.6	8.1
trans-$Pt(NH_3)_2Cl_2$	27.0	27.0	< 1.0
cis-$Pt(NH_2C_5H_4)_2Cl_2$	480.0	2.4	200
cis-$Pt(NH_2C_6H_5)_2Cl_2$	3200	12.0	267
mer-$Rh(NH_3)_3Cl_3$	235	86	2.6
mer-$Rh(NH_3)_3(NO_3)_3$	135	59	2.3
mer-$Ir(NH_3)_3Cl_3$	> 1500		

The following complexes[16] cause filamentation of *E. coli* at the given concentration levels ($\mu g\ cm^{-3}$): $Ru(NH_3)_3Cl_3$ (5), $(NH_4)_3[RuCl_6]$ (15–30), $K_2[Ru(NO)Cl_5]$ (40–100), $K_3[Rh(NO_2)_6]$ (20–60), *trans*-$[Rh(py)_4Cl_2]Cl*$ (2.5), $UO_2(acetate)_2$ (25–75).

*This complex is probably reduced to the Rh(I), d^8 species *in vivo* prior to biological activity.

leaving groups, and are neutral species. The ammine groups in *cis*-DDP are inert, and are stronger *trans*-directing ligands than the chloro groups. The chloro groups are therefore much more labile and so offer the possibility of ligand substitution at the platinum centre. In blood plasma the high concentration of chloride ions will result in very slow replacement of chloride by other ligands except when their nucleophilicity is high. Thus during transport the *cis*-$Pt(NH_3)_2Cl_2$ will probably remain unchanged. When the local chloride concentration has decreased such substitution will take place, with the result that the *cis* complex may bind strongly to a biological ligand, particularly a chelating one. Thus the behaviour of these Pt(II) compounds may be understood in terms of the tendency of a group to leave (the *trans* effect) and the nucleophilicity of the attacking ligand, as described in Chapter 2. The differing anti-tumour properties of *cis*- and *trans*-$Pt(NH_3)_2Cl_2$ have been rationalized in terms of the *trans* effect.[18] Substitution of one chloro group in these complexes by a heterocyclic nitrogen donor means that the ammine group in this *cis* complex now has *trans* to it a new ligand of greater *trans* effects strength than itself. Thus loss of ammonia will occur so that eventually the *cis* complex could lose all four of its original ligands giving a complex that is firmly bound to the DNA, and which cannot be removed by any intracellular repair mechanism. The *trans* complex cannot do this as replacement of one chloro group means that the new ligand only facilitates loss of the second chloro group, and the ammine groups are not labilized.

Other active platinum(II) complexes contain different anions. These may either be good leaving groups or may be lost from the complex by metabolic action, for example methyl malonate. Anionic ligands are required to give neutral, liposoluble complexes. The amine ligands also affect reactivity, but not usually by electronic effects as their basicity usually only changes slightly. Variation in amine affects ID_{90} and LD_{50} differently. LD_{50} is very sensitive to the nature of the amine, as the toxic side effects are reduced by increasing liposolubility. So values of LD_{50} increase rapidly with increase in the size of the amine ligand. The therapeutic index thus increases — but not because the potency of the drug towards the tumour has increased. Indeed the ID_{90} value may increase, but this is masked by the much greater change in LD_{50}. The complex Pt(en)OX, oxalatoethylenediamineplatinum(II) is of interest as this is extremely toxic to the neuromuscular system, a fact that may be of some mechanistic interest.

Binding of cis-$Pt(NH_3)_2Cl_2$ to DNA

Several lines of argument suggest[17] that DDP binds to the DNA molecule in the tumour cell, preventing replication of DNA. Such inhibition of DNA synthesis has been demonstrated for mammalian cells at therapeutic levels of DDP. This takes place before synthesis of RNA and protein is effected. Furthermore, the inhibition of DNA synthesis which occurs at very low levels of DDP persists even after removal of the reagent from the culture medium, thus suggesting that the action of DDP does not arise through inhibition of an enzyme vital to the synthesis. There is much direct and indirect evidence to support the postulate that DDP binds directly to the DNA, including the work on the filamentation of *E. coli*. Of particular significance is the fact[20] that *cis*-DDP is highly effective at inducing the production of phage particles from lysogenic strains of *E. coli*.

The binding of *cis*-Pt(NH$_3$)$_2$Cl$_2$ to DNA is presumed to arise from the loss of both chloro groups. Loss of one chloro group would not lead to distinguishable situations for the binding of *cis* and *trans* isomers. The *cis* isomer will bind to two groups about 3.3 Å apart while the *trans* isomer will bind to two groups about 5 Å apart and which approach in mutually *trans* positions. UV spectra of DNA confirm that both platinum complexes bind to it *via* the organic bases. Kinetic studies on their interaction with individual bases suggest that reaction with guanosine is most rapid and that this base in DNA will therefore bind platinum preferentially. Other anti-cancer agents, the alkylating agents, also react preferentially at guanosine. Binding of the *cis*-DDP complex to DNA could take place in at least three ways: *interstrand* in which the platinum bridges bases on opposite strands of DNA; *intrastrand* in which two adjacent bases on a single strand are bridged; and *bidentate chelation* to one base only. It has been shown experimentally[21] that interstrand cross-links result from only 1 in 400 reactions of DNA with *cis*-Pt(NH$_3$)$_2$Cl$_2$. Such interstrand linking is postulated to involve bridging of two 6-amino functions of adenosine groups on opposite chains.[17] The intrastrand link is probably at the N-7 positions on adjacent guanine bases, while bifunctional attack at a single base could involve chelation of the O-6 and N-7 functions of guanosine. There is no explanation at present as to why such binding is cytotoxic, although it is clear that the *trans* isomer could not bind in this way.

Toxicity and new drugs

As with all anti-cancer drugs there are considerable side effects with *cis*-DDP. Most patients suffer nausea, diarrhoea, a decrease in hemoglobin levels, and kidney damage. Kidney damage is now largely minimized by hydration of the patient prior to and during treatment. It would be of value if techniques could be developed to give constant low concentrations of the drug *in vivo*. Possibilities here involve *in vivo* reduction of Pt(IV) complexes, the breakdown of polymeric platinum compounds and the slow removal of inert blocking ligands. Other platinum complexes are being tested. Noteworthy here are the platinum blues formed from the hydrolysis products of *cis*-Pt(NH$_3$)$_2$Cl$_2$ (or PtCl$_2$) and pyrimidine bases or amides. The platinum blue formed from uracil appears to be effective and also has a high solubility in aqueous solution. Much work needs to be done on the structures and solution chemistry of these platinum blues.[18] Many are polymeric compounds and may function by liberation of active monomeric species.

Many current studies are concerned with the reaction of Pt(II) compounds with nucleosides, nucleotides, and DNA.[22-24] Circular dichroism is a particularly valuable tool in this context. The use of Pt(II) compounds with different numbers of leaving groups has allowed the characterization of monofunctional, and *cis* and *trans* bidentate binding of Pt(II) compounds to DNA.[23] It has also been shown that the initial binding of Pt complexes may involve an intercalation mode followed by rearrangement to coordinate binding.[24]

METAL ION TOXICITY AND CHELATION THERAPY

In this section it is proposed to summarize the occurrence and toxic effects of some environmentally and industrially important metal ions which are known to be involved in metal poisoning. The form of the metal ion is important in determining its toxicity, and so some attention will be paid to the phenomenon of biological methylation of metal ions. The formation of organo complexes increases liposolubility and toxicity. The classic example is the Minimata disaster, where inorganic mercury effluents were converted into methylmercury as a result of vitamin B_{12} activity in microorganisms. Finally the treatment of specific cases of metal poisoning by chelation therapy will be discussed in some detail.

Toxic metals and compounds

Metal ions of environmental concern are usually the soft, heavy metals and metalloids, which have minimal biological roles, and cause many problems when they appear in biological situations as a result of their very strong binding to biological macromolecules. They include compounds of arsenic (from coal burning, insecticides, and herbicides), cadmium (associated with zinc extraction and several industrial operations), mercury (industrial effluents and organomercurial fungicides), and lead (inorganic lead in industrial and domestic environments, and organic lead from antiknock agents in petrol). In addition iron compounds may cause siderosis, either from overtreatment with tonics etc. or in the well-known case of beer brewed in iron pots. These toxic effects include irritation of the mucosa, giddiness, diarrhoea and may cause cardiac collapse. Other metal ions that have received publicity at different times are Be^{2+} and Tl^+, while plutonium compounds are well known to be extremely toxic.

The soft metal ions will bind tightly to a range of biological sites, particularly those with sulphur ligands. They will often displace native metal ions from and inhibit metalloenzymes. This will be reflected in the metal ion content of the urine. Several general reviews of the biochemical toxicity of these cations are available.[25] These are reflected in behavioural disturbances (arsenic, cadmium, lead, mercury, methylmercury, thallium, tin, and vanadium), anaemia, alterations in membrane permeability, and inhibition of oxidative phosphorylation and protein synthesis. Heavy metal complexes also bind to nucleic acid bases and phosphate groups so affecting their structure. There is also much evidence of the carcinogenic effects of a wide range of metals and their compounds,[26] including chromium, nickel, arsenic, cadmium, beryllium, all of which are associated with cancer in humans. Cadmium is also notorious for its toxic effect on bone, as in 'itai-itai byo' disease in Japan. It is beyond the scope of this book to discuss environmental and biochemical aspects of this topic in detail but recent reviews are given for Pb[27], Hg[28], Cd[29], Pu[30] together with general reviews on metal toxicity[31].

Biomethylation

Methylation greatly influences the absorption, distribution, and toxicity of a number of metals and metalloids. The Minimata disaster in Japan arose from the consumption of fish contaminated by large amounts of methylmercury. The source of the mercury was traced to effluents containing inorganic mercury compounds. But metallic mercury and inorganic mercury compounds undergo methylation by enzymatic and non-enzymatic pathways in microorganisms in the river sediment. The methylmercury thus formed was taken up and accumulated by fish. Hg(II) compounds cannot cross the blood–brain barrier and so are not found in the brain. Their toxic effects are exerted elsewhere. In contrast the liposolubility of Ch_3Hg^+ allows it to accumulate in the brain where it causes irreversible damage through its strong binding. Early symptoms of methylmercury poisoning are loss of sensation at the extremities of the limbs. This is followed by loss of coordination in gait, slurred speech, deafness, blindness, coma, and death.[28] One interesting observation is that inorganic compounds of selenium decrease the toxicity of inorganic and organic mercury compounds, apparently by displacing these from their sites on proteins with subsequent changes in their distribution.[32] They must bind to new sites with resulting decrease in toxicity.

Biomethylation has also been demonstrated for arsenic,[33] lead,[34] thallium,[35] chromium, selenium, sulphur, platinum, and palladium. Some of these methylated species may be less toxic than the original metal ion. Thus $(CH_3)_2Tl^+$ is less toxic than Tl^+, because the toxicity of the thallium(I) cation arises from its varied and complex effects on the biochemistry of K^+. The dimethylthallium(III) cation is unlikely to function in this manner.

Mechanisms of biomethylation

These have been reviewed.[36,37] Biomethylation can occur by at least three mechanisms, and at least three major coenzymes have been implicated as methyl-transfer agents, namely S-adenosylmethionine, N^5-methyltetrahydrofolate derivatives, and vitamin B_{12} derivatives. We will only consider methylation *via* vitamin B_{12}. As noted in Chapter 3, methyl transfer from cobalt(III) methylcorrinoids can in principle involve transfer as the carbanion CH_3^-, leaving Co(III), the radical $CH_3^{\cdot}$, leaving Co(II), or the carbonium ion CH_3^+, leaving Co(I). Methylation by methylcorrinoid derivatives occurs *via* CH_3^- or $CH_3^{\cdot}$, while the other coenzymes transfer carbonium methyl groups, CH_3^+. Specific examples are:

$$\text{Hg}^{2+} + \text{CH}_3\text{Co(III)corrinoid} \longrightarrow \text{HgCH}_3^+ + \text{Co(III)corrinoid}$$
$$\text{Cr}^{2+} + \text{CH}_3\text{Co(III)corrinoid} \longrightarrow \text{Cr(III)CH}_3^{2+} + \text{Co(II)corrinoid}$$

Thus transfer of a methyl radical involves a one-electron oxidation of the methyl-accepting species.

It has been suggested that the division of cations into the two mechanistic classes illustrated above is dependent on the redox potential of the metal ion. Those with E_0' greater than 0.805 V involve carbanion transfer (e.g. Pb, Tl, Se, Pd, Hg) while those with lower redox potentials involve radicals (e.g. Sn, Cr, As).

Chelation therapy

Most treatments for metal poisoning use chelating agents, which it is hoped will form soluble, stable, and non-toxic complexes which are readily excreted. In practice only relatively few chelating agents have found clinical use, and a number of problems have been encountered with these. It should be stressed that metals and their compounds can enter the body in several ways and this will often determine the nature of the treatment. Metal dusts or fumes may be breathed into the lungs, or the metallic compound may enter the blood stream or under the skin in accidents. But most cases of metal poisoning occur as a result of ingestion, due to the presence of metals in food and drink. In such cases a large fraction of the ingested compound will pass straight through the body, the fraction crossing membranes increasing with lower charge on the compound and the presence of liposoluble groups. Only Na^+, K^+, Ca^{2+} are readily absorbed in the gastrointestinal tract.

Several criteria may be listed as tests of potential chelating drugs. The chelating agent must bind the metal strongly enough to compete for it with biological ligands and must do this selectively. If it is non-selective then there will be harmful side effects from the removal of other metals, particularly calcium and zinc, from the body. The ligand must also be non-toxic itself. Other questions relate to the way in which the chelating ligand attacks the deposit of heavy metal. The metal will be in a lipophilic environment in the body. If the ligand is sufficiently lipophilic to penetrate membranes to reach the deposits, then a lipophilic complex will be formed which will not be excreted, but may well be redistributed in the body with aggrevation of its toxicity. Either therefore the lipophilic/phobic nature of the complex must change to allow excretion, or the ligand stays in the blood stream and eliminates the metal slowly by disturbing the equilibrium distribution in the body, i.e. the metal stores are not attacked directly. To ensure that the ligand and its complex are always lipophobic, the ligand should have more ionizable groups than the metal ion has charges.

Ligands which are selective for a particular metal may be picked by careful analysis of formation constant data. From a more empirical point of view, ligands may be selected in terms of the ideas of hard and soft acids and bases, as outlined in Chapter 3. This topic has been reviewed recently by Jones and Vaughn.[31] Common chelating agents in current use include, in order of

TABLE 10.3 Some ligands used in chelate therapy

| *Aurinetricarboxylic acid* | *British Anti Lewisite* | *Penicillamine* |

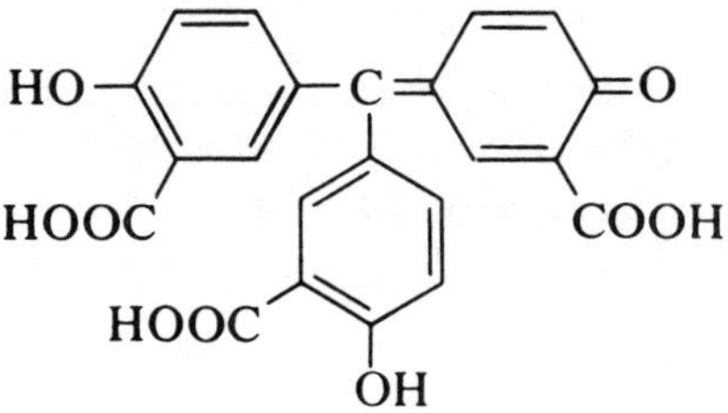

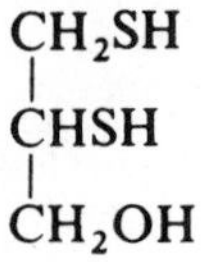

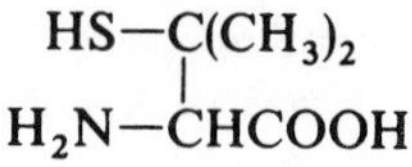

Cryptand

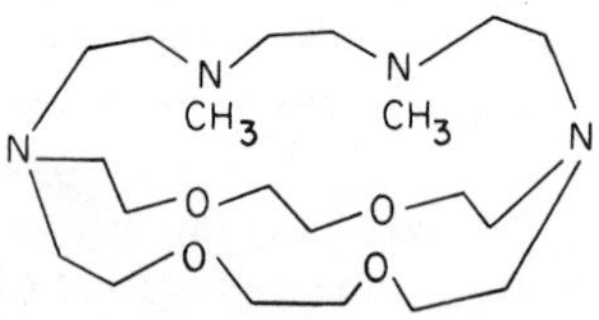

EDTA

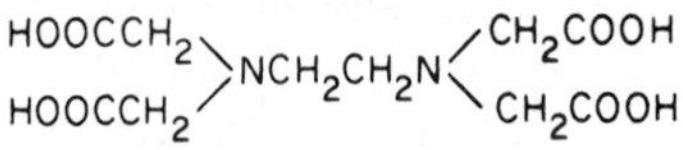

Desferrioxamine

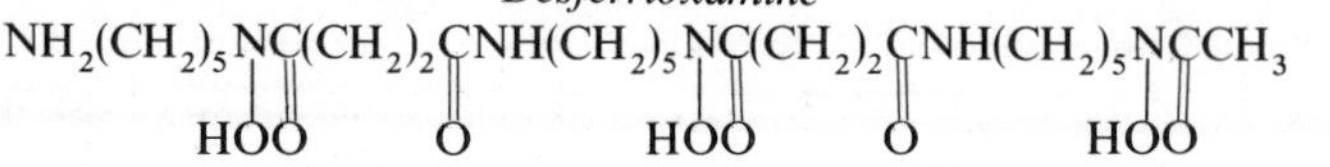

$$NH_2(CH_2)_5NC(CH_2)_2CNH(CH_2)_5NC(CH_2)_2CNH(CH_2)_5NCCH_3$$
$$HOO \quad O \quad HOO \quad O \quad HOO$$

decreasing softness, British Anti-Lewisite, BAL, (sulphur donors), D-penicillamine (N, S, O donors) and CaEDTA (N, O donors). These and other ligands will be discussed briefly and consideration given to the topic of mixed-ligand chelate therapy, in which the metal is removed by binding to two or more different chelates. Some ligands are shown in Table 10.3.

BAL (Dimercaprol)

This ligand was developed in the First World War to treat patients poisoned by the gas Lewisite, $ClCH=CHAsCl_2$. This poison gas functions by binding to the —SH groups of enzymes, but BAL binds particularly strongly to arsenic and so removes it.

BAL is used for treatment of poisoning by Hg, As, Cd, Au, Ti, Tl, and Bi and their compounds. The resulting complex is excreted. The ligand binds strongly to a range of metals and so must only be used in low concentration. A caution has been expressed over its use in the treatment of mercury compounds, as in the case of organomercury compounds this results in increased levels of mercury in the brain[38] in the early stages of poisoning. BAL is ineffective against lead.

Penicillamine

This ligand is particularly useful in promoting the urinary excretion of copper from patients with Wilson's disease. The copper complex is of interest. It will not form unless halide is present and a small amount is found in the complex. X-ray diffraction studies show the structure to involve a central halide surrounded by eight Cu(I) ions, bridged by sulphur donors which also bind to six Cu(II) ions, whose coordination sphere is completed by the amino group of the penicillamine ligand.[39]

Polyaminopolycarboxylic acids

EDTA has been much used in the treatment of metal ion poisoning, a notable example being lead. It is marginally effective against Cd^{2+}. It is used as the Ca^{2+} salt to prevent labilization of calcium from the skeleton. EDTA will bind Ca^{2+} and disturb the equilibrium distribution at the ultimate expense of the skeleton. EDTA at physiological pH values will be lipophobic and so treatment with EDTA is a slow process.

A number of related ligands have also been used. Thus a lipid has been linked to diethylenetriaminepentaacetic acid (DTPA). Presence of the lipid moiety results in DTPA being transported much more readily than DTPA itself across cell membranes into the interior of the cell. This has been used in the elimination of plutonium, and is much more successful than previous reagents. Two molecules of the DTPA-lipid system bind to each plutonium species and the chelated complex is ultimately excreted.

Miscellaneous ligands

Beryllium poisoning was once a problem in the light-alloys industries. It caused formation of large lumps on the body, and was very resistant to many chelating agents. Be^{2+} is a very hard cation and will bind to oxygen donors. One drug proved to be successful, aurinetricarboxylic acid. This formed a liposoluble complex so that the Be^{2+} was left in the cell as the complex. Fortunately this complex is not toxic, it is interesting that aurinetricarboxylic acid is also known as 'aluminon', as it is a spot test for aluminium. Beryllium and aluminium are in a diagonal relationship and so it is not surprising that aurinetricarboxylic acid also binds to beryllium.

A number of naturally occurring ligands have been used or suggested. Nature has solved the question of designing selective ligands, so the apoprotein of a metalloprotein would be ideally designed to treat poisoning by excess of that metal. Excess iron may be removed by 'desferrioxamine', which is an iron-binding compound found in certain microorganisms. 'Desferrioxamine' is a polyhydroxamic acid (Figure 10.3).

Another approach to this problem could well involve the use of crown ethers or, particularly, cryptates. Certainly the cryptates show extremely high selectivity for specific cations. Thus the cryptand shown in Figure 10.3 shows a very high selectivity for Cd^{2+} ($\sim 10^6$) with respect to Zn^{2+} and Ca^{2+}, and thus allows the complexation of the highly toxic Cd^{2+} without affecting Zn^{2+} and Ca^{2+}. Cryptand 2.2.2 is effective for removal of lead,[40] and for the removal of strontium-85 and radium-224 without complexing calcium.[41]

Mixed-ligand chelate therapy/synergistic chelation therapy

A mixture of two ligands is sometimes more effective in removing a toxic metal than either ligand operating singly. This has been attributed to the formation of a mixed-ligand chelate in which both ligands bind to the metal giving a complex of particular stability. Such mixed ligands are therefore assumed to be more effective in chelating with the toxic metal. However this view is now receiving less support, and indeed an earlier dramatic claim that the mixed ligands salicylic acid and diethylenetriaminepentaacetic acid were successful against poisoning by cadmium and plutonium has now been withdrawn.[42] It is unlikely that these particular ligands form mixed complexes with plutonium. An alternative explanation for any enhanced activity of mixed ligands in chelation therapy is that the two ligands bind to the toxic metal at different stages in its removal from the body. Thus one ligand may mobilize the metal while the other (of greater hydrophilic character) traps it in the plasma prior to excretion.[43]

DRUGS THAT APPEAR TO INVOLVE INTERACTION WITH METAL IONS

Many drugs are suspected of acting *via* chelation. A study of Table 10.4 indicates that many of the drugs listed there will act as good ligands *in vitro*. However it is difficult to relate the activity of a drug to its *in vitro* properties. Thus *in vivo* a potentially chelating drug has to compete for a metal ion with a wide range of other ligands and there is no guarantee that it will do this effectively. On the other hand, there are well-established cases where the medicinal behaviour of a compound may be attributed to the control of inhibition of enzyme activity *via* metal ion chelation. An example of this is disulfiram (tetraethylthiuram disulphide) which is used in the treatment of chronic alcoholism in cases where the patient wants to be cured. The drug inhibits the metalloenzyme aldehyde oxidase and so the metabolism of ethanol stops with the formation of acetaldehyde, so producing unpleasant symptoms and discouraging further indulgence. A second example involves metal complex formation to give a neutral, liposoluble product that will pass through cell membranes. Thus here the metal ion acts as a carrier for the drug.

Many questions have to be answered in the process of assessing the value and mode of action of new drugs. What are the side effects of the drug, or the products of its metabolism? How easily is the drug absorbed? How stable is it in *vivo*? How and where is the drug bound in its cycles around the body? The pH of the body solution and the pK_a of the drug will determine to what extent it is ionized *in vivo*. This is very important as in many cases biologically active species are weak acids or weak bases and only one form is active. Again, only unionized species are liposoluble. Sometimes the drug itself is not the active species, the pharmacologically active form is one that is metabolized from it. One drug may also interact with another. They may reinforce each other's activity, or they might have dangerous properties when together.

TABLE 10.4 Some drugs that may involve metal ions

Acetazolamide

(carbonic anhydrase inhibitor)

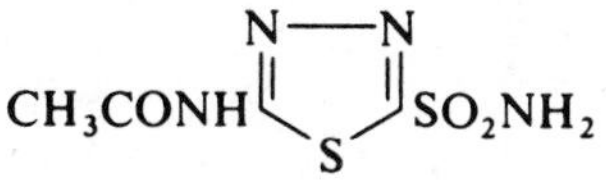

$$CH_3CONH - \underset{S}{\overset{N---N}{\bigsqcup}} - SO_2NH_2$$

Amphetamine

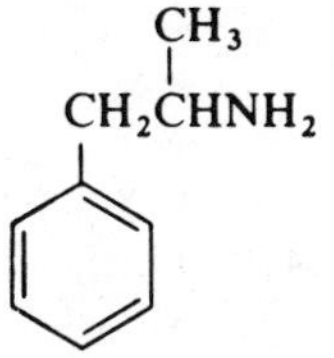

$$CH_3$$
$$CH_2CHNH_2$$

Aspirin

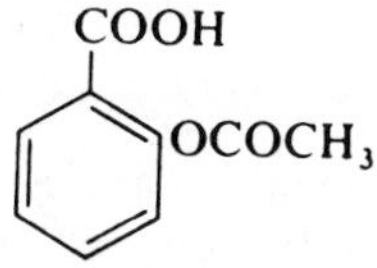

COOH
OCOCH$_3$

D.M.D.C.

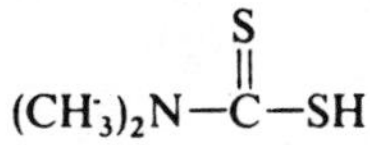

$$(CH_3)_2N - \overset{S}{\overset{\|}{C}} - SH$$

Ethambutol

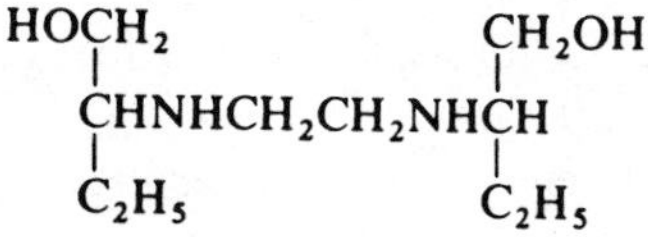

$$HOCH_2 \qquad\qquad CH_2OH$$
$$CHNHCH_2CH_2NHCH$$
$$C_2H_5 \qquad\qquad C_2H_5$$

Kojic acid

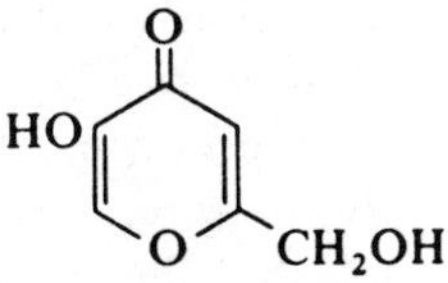

$$HO$$
$$O$$
$$O \qquad CH_2OH$$

(antibiotic)

Isoproniazid

$$H_3C$$
$$\diagdown$$
$$\qquad CHNHNHCO - \bigcirc N$$
$$\diagup$$
$$H_3C$$

Isonicotinic acid hydrazide

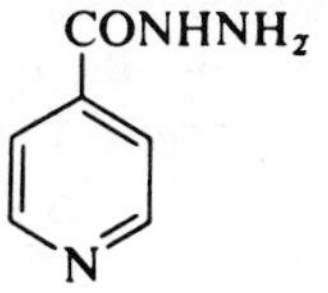

$$CONHNH_2$$

Phenacetin

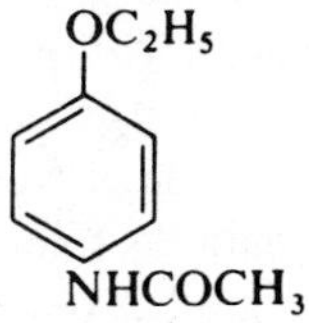

$$OC_2H_5$$

NHCOCH$_3$

Nialamide

$$\bigcirc - CH_2NHCOCH_2CH_2NHNHCO - \bigcirc N$$

---

disulfiram

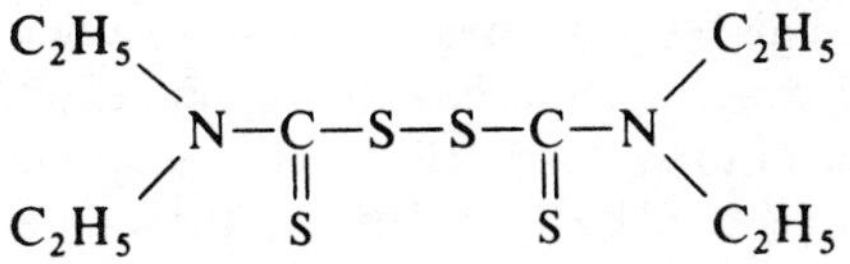

$$C_2H_5$$
$$\diagdown \qquad\qquad\qquad\qquad C_2H_5$$
$$N - C - S - S - C - N$$
$$\diagup \ \ \overset{\|}{S} \quad\ \overset{\|}{S} \ \diagdown$$
$$C_2H_5 \qquad\qquad\qquad\qquad C_2H_5$$

Thiacetazone

$$S$$
$$CH = NNHC^{\|}NH_2$$

NHCOCH$_3$

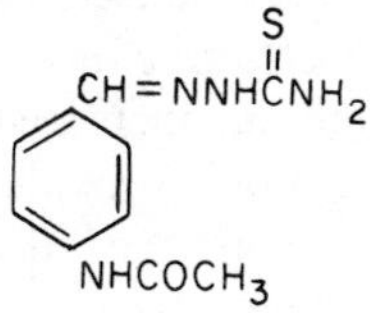

Normally the properties of a drug are associated with the shape of the molecule (or part of the molecule) and its ability to combine with a receptor site. Thus the role of the sulphonamides in controlling infections is associated with the presence of a p-NH$_2$ group which will fit in the bacterial enzyme receptor and so prevent the growth and reproduction of the bacteria. As noted earlier this has been exploited in the use of metal complexes, usually with substituted 1,10-phenanthroline, which are active against a range of microorganisms, including some which have a high resistance to normal antibiotics. Here the shape and charge distribution of the complex are such as to impart valuable pharmacological properties to the compound.

The cardiac glycosides

Digitalis and a number of closely related drugs have a specific and powerful action on the heart. Digitalis itself is the dried leaf of the foxglove plant, while its active constituents are the glycosides. They are found in a number of plants having digitalis action. Digitalis affects heart muscle, which is much more sensitive to it than are other muscles, causing an increase in the force of contraction, with a resulting slowing of the cardiac rate from the rapid, irregular and ineffective beat of the failing heart. Toxic doses of digatalis result in an excessive slowing of the heart rate.

It will be recalled that reference to the cardiac glycosides has been made in the context of the distribution of the alkali and alkaline earth metal cations in the cell. In particular it was noted that ouabain is an effective inhibitor of the active transport of ions across cell membranes by interfering with the transport ATPase. Studies on heart muscle preparations show that the activity of the cardiac glycosides is modified by altering the K^+ concentration, while toxic doses cause a loss of K^+ from the heart and a gain in Na^+. The symptoms resulting from an overdose of cardiac glycosides can be treated by increasing the plasma K^+ concentration. Similarly, depletion of plasma K^+ concentration in patients treated with digitalis may induce toxicity. There are a number of other parallels that indicate that the action of cardiac glycosides is due to inhibition of transport ATPase. Thus magnesium catalyses transport ATPase and decreases the toxic action of the cardiac glycosides. It is not possible at present fully to understand the action of the cardiac glycosides purely in terms of the inhibition of the cardiac transport ATPase system.

The contraction of muscle is dependent upon the presence of calcium ions in the muscle cells. It is known that ouabain produces an increase of calcium uptake into beating heart muscle; again the mechanism is uncertain. However, a decrease in calcium concentration diminishes the toxicity of the cardiac glycosides while an increase in calcium concentration has the reverse effect. Calcium and digitalis provide an example of the synergic effect. The symptoms produced by over-digitalization have been reduced by the injection of the sodium salt of EDTA due to the resulting decrease in plasma calcium concentration.

The urinary system

The diuretics are drugs that promote the formation of urine. The mercurial diuretics are particularly well known and are very effective. All are derivatives of mercuripropanol $RCH_2CH(OH)CH_2HgX$, where R is a polar, hydrophilic group. The mercurials are given by intramuscular injection, as they are too toxic for intravenous injection and too poorly absorbed to be given orally. The organo groups confer liposolubility on the mercury. The drug is taken up selectively in the tissues. All the organomercurial diuretics are broken down in the kidneys, giving mercuric ions, which may be the active agents. It is quite clear, however, that the mercurials operate by inhibiting the SH-containing enzymes. They are known, for example, to inhibit ATPase and succinic dehydrogenase in kidney slices. This same property of mercurials has resulted in their being used to control bacterial infections, where again they combine with —SH groups in the bacteria.

Acetazolamide (Table 10.4) and related drugs have a diuretic action probably by specifically inhibiting the action of carbonic anhydrase by binding to the zinc of that enzyme.

The monoamine oxidase inhibitors

These are used as stimulants. They are often derivatives of hydrazine. There is a correlation between *in vitro* monoamine oxidase inhibiting power and the clinical effectiveness of these compounds. That they are good ligands *in vivo* is demonstrated by one side effect, the decalcification of bone. It is thought therefore that these stimulants act by binding to the copper of monoamine oxidase.

Antimicrobial drugs

Many of these are obviously excellent ligands. Thus the use of tetracyclines affects the coagulation of blood due to their high affinity for Ca^{2+} and the resulting inhibition of the cascade mechanism described in Chapter 9. It is possible therefore that the activity of some of these antimicrobial drugs is dependent upon the complexing of metal ions. Indeed it is known is some cases that their efficiency is increased when coordinated. It may be that the metal complex is more liposoluble than the drug, and so the metal ion helps to transport it across the cell membrane. Alternatively, in other cases, the metal ion itself may be the toxic agent and the role of the coordinated antibiotic is one of carrier across the membrane.

Oxine (8-hydroxyquinoline) has antifungal and antibacterial properties when chelated to Fe(III). Albert has shown that the ligand and Fe(III) are both inactive when separated from each other, but are highly effective together, particularly when present in the 1:1 molar ratio. Albert has suggested that the 3:1 complex alone is capable of penetrating the cell membrane. Presumably the coordinatively unsaturated and charged 2:1 and 1:1 complexes bind too strongly to the membrane. Once the 3:1 complex is inside the cell it must break down to the 2:1 and 1:1 species. It appears that the metal is necessary for the entry of the oxine molecule into the cell. In support of the suggestion that liposolubility is an important effect in determining the toxicity of this complex, it should be noted that the introduction of lipophobic groups into the oxine results in a decrease in antibacterial action.

The antitubercular drug isoniazid (Table 10.4) acts by interfering with the metabolism of the tubercle organism. However, the chelated drug is more effective, again presumably due to the fact that the complex is more liposoluble than the drug itself. A similar reason probably accounts for the effectiveness of copper ions in enhancing the activity of the antitubercular drug thiacetazone (*p*-acetamidobenzaldehyde thiosemicarbazone).

Other examples where the antibacterial action of drugs is enhanced by metal ions are kojic acid, bacitracin, and sodium dimethyldithiocarbamate (DMDC). The last-named drug requires copper.

The possibility of metal ion involvement in the tetracyclines has attracted some attention. It has been shown that there is a correlation between the possession of antibacterial properties and the ability to form stable chelates with calcium ions. A similar correlation has been drawn between active tetracyclines and the ability to form 2 : 1 complexes with Cu^{2+}, Ni^{2+}, and Zn^{2+}, while the presence of excess metal ions inhibits the action of the tetracyclines.

The structure of the dipotassium salt of oxytetracycline shows[44] the chelation of K^+ by oxygen donors, a view confirmed by NMR studies in solution in DMSO. It is now generally accepted that the action of the tetracyclines is directed towards the ribosomes of the bacterial cells and hence results in inhibition of protein synthesis. So while metal ions may enhance the toxicity of tetracyclines by complex formation and transport of a neutral complex through the cell wall, it seems that the ultimate target of the tetracyclines is the Mg^{2+} which is necessary for the stabilization and function of the ribosomes. Thus the inhibition of the growth of *E. coli* by tetracyclines is reversed by the addition of high concentrations of Mg^{2+}. The TB drug ethambutol is also known to act by interference with a bacterial ribosome–Mg^{2+}–spermidine complex.

The glycopeptide antibiotic bleomycin is isolated from *Streptomyces verticillus* as the copper complex. Bleomycin has attracted attention because of use in the treatment of certain tumours in addition to its antimicrobial and antiviral properties. It is suggested that the bleomycins bind to and cleave DNA at G-T and G-C sequences,[45] in a reaction that is dependent on Fe(II) and dioxygen and which is postulated[46] to involve the oxidation of an Fe(II)–bleomycin–DNA complex. Its activity is inhibited by Cu(II), Zn(II), and Co(II) (which probably compete for the metal site) and chelating agents.

The binding of metals to bleomycin has been much studied by NMR techniques, (^{1}H and ^{13}C) which show that Zn(II) and Cu(II) are bound by a histidyl residue and a pyrimidine group.[47] Other ligands are probably an *a*-amino and a carbamoyl group. More direct NMR evidence for the ligand groups has been obtained by the study of Hg(II)-substituted bleomycin and the observation of ^{199}Hg–^{1}H coupling.[48] These results confirmed the binding *via* the histidyl group, the C2(H) and C4(H) resonances being doublets presumably due to the ^{199}Hg–^{1}H coupling ($I = \frac{1}{2}$ for ^{199}Hg). No evidence could be found for pyrimidine binding but this might reflect the preference of Hg(II) for linear coordination.

317

REFERENCES

1. M. J. Seven and L. A. Johnson (Eds.) *Metal Binding in Medicine*, Lippincott, Philadelphia, 1960; Trace elements in the metabolism of connective tissue, *Fed. proc.*, **30**, 983 (1971); *Trace Element Metabolism in Animals* (Eds. W. G. Hoekstra, J. W. Suttie, H. E. Ganther, and W. Mertz), University Park Press, Baltimore, 1974; E. J. Underwood, *Trace Elements in Human and Animal Nutrition*, (3rd edn.), Academic Press, New York 1977; Special Issue on Trace Elements in *Clin. Chem.*, **21**, 467 (1975).
2. J. J. R. Frausto da Silva and R. J. P. Williams, *Structure and Bonding*, **29**, 67 (1976).
3. P. M. May, P. W. Linder, and D. R. Williams, *J. C. S. Dalton*, **1977**, 588.
4. M. Chvapil, S. L. Elias, J. N. Ryan, and C. F. Zukoski, Pathophysiology of zinc, in *Neurobiology of the Trace Metals Zinc and Copper* (Ed. C. C. Pfeiffer), Academic Press, New York, 1972.
5. A. Albert, *Selective Toxicity*, (5th edn.), Methuen, London.
6. A. Shulman and F. P. Dwyer, in *Chelating Agents and Metal Chelates* (Eds. F. P. Dwyer and D. P. Mellor), Academic Press, New York, 1964, p. 383.
7. B. Rosenberg, Some Biological Effects of Platinum Compounds, *Platinum Metals Review*, **15**, 42 (1971).
8. P. J. Sadler, *Structure and Bonding*, **29**, 171 (1976).
9. A. A. Isab and P. J. Sadler, unpublished work.
10. N. A. Malik, P. J. Sadler, S. Neidle, and G. L. Taylor, *J. C. S. Chem. Comm.*, **1978**, 711.
11. M. T. Doig, M. G. Heyl, and D. F. Martin, *J. Chem. Ed.*, **50**, 343 (1973); F. N. Johnson (Ed.), *Lithium in Research and Therapy*, Academic Press, New York, 1975; N. J. Birch, *Inorganic Perspectives in Biology and Medicine*, **1**, 173 (1978).
12. N. Bagchi, T. R. Brown, and R. E. Mack, *Biochim. Biophys. Acta*, **542**, 163 (1978).
13. J. J. R. Frausto da Silva and R. J. P. Williams, *Nature*, **262**, 237 (1976).
14. R. Ebstein, R. Belmaker, L. Grunhaus, and R. Rimon, *Nature*, **259**, 411 (1976).
15. A. J. Thomson, R. J. P. Williams, and S. Reslova, *Structure and Bonding*, **11**, 1 (1972).
16. M. J. Cleare, *Coord. Chem. Revs.*, **12**, 349 (1974).
17. A. J. Thomson, *Platinum Metals Rev.*, **21**, 2 (1977).
18. S. J. Lippard, *Acc. Chem. Res.*, **11**, 211 (1978).
19. D. R. Williams, *Chem. Rev.*, **72**, 203 (1972).
20. S. Reslova, *Chem. Biol. Interactions*, **4**, 66 (1971).
21. J. M. Pascoe and J. J. Roberts, *Biochem. Pharmacol.*, **23**, 1345, 1359 (1974).
22. See for example R. S. Tobias, G. Y. H. Chu, and H. J. Peresie, *J. Amer. Chem. Soc.*, **97**, 5305 (1975); H. J. Peresie and A. D. Kelman, *Inorg. Chem. Acta*, **29**, L247 (1978); J.-P. Macquet and T. Theophanides, *Inorg. Chim. Acta*, **18**, 189 (1976).
23. J.-P. Macquet and J.-L. Butour, *Eur. J. Biochem.*, **83**, 375 (1978).
24. B. Norden, *FEBS Letters*, **94**, 204 (1978).
25. B. Weiss, *Fed. Proc.*, **37**, 22 (1978); B. L. Vallee and D. D. Ulmer, Biochemical effects of mercury, cadmium and lead, *Ann. Rev. Biochem.*, **41**, 91 (1972); A. Aitio, M. Ahotupa, and M. G. Parkii, *Biochem. Biophys. Res. Comm.*, **83**, 850 (1978).
26. F. W. Sunderman, *Fed. Proc.*, **37**, 40 (1978).
27. M. E. Hilburn, *Chem. Soc. Rev.*, **8**, 63 (1979).
28. *Trace Elements in the Environment*, Adv. Chem. Ser., 123, A. C. S.; D. L. Rabenstein, *Acc. Chem. Res.*, **11**, 100 (1978); *J. Chem. Ed.*, **55**, 292 (1978).
29. J. H. Mennear (Ed), *Cadmium Toxicity*, Dekker, 1979.
30. M. C. Thorne and J. Vennart, *Nature*, **263**, 555 (1976); D. Ramsden and T. F. Johns, *Nature*, **266**, 216 (1977); *The Toxicity of Plutonium*, H. M. S. O., London, 1975; *The Metabolism of Compounds of Plutonium and Other Actinides*, I. C. R. P. publication 19, Pergamon Press, Oxford, 1972.
31. T. W. Clarkson, *Fed. Proc.*, **36**, 1634 (1977); M. M. Jones and W. K. Vaughn, *J. Inorg. Nucl. Chem.*, **40**, 2081 (1978).

318

32. K. Sumino, R. Yamamoto, and S. Kitamura, *Nature*, **268**, 73 (1977).
33. J. M. Wood, *Science*, **183**, 1049 (1974).
34. P. T. S. Wong, Y. K. Chau, and P. L. Luxon, *Nature*, **253**, 263 (1975).
35. F. Huber and H. Kirchmann, *Inorg. Chem. Acta*, **29**, L249 (1978).
36. A. O. Summers and S. Silver, *Ann. Rev. Micro.*, **32**, 637 (1978).
37. W. P. Ridley, L. J. Dizikes, and J. M. Wood, *Science*, **197**, 329 (1977); J. M. Wood, A. Cheh, L. J. Dizikes, W. P. Ridley, S. Rakow, and J. R. Lakowicz, *Fed. Proc.*, **37**, 16 (1978).
38. A. J. Canty and R. Kishimoto, *Nature*, **253**, 123 (1975).
39. P. J. M. W. L. Birker and H. C. Freeman, *Chem. Comm.*, **1976**, 312; *J. Amer. Chem. Soc.*, **99**, 6890 (1977).
40. P. Baudot, M. Jacque, and M. Robin, *Toxicol. Appl. Pharm.*, **41**, 113 (1977).
41. W. H. Müller, *Naturwissenschaften*, **57**, 248 (1970); **61**, 455 (1974).
42. J. Schubert and S. K. Derr, *Nature*, **275**, 311 (1978); J. Schubert, *Nature*, **281**, 406 (1979); R. A. Bulman, F. E. H. Crawley, and D. A. Geden, *Nature*, **281**, 406 (1979).
43. P. M. May and D. R. Williams, *Nature*, **278**, 581 (1979).
44. K. H. Jogun and J. J. Stezowski, *J. Amer. Chem. Soc.*, **98**, 6018 (1976).
45. M. Takeshita, A. P. Grollman, E. Ohtsubo, and H. Ohtsubo, *Proc. Natl. Acad. Sci. U.S.A.*, **75**, 5983 (1978).
46. E. A. Sausville, J. Peisach, and S. B. Horwitz, *Biochemistry*, **17**, 2740, 2746 (1978).
47. A. E. G. Cass, A Galdes, H. A. O. Hill, and C. E. McClelland, *FEBS Letters*, **89**, 187 (1978); J. C. Dabrowiak, F. T. Greenaway, and R. Grulich, *Biochemistry*, **17**, 4090 (1978).
48. A. E. G. Cass, A. Galdes, H. A. O. Hill, C. E. McClelland, and C. B. Storm, *FEBS Letters*, **94**, 312 (1978).
49. D. H. Brown and W. E. Smith, *Chem. Soc. Rev.*, **9**, 217 (1980).

CHAPTER 11

A POSTSCRIPT ON NON-METALS AS TRACE ELEMENTS

Considerable attention has been paid to the role of metals in biology and medicine. Brief reference has also been made in Chapter 10 to the importance of some but not all of the essential trace metals in nutrition, and to the clinical implications of deficiencies or over-supply of these elements. However it is important to recall that certain non-metals are also essential elements. Hydrogen, carbon, nitrogen, oxygen, phosphorous, sulphur, and chlorine are all well-known bulk components of biological systems, although much of the biological chemistry of these elements is associated with organic compounds. In this chapter it is hoped to examine the biological role of some trace non-metals, including boron, fluorine, silicon, selenium, and iodine. Arsenic and bromine may also be essential trace elements. It should be recalled that, with the exception of the halides, these non-metallic elements form oxo anions, which may be polymerized.

The biological functions of some of these elements are not well characterized and often very little biochemical information is available. It should be stressed that beneficial and toxic effects are primarily a function of concentration, and so it is important that these elements are present at the correct concentration levels. The question of analytical techniques is an important one, and, while the sensitivity of techniques has improved considerably, problems may still arise from contamination and lack of specificity. Extreme care is required for work with some of these trace non-metals, such as silicon.

Discussion of several trace non-metals is contained in a number of reviews,[1-4] while some useful analytical data is available on the element composition of human tissues and body fluids.[5] Some general reviews are also cited in Chapter 10.

Boron

It has been known for many years[6] that boron is essential for higher plants. It is probable that boron is present as an anionic borate species. It has been suggested that boron may function by complexing with polyhydroxy compounds, as is well known for boric acid, and that in particular boron is able to bind to sugars and hence facilitate translocation of sugar across plant membranes. While this is not now accepted, it may be that boron exerts an indirect effect on sugar transport, such as a stimulation of the biosynthesis of auxins. Increased translocation of sugar results from auxin-stimulated growth.[7]

Several enzyme reactions are inhibited by borate, while the effect of borate on a wide range of plant processes has been explored.[6] Nevertheless it is difficult to bring together this information in a cohesive manner. It has been proposed that the primary effect of boron deficiency is one of membrane breakdown.

Silicon[8]

The demonstration[9] that silicon is an essential trace element was accomplished by careful nutritional studies with rats using all-plastic systems. Rats show considerable increase in growth rates on the addition of sodium metasilicate ($Na_2SiO_3.9H_2O$) to their diet (50 mg per 100 g). Silicon is thus essential for growth and for skeletal development in chicks and rats. Deficiencies produce defects in the development of bone structure and in connective tissue. Si appears to be associated with bone regions where active calcification is taking place, for example in the bone-forming cell, the osteoblast.[10] Concentrations of Si in connective tissue decrease with age, and so it is possible that this deficiency is associated with atherosclerosis.

Little is known about the uptake of Si into living systems. It will be present as silicic acid, and it is thought likely that cross-linking interactions occur with carbohydrates. Hyaluronic acid of the human umbilical cord appears to be the richest source of silicon in *homo sapiens*, containing 1.53 mg free silicon and 0.36 mg bound silicon per gram.[11]

SiO_2 is used for structural purposes in diatoms, some protozoa, some sponges, and one family of plants.

Selenium

Some specific examples are known which demonstrate the biochemical role of selenium. The molybdoenzyme formate dehydrogenase contains Se[12] as does glutathione reductase[13] and peroxidase. The uptake of selenium probably follows whatever uptake mechanism holds for sulphur, but at some stage these two elements must be distinguished between, in view of the specific requirement for selenium in these enzymes. Frausto da Silva and Williams[2] suggest that the answer may be found in the differing redox states of these two elements, and that selenium may be bound in an intermediate oxidation state. They also point out that selenium undergoes unique, fast, two-electron redox reactions in the region of 0.0 volts, in which oxygen atom reactions can be involved. This could account for the specific biological requirement for selenium.

Deficiency of selenium results in liver necrosis in rats. In young cattle, sheep, and rabbits selenium deficiency causes degenerative changes in muscle cell membranes, so that calcium accumulates inside the cell and enzymes leak out into the blood plasma. The death of muscle cells leads to the development of muscle weakening and heart failure. The biochemical explanation for this muscle disease invokes the enzyme glutathione peroxidase, which scavenges peroxides. Deficiency of selenium results in lower levels of this enzyme, and so the peroxides are able to oxidise the lipids and sulphur amino acids of the cell membranes. X-ray photoelectron spectroscopy on glutathione peroxidase shows Se $3d$-electron signals.[14]

In both these cases the susceptibility to selenium deficiency is influenced by the levels of vitamin E, which also protects cell membranes against oxidative damage. Addition, at low concentration levels, of naturally occurring selenium-binding compounds (factor 3) and sodium selenite, Na_2SeO_3, provides protection against necrosis.[15] Of interest is the fact that monoselenodicarboxylic acids, $HO_2C(CH_2)_nSe(CH_2)_nCOOH$, where n is an odd number (5, 7, 9, 11), are also effective.[16]

Toxicity

Several plants (such as *Astragalus racemosus*, 'locoweed') concentrate selenium when grown in selenium-rich soils. Livestock eat these plants and suffer from 'blind staggers', with disorders of the nervous system and loss of appetite. A rather different disease is produced as a result of animals feeding on material containing protein-bound selenium, as this is less soluble.

Degradation of these selenium compounds in animals results in the excretion of dimethylselenium, which has a strong garlic-like odour. The mechanism of this reaction has been elucidated. Selenium(IV) (H_2SeO_3) reacts with glutathione to give a —SSeS— containing species whch is then enzymatically reduced to H_2Se *via* a glutathioneselenopersulphide. This is then methylated, using *S*-adenosylmethionine as the methyl donor.[17]

$$H_2SeO_3 + 4GSH \longrightarrow GSSeSG + GSSG + 3H_2O$$
$$GSSeSG + NADPH + H^+ \xrightarrow[\text{reductase}]{GSH} GSSeH + GSH + NADP$$
$$GSSeH + NADPH + H^+ \longrightarrow H_2Se + GSH + NADP$$
$$H_2Se \longrightarrow Me_2Se + Me_3Se^+$$

Chemotherapy

It is well known that selenium can protect against mercury poisoning. Less well known is the fact that there appears to be a correlation between high Se levels in diets and low cancer mortality. This is a complex area but one that should be explored further.[18] Human dietary intake[3] of selenium is about 55–110 mg per year, and selenium concentration in the blood lies in the range 0.09–0.29 $\mu g\,cm^{-3}$. Of interest is the fact that selenium blood levels for New Zealand are lower than this (0.068 $\mu g\,cm^{-3}$), indicative of a selenium-deficient country! Selenium taken orally appears to be concentrated in the liver and kidney. A further example of the protection afforded by selenium against the toxic effect of soft metals is in the case of cadmium compounds.[19] As in protection against mercury compounds, selenium appears to divert the toxic metal to binding sites different from those at which it exerts its toxicity. This may reflect competition for these sites.

Arsenic

Although the toxic properties of this element and its compounds are well known, there is increasing evidence[20] to suggest that deficiency of arsenic is manifested in low birth rates and retardation of growth in rats, pigs, and goats. Supplementation of the diet with sodium arsenite resulted in an increase in growth rate.

Chloride and Bromide

The halogen anions differ from the non-metallic compounds considered so far in that they are simple anionic species instead of oxo anions. Chloride is distributed widely. It can pass through membranes and is important in the maintenance of osmotic balance. There is some doubt over the role of bromide as a trace element, although it is well known to exert a sedative effect.

Fluoride

Fluoride is essential in diet (at 2.5 p.p.m.) for optimal growth, and fluoride deficiency is associated with anaemia. It is interesting that CaF_2 is used by some molluscs. Most attention has been paid to fluoride metabolism in a dental context, because of the use of fluoride to prevent dental caries. Fluoride is readily taken up by calcium hydroxyapatite, $Ca_{10}(PO_4)_6(OH)_2$, the fluoride ions displacing hydroxide ions. Also fluoride enhances the precipitation of calcium phosphate and so may accelerate remineralization.

The formation of dental caries is well described.[21] It is initiated by the formation of plaque on the tooth surface. Acids produced by bacteria dissolve the dental enamel under the plaque, but strangely, not initially on the surface of the enamel. The outer surface of the enamel is often not affected until substantial damage is caused to the subsurface regions. It is thought that fluoride ion can facilitate the formation of apatite at this stage, before the outer surface is affected. Thus remineralization of the incipient lesion is accomplished, but the mechanism is unclear.

Fluoride is also used to stabilize the enamel mineral against decay, either by the use of fluoride in toothpaste, or by the treatment of teeth with fluorides directly. Sometimes teeth are treated with the acid phosphate fluoride prior to the administration of fluoride. Etching of the teeth is thought to increase fluoride penetration and uptake. Stannous fluoride has also been used, and produces increased surface hardness.[22] It has been suggested that the Sn(II) reacts with phosphate to give insoluble tin phosphates, but such compounds have not yet been identified.[23]

The concentration of fluoride in drinking water necessary for the prevention of caries is about $1 \, mg \, dm^{-3}$, but intake levels will also depend on other factors. Exposure to high levels of fluoride (e.g. above $8 \, mg \, dm^{-3}$) may upset the balance between mobilization and formation of bone and so affect skeletal tissue. Excessive absorption of fluoride results in fluorosis. This is usually an occupational hazard. Large amounts of fluoride are also emitted during the production of superphosphate fertilizer. Fluorosis results in thyroid problems, growth retardation, and eventually kidney damage. Long-term effects include increased mineralization in the body, probably due to enhanced precipitation of calcium phosphate. This results in deformed bones, which may actually be fused together, and calcified ligaments.

Iodide

The main physiological role of iodide is well known to be in the metabolism of the thyroid gland[24] and the associated hormones. The thyroid gland accumulates iodide, a property shown also by salivary glands, mammary glands, and other organs. It is thought at present, however, that iodide only plays a major role in the activity of the thyroid gland. Uptake is hindered by a number of anions.

The first step in the biosynthesis of thyroid hormones involves the iodination of tyrosine residues bound to the protein thyroglobin, giving 3-iodo-, and 3,5-diodotyrosine residues. These couple to form the hormones 3,5,3',5'-tetraiodothyronine (thyroxine), 3,5,3'-triiodothyronine, and 3,3',5'-triiodothyronine. Both iodination and coupling reactions are catalysed by the enzyme thyroid peroxidase,[25] while the latter reaction also appears to require iodide and hydrogen peroxide. The hormones are then cleaved from the thyroglobin. This biosynthetic scheme is represented below for thyroxine, which involves combination of the diiodotyrosine compounds. The other hormones result from the two alternative combinations of the iodo and diiodo species.

Thyroglobin (protein-NHCOCH(NH$_2$) CH$_2$C$_6$H$_4$OH)

I$^-$, thyroid peroxidase

3-Iodotyrosine   3,5-Diiodotyrosine   Thyroxine

Thyroid hormones have a number of effects on metabolism, including a role in the control of body temperature, while their synthesis and release is controlled by the pituitary gland through the thyroid-stimulating hormone.

Iodine deficiency results in some well-defined symptoms, including lack of energy, a feeling of cold and dry, yellowish skin. These may be controlled by treatment with thyroid hormones or iodine. Lack of thyroid hormones may lead to an enlargement of the thyroid gland in an attempt to compensate. Goitres may also result in rare cases from the presence of compounds that prevent the uptake of iodide (for example thiocyanate, and the antithyroid agent, goitrin, which is found in brassicas). Iodine deficiency is particularly serious in the case of children, since it results in retardation of physical and mental development. Indeed iodine deficiency in pregnancy may result in the birth of hypothyroid children (cretins).

Excess thyroid hormone (hyperthyroidism) leads to the over-use of metabolic reserves, with nervousness, tremors, weight loss, and heat intolerance. This is

associated with increased peroxidase activity and hence increased iodination of
the thyroglobulins. Hyperthyroidism may result from thyroid tumours. These
may be treated by the use of radioactive isotopes of iodine, which are readily
taken up by the thyroid cells.

REFERENCES

1. E. J. Underwood, *Trace Elements in Human and Animal Nutrition*, (4th edn.), New York, Academic Press, 1977.
2. J. J. R. Frausto da Silva and R. J. P. Williams, *Structure and Bonding*, **29**, 67 (1976).
3. N. J. Birch and P. J. Sadler, *Inorganic Biochemistry*, Specialist Periodical Report of the Chemical Society, (Ed. H. A. O. Hill), **1**, 356 (1979).
4. H. J. M. Bowen, *Trace Elements in Biochemistry*, Academic press, New York, 1966.
5. G. V. Iyengar, W. E. Kollmer, and H. J. M. Bowen, *Elemental Composition of Human Tissues and Body Fluids*, Verlag Chemie, 1978.
6. W. M. Dugger, *Adv. Chem. Ser.*, **123**, 112 (1973).
7. J. J. Dyar and K. L. Webb, *Plant Physiol.*, **36**, 674 (1961).
8. J. D. Birchall, in *New Trends in Bioinorganic Chemistry* (Eds. R. J. P. Williams and J. J. R. Frausto da Silva), Academic Press, London, 1978.
9. K. Schwarz and D. B. Milne, *Nature*, **239**, 333 (1972).
10. E. Carlisle, *Fed. Proc.*, **34**, 927 (1975).
11. K. Schwarz, *Proc. Natl. Acad. Sci. U.S.A.*, **70**, 1608 (1973).
12. L. G. Ljungdahl and J. R. Andreesen, *FEBS Letters*, **54**, 279 (1975).
13. J. T. Rotruck, A. L. Pope, H. E. Ganther, A. B. Swanson, D. G. Hafeman, and W. G. Hockstra, *Science*, **179**, 588 (1972).
14. D. Chiu, A. L. Tappel and M. M. Millard, *Arch. Biochem. Biophys.*, **184**, 209 (1977).
15. K. Schwarz and C. M. Foltz, *J. Amer. Chem. Soc.*, **79**, 3292 (1957); K. Schwarz, *Med. Clin. North Amer.*, **60**, 745 (1976).
16. K. Schwarz and A. Fredga, *Bioinorg. Chem.*, **4**, 235 (1975).
17. H. E. Ganther, *Chem. Scripta*, **8A**, 79 (1975); J. R. Prohaska and H. E. Ganther, *Biochem. Biophys. Res. Comm.*, **76**, 437 (1977).
18. G. N. Schrauzer, D. A. White, and C. J. Schneider, *Bioinorg. Chem.*, **7**, 23 (1977); R. J. Shamberger, S. A. Tytko, and C. E. Willis, *Arch. Environ. Health*, **1976**, 231.
19. J. Parizek, J. Kalonskova, A. Babicky, J. Benes, and L. Pavlik, in *Trace Element Metabolism in Animals* (Ed. W. G. Hoekstra), University Park Press, Baltimore, 1974, p. 119.
20. F. H. Nielsen, *Fed. Proc.*, **34**, 923 (1975).
21. D. B. Scott, J. W. Simmelink, and V. Nygaard, *J. Dent. Res.*, **53**, (Suppl. 2), 65 (1974). (Several other useful articles are available in this supplement.)
22. D. J. Purdell-Lewis, J. Arends and A. Groeneveld, *Caries Res.*, **12**, 43 (1978).
23. P. Grøn, *Caries Res.*, **11**, (Suppl.) 172 (1977).
24. W. A. Harland and J. S. Orr, *Thyroid Hormone Metabolism*, Academic Press, London, 1975.
25. D. Deme, J. M. Gavaret, J. Pommier, and J. Nunez, *Eur. J. Biochem.*, **70**, 435 (1976).

INDEX

330

338